Engineering
Composite Materials

Engineering
Composite Materials

SECOND EDITION

Bryan Harris

Book 673
First published in 1999 by
IOM Communications Ltd.
1 Carlton House Terrace
London SW1Y 5DB

ISBN 1-86125-032-0

IOM Communications Ltd.
is a wholly-owned subsidiary of
The Institute of Materials

Typeset in India by Emptek Inc.
Printed and bound in the U.K. at
The University Press, Cambridge

To Margaret

In order to arrive at what you do not know
  You must go by a way which is the way of ignorance.
In order to possess what you do not possess
  You must go by the way of dispossession.
In order to arrive at what you are not
  You must go through the way in which you are not.
And what you do not know is the only thing you know
And what you own is what you do not own.
And where you are is where you are not.

T. S. Eliot, 'East Coker'
(*The Four Quartets,* Faber and Faber, London, 1940).

Contents

Preface to the Second Edition

What has been, will be again, what has been done, will be done again, and there is nothing new under the sun! Take anything which people acclaim as being new: it existed in the centuries preceding us. No memory remains of the past, and so it will be for the centuries to come — they will not be remembered by their successors.
Ecclesiastes, 1: 9-11, The New Jerusalem Bible, 1985, (Darton, Longman and Todd, London).

It is a salutary thing to be faced with revising the first edition of a text-book some ten years after the original publication. In a fast-moving field like composite materials, a great deal happens in ten years and what seemed definite when the first Preface was being written can appear to have been ephemeral in retrospect. The experience could lead one to renounce completely the practice of writing Prefaces, but I suppose that even scientific authors are sufficiently egotistic to be prepared to risk making the same mistakes again.

The changes that have taken place in the subject and in the associated industrial scene are quite remarkable, and not at all as might have been predicted. The current level of interest in the subject might be judged from the large numbers of books and journal papers that appear — the number of scientific journals dealing exclusively with composites has tripled or quadrupled in ten years — and the numbers of expensive international conferences that are held, year by year, and even in the same year, beggars belief. The picture of academic activity therefore suggests a flourishing area of research and development: there has been a world-wide proliferation of academic centres for the study of composite materials. In the U.K. the teaching of composite materials in engineering and materials courses has also become common, compared with the situation in the early 1980s. Many new specialist books on MMCs, CMCs and carbon/carbon composites have appeared, and others are in preparation.

And yet the obverse side of the coin — the industrial scene — looks rather different, depending on one's national viewpoint. Many of the U.K. companies which became deeply involved in composites research (and advanced materials research in general) in the seventies and eighties have now renounced all or most of their interest in composite materials — ICI, GKN, BP, Courtaulds. Others retain only a low-key interest. Laboratories and centres of expertise built up at great expense have been dispersed, profitable commercial aspects have been sold off, frequently to the U.S., in the rush to return to 'core business,' *e.g.,* oil and textiles, and the country which gave birth to high-performance carbon fibres has thrown away the initiative, as so often in the past. There is now no longer any manufacturing of high-performance fibres in the U.K. But although one French oil company also pulled out of composite materials, the composites industry in the rest of Europe has not shown the same state of decline as that in the U.K. Japanese and U.S. interests in composites also appear still to be thriving.

As for materials, of the three major groups of composite materials, polymer composites continue to dominate the information and publicity scenes. Market penetration has increased substantially, and a wide spectrum of materials, from modest moulding compounds to high-performance carbon-fibre-reinforced plastics, are commonly available, even at a price. Designers are increasingly aware of the potential value of using them and increasingly conscious of the scope they offer for re-evaluation of function and for holistic design instead of piece-meal substitution, as in the past. The results of the tremendous amount of analytical activity that the decade has seen are finding applications now that sophisticated (or even just tedious) mathematical solutions to design problems can be carried out very rapidly on the computers which now sit on most desk-tops.

The metal-matrix composites 'fight-back' referred to in the Preface to the First Edition has continued, but with the slightly unexpected difference that there has been almost as much interest in particulate MMCs as in fibre-reinforced metals for some applications. At the time of writing, there are still far more potential applications than actual applications, but the research continues, particularly for high-temperature applications. Research on CMCs follows an erratic path as potential users try to decide whether or not they can live with the cracks which they inevitably contain. The fact that such materials are commercially available, even though they are extremely costly, suggests that some are prepared to take the risk.

I should like to correct an impression about this book that was apparently gained by some of the reviewers of the First Edition. It was never intended to be an authoritative treatise on composite mechanics, as at least one reviewer assumed. Even in 1986, there were already several classical treatments — R.M. Jones's book *Mechanics of Composite Materials*, for example, and the Ashton, Halpin, Tsai *Primer on Composite Materials*. Now, in 1997, there is an abundance of such books. My intention was never, and is not now, to compete with these standard works for advanced users of composites. In the U.K., undergraduate courses in Engineering and in Materials Science frequently have a need for an input on composite materials that covers the subject selectively but sufficiently widely to make new graduates aware of the nature and performance of this class of modern engineering materials. And postgraduates beginning work on composites research projects need an introductory text that lays the foundations and points the way forward. This is how I have approached the revision. Reviewers of the First Edition also commented that the book was not up to date. They will therefore be torn if they read the new Edition, because although there is abundant reference to new developments and new publications, I have resisted throwing out many references to earlier material for the reason enshrined in the quotation from Ecclesiastes at the head of this Preface. Most people recognise and use the quotation "*there is nothing new under the sun*" either without appreciating the full context or without realising just how true the statement may be. When one has worked in a field for thirty years or more, one begins to recognise just how often the wheel is indeed reinvented.

In conclusion, I should like to add a comment to provoke thought for the future. In *his Introductory Remarks* to a 1970 Royal Society *Discussion Meeting on Strong*

Fibrous Solids*** Sir Alan Cottrell, then Chief Scientist at the Cabinet Office, pointed out that in relation to composites the term 'consumer durables' could carry an uncomfortable feel about it in the future. He foresaw that there would be little incentive to do anything other than dump old composites components because, unlike scrap metals which have a recovery value, the value of composites lies mainly in their form rather than in their substance. He concluded, "I hope that our enthusiasm for these exciting new materials will not lead us to overlook this problem." The fact that 27 years later so few groups are working on recycling and recovery reveals the prophetic nature of these words.

Bryan Harris,
Department of Materials Science & Engineering, University of Bath
July 1998

---

*Published as Proceedings of Royal Society, (London), **A319**, 1970, 1-43.

Preface to the First Edition

Composite materials is an exciting field to work in, perhaps because it has brought together a range of inter-linked disciplines. It has provided plenty of analytical problems for theoreticians and experimental programs for academic researchers as well as opening up new possibilities for designers. The practical history of the subject goes back, perhaps, to the second world war when some of the early development work on glass-reinforced plastics was done and James Gordon was experimenting with flax as a reinforcing fibre, although few apologists attempting to establish venerable antiquity for the subject can resist references to the biblical Israelites-versus-Egyptians restrictive-practices dispute (**Exodus** v. 7 *et seq*) or the Romano-Egyptian utilisation of reinforced concrete.

Research and development in composites in post-war years were relatively slow, perhaps because physical metallurgy was riding high on the elucidation of the behaviour of metallic alloy systems in terms of dislocation theory, and alloy development was pushing sufficiently well ahead to produce enough new materials with improved properties to satisfy the engineering demands of the period. Certainly, polymer development was still firmly in the hands of the polymer chemists. No disrespect is intended here: this is a matter of historical development. Engineers had not yet foreseen what composites had to offer. My own recollection of life at ICI Metals Division Research Department in the mid-fifties was that a tiny effort only was devoted to these odd plastic laminates by a small Chemistry Group, while the rest of us got on with the more respectable job of alloy development.

By the late fifties, however, engineers were becoming more exigent, and early materials scientists — in reality, a cosmopolitan group of metallurgists, physicists and chemists — were scanning the Periodic Table to see what the possibilities were for developing new ultra-high-strength materials. The old British National Research and Development Council, then Directed by John Duckham, ran a brain-storm session in Cambridge, when leading U.K. scientists (and a few obscure research students, of whom I, somehow, was one) made all kinds of outrageous suggestions for the development of newer and better engineering materials. Rash statements made then about things like ultra-fine particulate materials, whiskers, and very strong filaments (I seem to recall Tony Kelly suggesting carbonising a fine metallic filament to produce a continuous thread of titanium carbide, for example) are now everyday concepts, many of them translated into practical reality. Then, it seemed very much like reaching for the moon.

By the time I went to the U.S.A. in 1962, Morgan in the U.K. were employing new graduates to work on whiskers. The Thermokinetics company in the U.S. was selling alumina whiskers, in gram quantities, for something of the order of $100/g, and Texaco were experimenting with pyrolytic formation of continuous boron filaments. But a different kind of development that was going on simultaneously was the laying down of the foundations of the theory of metal-matrix composites, principally by metallurgists. By the mid 1960s, the subject was already a discipline, with a rapidly growing publication rate, and into this ferment of ideas was released the catalyst of practical reality — the development of continuous high-performance carbon fibres by Leslie Phillips and his colleagues at the Royal Aircraft Establishment, Farnborough. The availability of a fibre with such good properties that it could be handled and used in reasonably simple manufacturing operations gave aerospace designers their first sniff of something usable and initiated a new phase of reinforced plastics development. Applications were announced: new analytical work on the accurate prediction of the properties of complex laminates began to lay down ground-rules for design: interest in the old GRP was re-awakened, realisation having dawned that there were plenty of things in heaven and earth still not dreamt of in the old philosophy and the search was on for newer and better fibres.

Concurrently, interest in metal-matrix composites waned as it became increasingly evident that none of the currently available fibres was going to be any good for reinforcing metals for high-temperature application, while for low-temperature work reinforced plastics were then offering all that was needed. Interest waned, but was clearly not eclipsed, despite the fact that there were few publications in the open literature on fibre-reinforced metal-matrix systems during the seventies. In the eighties, however, the weather-vane has moved back to the old quarter: new reinforcing filaments are available, and the level of interest in metal-matrix composites indicated by the numbers of presentations at recent conferences in Europe and the U.S. (at least those which do not have closed sessions from which non-U.S. citizens are excluded) reflects an almost feverish re-awakening of interest in these materials. As an old-stager I find it amusing to see that the new generation of metal-matrix researchers is having to dig over a little old ground. But then, there really is nothing new under the sun: textile and paper physicists from way back have always maintained that they had already solved all the important problems that were subsequently re-solved by composites people. This is why composites is an exciting field, and I hope some who read this book might be persuaded to work in it.

I should perhaps admit that the book, like this Preface, reflects something of a personal view of the subject and its development. The book has grown out of a series of lectures that I offer to Materials Science students at Bath and has appeared in an edited form in presentations which I made at a Summer School on **High-Modulus Polymers and Composites** at the Chinese University of Hong Kong in 1984. In a monograph of limited size it is necessary to be selective, but I hope that I have provided sufficient general reference material to permit a reader intent on deeper involvement to acquire a more balanced view.

Bryan Harris,

May 1986

Acknowledgements

In writing this book I have made much use of the work of a generation of scientists and engineers, those who laid the foundations of this subject of composite materials and those who have built on those foundations. To all those to whose work I have made reference here I am greatly beholden, and I hope that they will accept that in using their work I am acknowledging their contributions with thanks. I am particularly indebted to that long succession of research students, research assistants, and colleagues, at the universities of Bath and Sussex, with whom I have shared the fun (not too inappropriate a word) of discovering how composites behave. The first edition was produced in the days before an author had total control over the manuscript in the environment of a desktop computer. And if there is now no-one to thank for the thankless task of copy-typing and retyping successive drafts, the author alone is to blame if errors remain. I am, however, very grateful to Dr. Jason Chant of British Gypsum for reading and commenting on several of the chapters. I should also like to acknowledge the anonymous software writers who have made it possible for me to produce text, equations and graphics without leaving my office. I have used Microsoft's *Word* and *Equation Editor* for the text and equations, MicroCal's *Origin* software for producing graphs of data, and *Professional Draw* by Gold Disc Inc. for the schematic diagrams and illustrations. Re-use of old graphs and other people's data was made possible by digitising scanned plots with BioSoft's *Ungraph* program and modelling was done with the aid of MathSoft's excellent *Mathcad*. The use of these tools has transformed for me the process of making a book in a way that I would never have believed possible.

1. The Nature of Composite Materials

1.1 WHAT ARE COMPOSITES?

Composite materials are extending the horizons of designers in all branches of engineering, and yet the degree to which this is happening can easily pass unperceived. The eye, after all, does not see beyond the glossy exterior or the race performance of a GRP[1] yacht, nor does it sense the complexity of the structure of a composite helicopter rotor blade or of a modern CFRP[2] tennis racket. Nevertheless, this family of synthesized materials offers the possibility of exciting new solutions to difficult engineering problems.

In composites, materials are combined in such a way as to enable us to make better use of their virtues while minimising to some extent the effects of their deficiencies. This process of optimization can release a designer from the constraints associated with the selection and manufacture of conventional materials. He can make use of tougher and lighter materials, with properties that can be tailored to suit particular design requirements. And because of the ease with which complex shapes can be manufactured, the complete rethinking of an established design in terms of composites can often lead to both cheaper and better solutions.

The 'composites' concept is not a human invention. Wood is a natural composite material consisting of one species of polymer — cellulose fibres with good strength and stiffness — in a resinous matrix of another polymer, the polysaccharide lignin. Nature makes a much better job of design and manufacture than we do, although Man was able to recognize that the way of overcoming two major disadvantages of natural wood — that of size (a tree has a limited transverse dimension), and that of anisotropy (properties are markedly different in the axial and radial directions) — was to make the composite material that we call plywood. Bone, teeth and mollusc shells are other natural composites, combining hard ceramic reinforcing phases in natural organic polymer matrices. Man was aware, even from the earliest times, of the concept that combining materials could be advantageous, and the down-to-earth procedures of wattle-and-daub (mud and straw) and 'pide' (heather incorporated in hard-rammed earth) building construction, still in use today, predate the use of reinforced concrete by the Romans which foreshadowed the pre-tensioned and post-tensioned reinforced concretes of our own era. But it is only in the last half century that the science and technology of composite materials have

---

[1]GRP is the usual abbreviation for glass-fibre-reinforced plastics
[2]CFRP is carbon-fibre-reinforced plastic

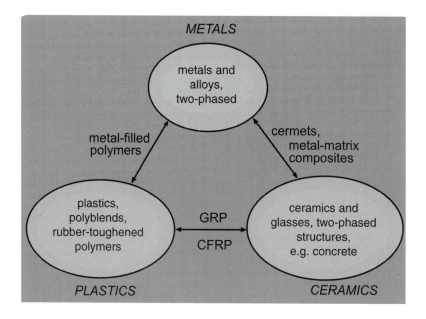

Fig. 1.1 Relationships between classes of engineering materials, showing the evolution of composites.

developed to provide the engineer with a novel class of materials and the necessary tools to enable him to use them advantageously.

The simple term 'composites' gives little indication of the vast range of individual combinations that are included in this class of materials. We have mentioned some of the more familiar ones, but the diagram Figure 1.1 gives a clearer idea of the scope for ingenuity which is available to the Materials Scientist and his customer, the Design Engineer. First, within each group of materials — metallic, ceramic and polymeric — there are already certain familiar materials which can be described as composites. Many members of the commonest and largest group of engineering materials, the family of steels, consist of combinations of particles of hard ceramic compounds in a softer metallic matrix. These particles are sometimes plate-like, sometimes needle-shaped, and sometimes spherical or polygonal. Polymers, too, are often two-phased, consisting of a matrix of one polymer with distributions of harder or softer particles contained within it; wood is a perfect example of this, as we have seen. And concrete is a classic example of a ceramic/ceramic composite, with particles of sand and aggregate of graded sizes in a matrix of hydrated Portland cement. These materials have been well known for many years, and Materials Scientists have learned to control their properties by controlling their microstructures; that is to say, the quantity, the form, and the distribution of what we might refer to as the 'reinforcing phase'. The idea of mixing components across the materials class boundaries is a natural extension of this idea. Making additions of hard, or fire-resistant, or simply cheap, ceramic powders to plastics to make filled polymers; and making additions of very hard, or abrasive, or thermally stable ceramic particles to metals to make the class of materials known as 'cermets' to produce machine tool tips capable of cutting hard metals at high speeds or high temperatures; are only two examples of important developments in our exploitation of these materials. But even more significant is the extension of this principle to incorporate filamentary metals, ceramics and polymers into the bulk forms of any of these three classes of materials to make fibre composites — reinforced plastics, like CFRP and GRP, metal-matrix composites (MMCs) like silicon-carbide-fibre-reinforced aluminium, and ceramic-matrix composites (CMCs) like carbon-fibre-reinforced glass.

Ideally, the properties of engineering materials should be reproducible and accurately known. And since satisfactory exploitation of the composite principle depends on the design flexibility that results from tailoring the properties of a combination of materials to suit a particular requirement, we also need to be able to predict those properties successfully. At present, some of the important engineering properties of composites can be well predicted on the basis of mathematical models, but many cannot.

1.2 CONVENTIONAL MATERIALS AND THEIR LIMITATIONS

It is difficult to draw up a table of materials characteristics in order to assess the relative strengths and weaknesses of metals, plastics and ceramics because each of these terms covers whole families of materials within which the range of properties is often as broad as the differences between the three classes. A comparison in general terms, however, can identify some of the more obvious advantages and disadvantages of the different types of material. At a simplistic level, then:

- *Plastics* are of low density. They have good short-term chemical resistance but they lack thermal stability and have only moderate resistance to environmental degradation (especially that caused by the photochemical effects of sunlight). They have poor mechanical properties, but are easily fabricated and joined.
- *Ceramics* may be of low density (although some are very dense). They have great thermal stability and are resistant to most forms of attack (abrasion, wear, corrosion). Although intrinsically very rigid and strong because of their chemical bonding, they are all brittle and can be formed and shaped only with difficulty.
- *Metals* are mostly of medium to high density — only magnesium, aluminium and beryllium can compete with plastics in this respect. Many have good thermal stability and may be made corrosion-resistant by alloying. They have useful mechanical properties and high toughness, and they are moderately easy to shape and join. It is largely a consequence of their ductility and resistance to cracking that metals, as a class, became (and remain) the preferred engineering materials.

On the basis of even so superficial a comparison it can be seen that each class has certain intrinsic advantages and weaknesses, although metals pose fewer problems for the designer than either plastics or ceramics.

1.3 STRONG FIBRES

The engineer who uses materials for structural or load-bearing purposes is quickly aware of an important feature of engineering solids, which is that they are never as strong as we would expect them to be from our knowledge of the strengths of the chemical bonds which hold them together. The reason for this is that all materials contain defects of various kinds which can never be entirely eliminated in practical manufacturing operations. For example, the strength of bulk glass and other ceramics is determined not by their strong covalent or ionic bonds, but by the many tiny pores or sharp cracks that exist either on the surface or in the interior. The most highly polished and dense bulk ceramic will rarely have a strength that exceeds one thousandth of the theoretically predicted strength. Similarly, metals contain faults in the stacking of atoms in their crystalline arrays, and the most damaging of these imperfections — dislocations — cause most bulk metal samples to deform plastically at loads which are, again, perhaps a thousandth or less of the theoretically calculated shear strength.

The strength of any sample of a glass or ceramic is actually determined by the size of the largest defect, or crack, which it happens to contain. Roughly, the strength is proportional to the inverse square root of the length of the largest flaw, a relationship developed in a thermodynamic argument by Griffith in the 1920s. The relationship in its simplest form is:

$$\sigma_{max} = \sqrt{\frac{2E\gamma_F}{\pi a}} \tag{1.1}$$

where σ_{max} is the strength of the material, E is its elastic stiffness (Young's modulus), γ_F is the work required to fracture the sample (a slight modification of Griffith's original idea that it represented the surface free energy of the material), and a is the flaw size. Since materials of this kind inevitably contain cracks and other defects with a wide range of sizes, one of the unfortunate consequences of the Griffith model is that the measured strengths of a batch of supposedly identical samples of a brittle material will be very variable simply because of the spread of flaw sizes.

If the flaw sizes can be reduced by careful control of the manufacturing process, the general level of strength of the material will be raised and its variability reduced. Glass fibres, for example, are manufactured by drawing molten glass very rapidly down to form fine filaments of the order of only ten microns (10 μm) in diameter. The resulting freshly formed surface is free of macroscopic defects and the fibre itself is too fine to contain any defects of the size that are found in bulk glass. Careful measurements of the strengths of freshly drawn glass fibres shows them to be up to 5 GPa by comparison with the very modest 100 MPa or so of bulk glass. The fresh fibre needs to be protected from abrasion by contact with hard objects and from corrosion by moisture in the atmosphere, both of which would result in the reintroduction of surface flaws and loss of the high strength achieved by drawing.

Similarly, the strengths of polymeric filaments, such as the polyacrylonitrile fibres used to make 'acrylic' textiles, are limited by the weak chemical bonding between the molecular chains which make up the filament and by the defects caused by the manufacturing process. By subjecting such filaments to carefully controlled stretching, oxidation and carbonisation processes, however, the polymer can be converted into fibres which are chemically almost entirely carbon, and with a crystal structure that is approximately that of graphite, in which the proportion of strong C—C bonds that lie in the direction of the fibre axis is very great. The load-bearing ability of these filaments, which are about 7 μm in diameter, is high, and the sizes of the residual defects are very small. As a result, each filament, finer than a human hair, is capable of supporting a load of 10-20 g without breaking, which is equivalent to strengths of up to 5 GPa. And so it is with the many other types of fibres, organic and inorganic, that are used in composite materials. The finer the filament that can be made from a given solid, the stronger it will be.

The theoretical strength of a given type of solid is determined by the strengths of the atomic or molecular bonds that hold the solid together. And although the practical strengths of solids are determined by the defects which they contain, it is nonetheless necessary to seek materials with the strongest chemical bonds if we are to have the best chance of exploiting the principle of composite materials construction. An appreciation of the nature and magnitude of chemical bonding leads us to scan the Periodic Table for elements in which there is a high density of covalent or mixed covalent/ionic bonds, and we naturally alight on the area containing light elements with directional bonds. The elements most likely to provide what we seek are those like carbon, boron and silicon. In the 1950s, when ideas about composite materials were first being formulated, the plain elements would not have been considered to be particularly promising as high-performance structural materials, but the invention of carbon fibres at the Royal Aircraft Establishment, Farnborough, in the U.K., and the simultaneous development of high-strength boron fibres at Texaco in the U.S.A., resulted in a very rapid expansion of research and development programs to exploit these new filamentary materials in engineering composites.

Silicon also appears unpromising as a structural material, but in combination with another 'light' element, oxygen, as the compound silica (SiO_2), it was long familiar as a high-strength filament in the form of melt-drawn glass or quartz fibres. And the list of other combinations of light-weight elements that have been successfully exploited in the search for high-performance reinforcing filaments includes alumina (Al_2O_3), silicon carbide (SiC), and silicon nitride (Si_3N_4) to name but three of the more important ceramic fibres. Polymeric fibres in which the strength and stiffness are due to the strong C—C bond are also utilized in fibre composites, and perhaps the two most important are highly drawn forms of the humble polyethylene and the more exotic aromatic polyamides (related to Nylon, but possessing much greater thermal stability). Table 1.1 gives a rough comparison of the mechanical properties of some of the more important commercial reinforcing fibres. Taking steel, in the form of piano wire, as a basis for comparison, it is clear why some of the new fibrous materials are of such interest to engineers, especially those concerned with light-weight structures. On a strength/density or stiffness/density basis, the comparison is almost startling.

Table 1.1 Typical Properties of Some Familiar Reinforcing Fibres

Material	Trade Name	Density, 10^3 kg.m^{-3}	Fibre Diameter, μm	Young's Modulus, GPa	Tensile Strength, GPa
High-carbon steel wire	e.g., piano wire	7.8	250	210	2.8
Short Fibres:					
α-Al$_2$O$_3$ (whisker crystals)		3.96	1 - 10	450	20
δ-Al$_2$O$_3$ + SiO$_2$ (discontinuous)	Saffil (U.K.)	2.1	3	280	1.5
Continuous Fibres: (inorganic)					
α-Al$_2$O$_3$	FP (U.S.A.)	3.9	20	385	1.8
Al$_2$O$_3$ + SiO$_2$ + B$_2$O$_3$ (Mullite)	Nextel 480 (U.S.A.)	3.05	11	224	2.3
Al$_2$O$_3$ + SiO$_2$	Altex (Japan)	3.3	10 - 15	210	2.0
Boron (CVD on tungsten)	VMC (Japan)	2.6	140	410	4.0
Carbon (PAN precursor)	T300 (Japan)	1.8	7	230	3.5
Carbon (PAN precursor)	T800 (Japan)	1.8	5.5	295	5.6
Carbon (pitch precursor)	Thornel P755 (U.S.A.)	2.06	10	517	2.1
SiC (+O)	Nicalon (Japan)	2.6	15	190	2.5 - 3.3
SiC (low O)	Hi-Nicalon (Japan)	2.74	14	270	2.8
SiC (+O+Ti)	Tyranno (Japan)	2.4	9	200	2.8
SiC (monofilament)	Sigma (U.K.)	3.1	100	400	3.5
Silica (E glass)		2.5	10	70	1.5 - 2.0
Silica (S or R glass)		2.6	10	90	4.6
Silica (quartz)		2.2	3 - 15	80	3.5
Continuous Fibres: (organic)					
Aromatic polyamide	Kevlar 49 (U.S.A.)	1.5	12	130	3.6
Polyethylene (UHMW)	Spectra 1000 (U.S.A.)	0.97	38	175	3.0

Since it is the reinforcing fibres that provide the means of creating composite materials of high strength and stiffness, combined with low density, it is worthwhile examining in a little more detail the nature of these fibres and their origins. Further information can be obtained on these and other fibres from the publications edited by Bunsell (1988) and Mai (1994).

1.3.1 GLASS FIBRES

Glass fibres are manufactured by drawing molten glass into very fine threads and then immediately protecting them from contact with the atmosphere or with hard surfaces in order to preserve the defect-free structure that is created by the drawing process. Glass fibres are as strong as any of the newer inorganic fibres but they lack rigidity on account of their molecular structure. The properties of glasses can be modified to a limited extent by changing the chemical composition of the glass, but the only glass used to any great extent in composite materials is ordinary borosilicate glass, known as E-glass. The largest volume usage of composite materials involves E-glass as the reinforcement. S-glass (called R-glass in France) has somewhat better properties than E-glass, including higher thermal stability, but its higher cost has limited the extent of its use. Wallenberger and Brown (1994) have recently described the properties of experimental calcium aluminate glass fibres with stiffnesses as high as 180 GPa.

1.3.2 CARBON FIBRES

By oxidising and pyrolysing a highly drawn textile fibre such as polyacrylonitrile (PAN), preventing it from shrinking in the early stages of the degradation process, and subsequently hot-stretching it, it is possible to convert it to a carbon filament with an elastic modulus that approaches the value we would predict from a consideration of the crystal structure of graphite, although the final strength is usually well below the theoretical strength of the carbon-carbon chain (Watt, 1970). The influence of strength-limiting defects is considerable, and clean-room methods of production can result in substantial increases in the tensile strength of commercial materials. Prior to sale, fibres are usually surface-treated by chemical or electrolytic oxidation methods in order to improve the quality of adhesion between the fibre and the matrix in a composite. Depending on processing conditions, a wide range of mechanical properties (controlled by structural variation) can be obtained, and fibres can therefore be chosen from this range so as to give the desired composite properties. Although the fibre is highly organized and graphite-like, the structure is not identical with that of graphite and the fibres should not, strictly speaking, be referred to by that name, although this is common in the U.S. (and in U.K. advertising jargon for sports equipment!). Recent developments in this field have led to the use of pitch as a precursor in place of textile fibres, and these newer materials have extremely high stiffnesses, compared to PAN-based fibres, but rather lower strengths (Fitzer and Heine, 1988).

Carbon fibres are inherently expensive, since they are produced from relatively costly precursor filaments. Since their introduction, however, their price has fallen from several hundred £/kg to tens of £/kg, depending upon quality, and although carbon is still an expensive reinforcement, more so than glass, for example, it is now much cheaper than most other common reinforcing fibres. The costs of the fibres listed in Table 1.1 vary from about £30 per kg for PAN-based carbon to over £5,000 per kg for boron.

1.3.3 SILICON CARBIDE

Continuous SiC monofilaments were first produced by pyrolytic decomposition of gaseous silanes onto fine filaments of carbon or tungsten. These are thick fibres, of the order of 100 μm in diameter, which continue to be of major interest to manufacturers of metal- and ceramic-matrix composites. The alternative production method, analogous to that for carbon described earlier, is based on the controlled thermal degradation of a polymer precursor (Yajima, 1980). This process typically takes a precursor such as some mixture of dimethyl dichlorosilane and diphenyl dichlorosilane and converts it in an autoclave to a polycarbosilane from which a continuous fibre is made by melt-spinning. The fibre is then converted by pyrolysis at 1300°C into a fibre consisting mainly of β-SiC of about 15 μm diameter. The characteristic commercial fibre of this type is that known as 'Nicalon' which is marketed by the Nippon Carbon Company. It has a rough surface, making for good fibre/matrix adhesion, but is somewhat reactive towards oxygen. It is well wetted by molten metals and is reasonably stable as a reinforcement for MMCs based on aluminium and copper although it lacks long-term thermal stability. There have been various attempts to improve this feature of the fibre, for example by reducing the oxygen content (Hi-Nicalon; see Toreki et al., 1994) and by adding titanium (Tyranno fibre).

1.3.4 ALUMINA AND ALUMINA/SILICA COMPOUNDS

The earliest work involving Al_2O_3 as a reinforcement for composite materials concerned the use of tiny filamentary crystals, of the order of 50 μm long and a few microns in diameter, which could be grown from the vapour phase in a highly perfect state and which, in consequence, had great strength and stiffness. These crystals, known as 'whiskers', proved to be very expensive and difficult to handle in processing operations, and their potential could not be satisfactorily realized. The DuPont Company in the U.S.A. subsequently developed an improved alumina fibre in the form of a polycrystalline yarn known as FP fibre (Dhingra, 1980). These filaments, which were manufactured by a sol-gel process, had a modulus comparable with those of boron and carbon, and a strength of the order of 1.8 GPa. Like SiC, this fibre also had a rough surface and was potentially well-suited for the reinforcement of metals such as aluminium and magnesium on account of its chemical inertness, high-temperature stability, and its ability to form a

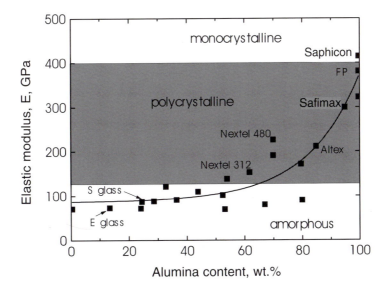

Fig. 1.2 Elastic modulus of SiO_2/Al_2O_3 fibres as a function of composition (after Wallenberger and Brown, 1994).

good bond with matrix alloys. As far as can be ascertained at present, however, FP has never been an economic prospect for its manufacturer, and alternative forms of alumina, usually containing other constituents, like Sumitomo's Altex fibre, for example, are currently preferred. Many of the oxide-based reinforcing filaments available at present are compounds of alumina and silica, and Wallenberger and Brown (1994) have given a useful graphical plot, drawn in modified form in Figure 1.2, showing how the tensile stiffness of this particular family of materials varies with composition. Some of the points in the diagram are labelled with their familiar trade names.

An alternative form of alumina-based reinforcement known as Saffil, a $\delta\text{-}Al_2O_3$ containing certain impurities which were advantageous for processing MMCs, was introduced in 1979 by ICI in response to the need for a reinforcing filament that was cheaper than those currently available. This discontinuous fibre was manufactured in the form of a continuous mat which could be used as layers or preforms, or as a shredded material, with controlled fibre aspect ratios (the ratio of length to diameter) ranging from 50:1 to 500:1. The material combined ease of handling with low cost (up to one tenth the cost of continuous reinforcing fibres), and unlike competing fibres was produced on a large-scale. Its intrinsic properties were as good as those of some competing fibres, but the fact that it was produced as a random mat meant that the potential strength of composites containing it was more limited. It has been used for the selective reinforcement of MMC components, but not on any large scale. The manufacturers have since relinquished their interest in the material.

1.3.5 ORGANIC FIBRES

Bulk polymers have elastic moduli no greater than 100 MPa, but if the polymer is spun into fibres and cold drawn so as to develop a high degree of molecular orientation, substantial improvements in both strength and rigidity can be achieved. In particular, polyolefin fibres with an exceptionally highly oriented, extended molecular chain structure can be achieved by super drawing in the solid state (Ward, 1980). In this process, both the crystalline and non-crystalline phases of the initially isotropic polymer are stretched out and aligned and there is an increase in crystal continuity. Such fibres have high strengths, and their elastic moduli are similar to those of glass and aluminium. Apart from their excellent mechanical properties such fibres have the important advantage over inorganic fibres that they are not brittle.

The other major development in organic fibres over the last three decades has been the production by DuPont of aromatic polyamide fibres, collectively known as aramids, of which the best known for many years in the composites industry has been Kevlar-49 (Yang, 1993). These polymers are based on p-oriented diamine and dibasic acid intermediates which yield liquid crystalline solutions in amide and acid solvents. These solutions contain highly-ordered extended-chain domains which are randomly-oriented in the absence of force, but which may be oriented by inducing shear forces in the liquid. Highly-oriented fibres can therefore be produced by wet-spinning these solutions, and poly(paraphenylene terephthalamide) fibres such as Kevlar and Twaron have strengths of the order of 2.6 GPa and moduli up to 130 GPa, depending on the degree of alignment of the polymer chains. Having properties intermediate between those of carbon and glass, aramids offer an extra degree of flexibility in composite design. One important characteristic of the fibre is that it is extremely difficult to cut on account of its fibrillar structure. Laser and water-jet methods are essential for trimming bare fabrics and composites containing them. Kevlar/resin composites are noted for their extremely high levels of toughness and resistance to impact damage. One disadvantage of aramids, by comparison with carbon fibres, for example, is that they are sensitive to moisture.

1.3.6 STYLES OF REINFORCEMENT

Many reinforcing fibres are marketed as wide, semi-continuous sheets of 'prepreg' consisting of single layers of fibre tows impregnated with the required matrix resin and flattened between paper carrier sheets. These are then stacked, as discussed in chapter 3, the orientations of each 'ply' being arranged in accordance with design requirements, and hot pressed to consolidate the laminate. This process is able to cope with curved surfaces, provided the degree of curvature is not too great, but there may be a possibility of local wrinkling of the fibres when prepregs are pressed into doubly curved shapes. One means of overcoming this problem is to use the reinforcement in the form of a woven cloth since textile materials can readily be 'draped' over quite complex formers. Many of the fine filamentary reinforcing fibres like glass, carbon and SiC can be readily woven into many kinds of cloths and braids, the fibres being effectively placed by the weaving process in the directions required by the designer of the final composite structure. In simple designs, this may call for nothing more elaborate than an ordinary plain weave or satin weave, with fibres running in a variety of patterns but in only two directions, say 0° and 90°, but weaving processes to produce cloth with fibres in several directions in the plane of the cloth are all readily available. Fibres of different types may also be intermingled during the weaving processes to produce mixed-fibre cloths for the manufacture of some of the 'hybrid' composites that will be discussed later.

Most of the continuous fibres that we have considered are expensive raw materials, and it is often only the fact that the overall cost of a manufactured composite product may nevertheless be lower than a competing product made from cheaper, conventional materials by more costly processes that makes a composites design solution an attractive alternative. Thus, although large quantities of glass fibres are supplied in chopped form for compounding with both thermoplastic and thermosetting matrix polymers, it may not seem economical to chop the more expensive types of reinforcement. Nevertheless, there are some advantages in using even these fibres in chopped form, provided they can be arranged in the composite in such a way as to make good use of their intrinsically high strengths and stiffnesses. Parratt and Potter (1980) described a process for producing both chopped fibres, like glass and carbon, and naturally short filaments, like whiskers or asbestos fibres, in the form of prepreg sheets with fibres that were very well aligned in either unidirectional or poly-directional patterns. These prepregs also have excellent 'drapability' and can be used to form complex shapes, as discussed by Tsuki et al. (1997). As will be seen in chapter 4, provided the short fibres are well above some critical length, which for carbon, for example, may be of the order of only a millimetre, they are able to contribute a high fraction of their intrinsic properties to the composite without the loss that occurs with woven reinforcements as a result of the out-of-plane curvature of the fibres.

1.4 THE SCOPE FOR REINFORCEMENT OF CONVENTIONAL
MATERIALS

The composite matrix is required to fulfil several functions, most of which are vital to the performance of the material. Bundles of fibres are, in themselves, of little value to an engineer, and it is only the presence of a matrix or binder that enables us to make use of them. The rôles of the matrix in fibre-reinforced and particulate composites are quite different. The binder for a particulate aggregate simply serves to retain the composite mass in a solid form, but the matrix in a fibre composite performs a variety of other functions which must be appreciated if we are to understand the true composite action which determines the mechanical behaviour of a reinforced material. We shall therefore consider these functions in some detail.

1.4.1 FUNCTIONS OF THE MATRIX

- The matrix binds the fibres together, holding them aligned in the important stressed directions. Loads applied to the composite are then transferred into the fibres, the principal load-bearing component, through the matrix, enabling the composite to withstand compression, flexural and shear forces as well as tensile loads. The ability of composites reinforced with short fibres to support loads of any kind is dependent on the presence of the matrix as the load-transfer medium, and the efficiency of this load transfer is directly related to the quality of the fibre/matrix bond.
- The matrix must also isolate the fibres from each other so that they can act as separate entities. Many reinforcing fibres are brittle solids with highly variable strengths. When such materials are used in the form of fine fibres, not only are the fibres stronger than the monolithic form of the same solid, but there is the additional benefit that the fibre aggregate does not fail catastrophically. Moreover, the fibre bundle strength is less variable than that of a monolithic rod of equivalent load-bearing ability. But these advantages of the fibre aggregate can only be realized if the matrix separates the fibres from each other so that cracks are unable to pass unimpeded through sequences of fibres in contact, which would result in completely brittle composites.
- The matrix should protect the reinforcing filaments from mechanical damage (*e.g.* abrasion) and from environmental attack. Since many of the resins which are used as matrices for glass fibres permit diffusion of water, this function is often not fulfilled in many GRP materials and the environmental damage that results is aggravated by stress. In cement the alkaline nature of the matrix itself is damaging to ordinary glass fibres and alkali-resistant glasses containing zirconium have been developed (Proctor & Yale, 1980) in an effort to counter this. For composites like MMCs or CMCs operating at elevated temperature, the matrix would need to protect the fibres from oxidative attack.
- A ductile matrix will provide a means of slowing down or stopping cracks that might have originated at broken fibres: conversely, a brittle matrix may depend upon the fibres to act as matrix crack stoppers.
- Through the quality of its 'grip' on the fibres (the interfacial bond strength), the matrix can also be an important means of increasing the toughness of the composite.
- By comparison with the common reinforcing filaments most matrix materials are weak and flexible and their strengths and moduli are often neglected in calculating composite properties. But metals are structural materials in their own right and in MMCs their inherent shear stiffness and compressional rigidity are important in determining the behaviour of the composite in shear and compression.

The potential for reinforcing any given material will depend to some extent on its ability to carry out some or all of these matrix functions, but there are often other considerations. We consider now the likely qualities of various classes of matrix materials.

1.4.2 METALS

Metals owe their versatility as engineering materials to the fact that they can be plastically deformed and can be strengthened by a variety of methods which, by and large, act by inhibiting the motion of dislocations. As a consequence of the non-directional nature of the metallic bond, dislocations are highly mobile in pure metals which are therefore very soft. But by controlling the number and distribution of dislocations the materials scientist can adjust the properties of a metal or alloy system to suit specific requirements. There are limitations, however. Increases in strength can usually be achieved only at the expense of the capacity for plastic deformation, with the consequence that the strongest alloys often lack tolerance of defects or other stress-concentrators. Since brittleness is a drawback no designer dares underestimate, this leads to the use of large safety factors which, in turn, means that the full potential of high-strength alloys is often not utilisable in practice.

Many solid-state hardening methods used in alloys involve producing a material in a metastable state which may subsequently revert to a more stable but unstrengthened condition if sufficient thermal energy is provided. Alloys strengthened by precipitation hardening, such as the strong aluminium alloys, alloys depending on phase transformations of the martensitic type, such as steels, and heavily cold-worked metals which depend simply on the presence of a high dislocation density, as in piano wires and lamp filaments, will all soften at elevated temperatures. The strongest aluminium alloys lose their strength at temperatures little over 150°C, for example.

Many conventional metallic materials are relatively heavy. For land-based engineering projects this may be of no consequence, but economic arguments relating to pay-loads (in civil aircraft) and tactical arguments relating to manoeuvrability (in military aircraft) have always been a powerful incentive for the use of low-density materials in aerospace engineering, and in these energy-conscious times the economic incentive for lightening automobiles has considerably influenced the motor car designer.

For structural applications involving compression or flexural rather than tensile loads, the relevant structural stiffness index is not simply Young's modulus, E, or the modulus/density ratio, E/ρ, but may instead be either of the ratios E/ρ^2 or E/ρ^3 (Gordon, 1968 and 1978, Ashby and Jones, 1980). For aerospace applications modifications to materials that result in lowered density may therefore be more profitable than attempts simply to improve their strength or stiffness.

These features show why it is worthwhile to attempt to use light, strong, stable fibres to reinforce some of the lighter engineering metals and alloys.

1.4.3 POLYMERIC MATERIALS

Few polymers are thermally stable by comparison with metals or ceramics and even the most stable, like the polyimides, or poly(ether ether ketone) (known as PEEK), are degraded by exposure to temperatures above about 300°C, as illustrated by some of the

Table 1.2 Thermal Stability of Some Matrix Polymers

Type and Polymer	Symbol	Crystallinity	Glass Transition Temp., T_g, °C	Max Use Temp., °C
Thermosets:				
Polyester	PE	No	80 - 100	50
Epoxy	Ep	No	120 - 180	150
Phenolic	Ph	No	130 - 180	200
Bismaleimide	BMI	No	180 - 200	220
Polyimide	PI	No	300 - 330	280
Thermoplasts:				
Polyamide (Nylon)	PA	Yes	80	125
Poly(phenylene sulphide)	PPS	Yes	100	260
Poly(ether ether ketone	PEEK	Yes	143	250
Polycarbonate	PC	No	145	125
Polysulphone	PS	No	190	150
Poly(ether imide)	PEI	No	210	170
Poly(ether sulphone)	PES	No	230	180
Thermoplastic polyimide	TPI	No	270	240

data in Table 1.2. There is nothing that reinforcement can do to combat chemical degradation, but the associated fall in strength and increase in time-dependent (creep or visco-elastic) deformation, a feature common to all polymers, though less serious in cross-linked resin systems than in thermoplastics, can be delayed by fibre reinforcement.

A more serious problem in polymers is their very low mechanical strength and stiffness in bulk form: and, like metals, the weakest plastics tend to be ductile but the strongest tend to be brittle, although there are exceptions.

Polymers are traditionally insulators and in their application as such strength is usually a secondary consideration. The electrical conductivity of plastics reinforced with carbon fibres is of importance in many aeronautical applications, however, where protection of avionics systems from external electrical activity (*e.g.*, lightning strike) is of importance. Most polymers are already low-density materials, and the addition of fibres confers no density advantage.

It is in the reinforcing of polymers that most developments have been made so far, and it is likely that there is still scope for improvement. There is a thriving, international reinforced plastics Industry, for which both the science and technology are highly advanced. And although the level of awareness of the merits of reinforced plastics on the part of designers in general engineering seems to be low, the same is not true of the aerospace industry for which the potential benefits of very high strength and stiffness, combined with low density, are easily recognized.

1.4.4 GLASSES, CERAMICS AND CEMENT

Glasses have high chemical stability, but many lose their mechanical strength at relatively low temperatures as they pass through the glass transition. Special glasses have been developed with high transition temperatures (T_g), however, and many are at least as resistant as some of the less stable steels. The principal problem with glassy materials is that they are always brittle and their measured strengths are very variable at ordinary temperatures. They are not able to relieve stress concentrations at crack tips by plastic deformation and they are not, therefore, fail-safe. A constant disadvantage, aggravated by low thermal conductivity, is that glass usually has poor thermal shock resistance unless, like borosilicate glass, it also has a low thermal expansion coefficient. Many of these difficulties can be overcome by reinforcement with fibres like carbon, and with the added bonus of a saving in weight (Phillips *et al.*, 1972).

Most ceramics suffer from the same defects as glasses in that, although they are potentially high-strength solids, they are also brittle and notch-sensitive. Most ceramics retain their strength to very high temperatures, however, unlike glasses, and several have good resistance to thermal shock. Improving toughness and reducing notch sensitivity are the main reasons for attempting to reinforce ceramics since their stiffnesses are not very different from those of the best reinforcing fibres. For high-temperature stability we might expect that the best possibility for reinforced ceramics would be an equilibrium system in which whisker-like crystals of a reinforcing phase were grown within the polycrystalline matrix of a chemically-related ceramic. Most recent work has followed the more established route, however, in attempting to reinforce a variety of glasses (such as borosilicate) and glass-ceramics (such as calcium aluminosilicate (CAS) and lithium aluminosilicate (LAS)) with fibres like carbon, silicon carbide and alumina (Prewo and Brennan, 1980; Phillips *et al.*, 1972), the fibres being impregnated by a slurry of fine glass powder and subsequently hot-pressed (Sambell *et al.*, 1970, 1972). Critical control of manufacturing conditions is needed in order to produce appropriate interfacial conditions for the optimum combination of strength and toughness. Once control of the glass-making technique is established, development of the process to include a glass-ceramic transformation is the logical next step in the production of a continuous-fibre composite. The advantage of the glass-ceramic route is that relatively modest processing temperatures are involved. For high-temperature exposure of such composites, however, or for work involving the production of CMCs by conventional pressing methods, there remains the problem of compatibility. At the time of writing, the concept of all-oxide CMCs is being strongly argued, although we already have comparative examples of stable pairings in the French SiC/SiC materials (Frety and Boussuge, 1990). After a period of intense activity in fibre-reinforced glass and glass-ceramic research with an eye on the gas-turbine industry, it has still not been possible for engineers to come to terms with the concept, already explored in the 1970s (Aveston *et al.*, 1971), that in these materials the matrix will crack at worryingly low stress levels (Harris *et al.*, 1992) leaving the reinforcing filaments exposed to any hostile environment.

Concrete is a particulate ceramic composite in which aggregate particles of graded sizes are embedded in a glassy or microcrystalline silicate matrix. Like most ceramics and glasses it is brittle and exhibits a very low tensile failure strain which engineers cope with by using it in compression or with macroscopic reinforcing bars to carry tensile loads. In design, the tensile load-bearing ability of the concrete is ignored. Reinforcement of concrete in the composite sense has been widely studied, with attempts to produce stronger, stiffer and tougher structural materials by adding fibres of asbestos, glass, steel, polymeric materials, and carbon. Improved properties can be obtained, although usually only at the expense of a severe economic penalty, given the fact that concrete is the cheapest structural material available. The addition of even a few percent of the cheapest reinforcing fibres may still raise the price sufficiently to lead the engineer to use a thicker section of plain concrete in preference to a more expensive thinner section of reinforced concrete. Such small additions do little to increase the cracking stress or elastic modulus of cement, but by introducing complex cracking and failure modes they can substantially improve the work of fracture. The scope for improvement of bulk poured concrete is probably negligible: it is much more likely that the present trend towards changes in design philosophy and the use of speciality products in fibre-reinforced concrete will continue. High volume fractions of well-aligned fibres can be obtained by vacuum dewatering and filament-winding and the products of these processes can be used in much thinner sections than is normal with concrete. Traditional attitudes and design procedures are of course no longer applicable for such materials.

1.4.5 CARBON

Carbon is a unique material with many attractive engineering qualities. It can be prepared in a variety of forms — conventional hot-pressed carbons and graphites, densified impermeable graphite, pyrolitic graphite, and vitreous carbon — with a wide range of engineering properties. It is valuable for its lubricating properties, its electrical properties, its nuclear characteristics and, in the pyrolytic and vitreous forms, for high strength and resistance to oxidative and chemical attack (Cahn and Harris, 1969). The opportunity to improve the mechanical properties of such an important material and reduce its brittleness somewhat has been the driving force for the development of an invaluable material, carbon-fibre-reinforced carbon (Savage, 1993) that has been used for rocket nozzles, aerospace components (including the ablative shields of space vehicles), and surgical implants.

1.5 GENERAL BIBLIOGRAPHY AND REFERENCES

M.F. ASHBY and D.R.H. JONES: *Engineering Materials*, Pergamon, Oxford, 1980.

B. ATZORI, M. QUARESIMIN and G. TRATTENERO: *Proceedings of 2ⁿᵈ International Seminar on Experimental Techniques and Design in Composite Materials*, M.S. Found, ed., Sheffield, Sheffield Academic Press, Sheffield, 1994, 193-211.

J. AVESTON, G.A. COOPER and A. KELLY: *Proceedings of NPL Conference on The Properties of Fibre Composites*, IPC Science and Technology Press Ltd, Guildford, 1971, 15-26.

R.W. CAHN and B. HARRIS: *Nature*, **221**, 1969, 132-141.

A.K. DHINGRA: '*New Fibres and their Composites*,' W. Watt, B. Harris and A.C. Ham, eds., Royal Society of London, *Phil. Trans. Royal Society, London*, **A294**, 1980, 411-417.

E. FITZER and M. HEINE: *Fibre Reinforcements for Composite Materials*, Composite Materials Series, A.R. Bunsell, ed., Elsevier, Amsterdam, **2**, 1988, 73-148.

N. Frety and M. Boussuge: *Ceramic-Matrix Composites: Components, Preparation, Microstructure and Properties*, R. Naslain and B. Harris, eds., Special Issue of *Composites Science and Technology*, **37**, 1990, 177-190.

J.E. Gordon: *Structures,* Penguin, Harmondsworth, Middlesex, 1968.

J.E. Gordon: *The New Science of Strong Materials*, Penguin, Harmondsworth, Middlesex, 1978.

A.A. Griffith: *Phil. Trans. Royal Society, London*, **A221**, 1920, 163-198.

B. Harris, F.A. Habib and R.G. Cooke: *Proceedings of Royal Society*, London, **A437**, 1992, 109-131.

N.J. Parratt and K.D. Potter: *Advances in Composite Materials*, Proc. ICCM3, A.R. Bunsell, C. Bathias, A. Martrenchar, D. Menkes, and G. Verchery, eds., Pergamon, Oxford, **1**, 1980, 313-326.

D.C. Phillips, R. Sambell and D.H. Bowen: *Journal of Materials Science*, **7**, 1972, 1454-1464,

K.M. Prewo and J.J. Brennan: *Journal of Materials Science*, **15**, 1980, 463-468.

B. Proctor and B. Yale: '*New Fibres and their Composites*,' W. Watt, B. Harris and A.C. Ham, eds., Royal Society of London, *Phil. Trans. Royal Society, London*, **A294**, 1980, 427-436.

R. Sambell: *Composites*, **1**, 1970, 276-285.

R. Sambell, A. Briggs, D.C. Phillips and D.H. Bowen: *Journal of Materials Science*, **7**, 1972, 676-681.

G. Savage: *Carbon-Carbon Composites*, Chapman and Hall, London, 1993.

W. Toreki, C.D. Batich, M.D. Sacks, M. Saleem, G.J. Choi and A. Morrone: *Composites Science and Technology*, **51**, 1997, 145-159.

N. Tsuji, G.S. Springer and I. Hegedus: *Journal of Composite Materials*, **31**, 1997, 428-527.

F.T. Wallenberger and S.D. Brown: *Composites Science and Technology*, **51**, 1994, 243-264.

I.M. Ward: *'New Fibres and their Composites,'* W. Watt, B. Harris and A.C. Ham, eds., Royal Society of London, *Phil. Trans. Royal Society, London*, **A294**, 1980, 473-482.

W. Watt: Discussion Meeting on *Strong Fibrous Solids*, *Proceedings of Royal Society*, London, **A319**, 1970, 5-15.

S. Yajima: *'New Fibres and their Composites,'* W. Watt, B. Harris and A.C. Ham, eds., Royal Society of London, *Phil. Trans. Royal Society, London*, **A294**, 1980, 419-426.

H.H. Yang: *Kevlar Aramid Fibre*, John Wiley, Chichester, 1993.

1.5.1 GENERAL BIBLIOGRAPHY

B.D. Agarwal and L.J. Broutman: *Analysis and Performance of Fibre Composites*, Second Edition, John Wiley, New York, 1990.

B.T. Astrom: *Manufacturing of Polymer Composites*, Chapman & Hall, London, 1997.

A.R. Bunsell, ed.: *Fibre Reinforcements for Composite Materials*, *Composite Materials Series*, Elsevier, Amsterdam, **2**, 1988.

K.K. Chawla: *Ceramic Matrix Composites*, Chapman & Hall, London, 1993.

T.W. Chou: *Microstructural Design of Fibre Composites*, Cambridge University Press, 1992.

T.W. Clyne and P.J. Withers: *An Introduction to Metal-Matrix Composites*, Cambridge University Press, 1993.

G. Eckold: *Design and Manufacture of Composites*, Woodhead, Abington, U.K, 1994.

R.K. Everett and R.J. Arsenault, eds.: *Metal-Matrix Composites, Processing and Interfaces, Mechanisms and Properties*, Academic Press, Boston, **1** and **2**, 1991.

D. Hull: *An Introduction to Composite Materials*, Cambridge University Press, 1981.

R.M. Jones: *Mechanics of Composite Materials*, Scripta Books, Washington DC, 1975.

A. Kelly and N.H. McMillan: *Strong Solids,* 3rd Edition, Clarendon Press, Oxford, 1986.

Y.W. Mai, ed.: *Advances in Inorganic Fibre Technology*, Special Issue of *Composites Science and Technology*, **51**, 1994, 123-296.

F.L. Matthews and R.D. Rawlings: *Composite Materials: Engineering and Science*, Chapman & Hall, London, 1994.

R. Naslain and B. Harris, eds.: *Ceramic-Matrix Composites*, Elsevier Science Publishers, London and New York, 1990.

M.R. Piggott: *Load-Bearing Fibre Composites*, Pergamon Press, Oxford, 1980.

P.C. Powell: *Engineering with Fibre-Polymer Laminates*, Chapman and Hall, London, 1994.

R. Warren, ed.: *Ceramic Matrix Composites*, Blackie, Glasgow, 1992.

W. Watt, B. Harris and A.C. Ham, eds.: *'New Fibres and their Composites,'* Phil. *Trans. Royal Society, London,* **A294**, 1980.

1.5.2 SOURCES OF DATA

N.L. Hancox and R.M. Meyer: *Design Data for Reinforced Plastics*, Chapman and Hall, London, 1994.

A. Kelly, ed.: *Concise Encyclopedia of Composite Materials*, Pergamon, Oxford, 1989.

G. Lubin, ed.: *Handbook of Composites*, Van Nostrand Reinhold, New York, 1982.

2. Making Composite Materials

2.1 THE COMBINING OF MATERIALS

In considering the scope for reinforcing conventional materials, we have already examined the functions of the matrix in a composite and the general nature of available reinforcing filaments. The problem facing the manufacturer of composites is to develop suitable methods for combining the matrix and the reinforcement so as to obtain the required shape of component with properties appropriate to the design requirements. In the early days of the subject, attention was focused on first producing a 'piece' of the composite — a sheet, or rod, or billet, with the fibres arranged in one or more directions — and then using that intermediate product to construct a finished 'component' in much the same way as steel castings were bolted together to build a car engine or steel plates were welded together to make a pressure vessel. Early use of composites in practical applications was on the basis of piecemeal substitution of the composite for a metallic part of identical shape without any consideration of the special nature of the composite, notably its anisotropic mechanical properties. The idea of bolting GRP 'top-hat' stiffeners to the underside of GRP ship decking seems ridiculous now, but the procedure was certainly investigated in the 1970s.

It is essentially the reinforcement 'architecture' that determines the load-bearing characteristics of a fibre composite, and the beauty of modern composites design and manufacturing procedures is that in many cases the composite material and the final component can be created and finished in a single operation. A typical example is in the process of resin-transfer moulding (RTM) which we shall discuss later. A fibre preform — a skeleton of the approximate shape of the finished component with the fibres arranged in the directions determined by the design requirements — is placed into a closed mould, and pre-catalysed resin is injected into the mould, which is often pre-heated. The resin cures in a short time, and the finished component, requiring only final cleaning up, is removed from the mould, the composite material and the finished article being formed in a single operation. The rate-controlling step in this near-net-shape manufacturing process is probably the manufacture of the fibre preform. In the filament-winding of hollow containers, fibres wetted by catalysed resin are wrapped onto a mandrel and, again, as the resin cures the component and the material are formed simultaneously. In this case, the locations of the fibres may have been determined by a computer programme and the same computer code can then be used to control an automatic filament-winding machine — a good example of CAD/CAM.

The range of processes used for manufacturing composites is now wide. Almost all of the basic metal, polymer and ceramic forming processes have been adapted, with greater or less success, and for each class of composites — fibre-reinforced plastics (FRPs), MMCs and CMCs — there is a range of preferred techniques, depending on the required product. In many cases, there has been technology transfer from one field to another, so that similar processes, suitably modified, may be used for different types of composite. The RTM process for FRPs is clearly related to the liquid-metal infiltration process which is used for MMCs and to the sol-gel process for CMCs. Almost all successful processes involve liquid-phase reactions — molten metal, liquid glass, ceramic sol — although many processes for ceramic composites are gas-phase processes, as in the chemical vapour deposition (CVD) processes for preparing carbon/carbon composites and some of the monofilament-reinforced ceramics like the SiC/SiC composites made by SEP in France.

What is feasible for the manufacturer is often limited by the available forms of the reinforcing filaments. The thick mono-filamentary reinforcements like boron and SiC are produced in continuous single filaments and because of their diameters they cannot easily be handled except as regular arrays of continuous fibres. But the majority of the fine, continuous reinforcements are manufactured as continuous tows, often containing thousands of individual fibres, and these can be manipulated by well-established textile handling techniques. They can be chopped into short lengths for mechanical blending with matrix resins and fillers, for example, or they can be woven into many kinds of cloths and braids, the fibres being effectively placed by the weaving process in the directions required by the designer. In simple designs, this may call for nothing more elaborate than an ordinary plain weave or satin weave, with fibres running in a variety of patterns but in only two directions, say 0° and 90°. But weaving processes to produce cloths with fibres in several directions in the plane of the cloth — *e.g.,* triaxial weaves with three groups of fibres at 120° to each other, and so-called 4D and 5D fabrics — are all readily available. The study of textile-based composites is currently a much researched topic (see, for example, Masters and Ko, 1996). Fibres of different types may also be intermingled during the weaving processes to produce mixed-fibre cloths for the manufacture of some of the 'hybrid' composites that will be discussed later.

2.2 THE INTERFACE

In materials where the mechanical response depends on loads being shared between two or more separate constituents or 'phases' and where paths for the propagation of cracks will be affected by the different mechanical properties of the components, the manner in which these adhere to each other becomes an important consideration. When we come to discuss the strength and toughness of composites, we shall see that the strength of the interfacial bond between the fibres and the matrix may make all the difference between a satisfactory material and an inadequate one. The problem is that the ideal situation for load sharing — a perfect bond between the fibre and the matrix, forcing them to deform as one, which is the first assumption that we shall make in attempting to calculate the elastic modulus of a fibre composite — often results in a lower tensile strength than expected because of the way cracks run in such a composite. In the early days of carbon-fibre technology, it was quickly established that the greater the degree of order in the graphitized fibre structure the poorer was the adhesive bond between the new high-modulus fibres and the existing polyester and epoxy resins. As a consequence, the higher the fibre stiffness the poorer the extent to which the composite was able to resist shear forces acting in the plane of a laminate *i.e.*, the lower the inter-laminar shear strength (ILSS). Hasty attempts were made to improve the ILSS by a variety of surface treatments, including coating with polymers, high-temperature and electrolytic oxidation, and vapour deposition of other compounds onto the fibre surface. Electrolytic treatment became the established method, but the need for control over the extent of the treatment was not appreciated and some early fibres were supplied by manufacturers with heavily oxidised surfaces that were very well wetted by the resins. This facilitated manufacture, but resulted in composites that were disastrously brittle, causing a serious set-back for aeronautical use of carbon fibres. In early work on the use of carbon fibres as reinforcements for aluminium alloys, the choice of an aluminium/silicon alloy as matrix (on account of the well-known fluidity of these alloys) led to a rapid reaction between the silicon and the carbon fibre, with the result that a very strong bond was created but the fibre properties were ruined. This kind of interaction between the fibre surface characteristics and the fibre/matrix bond, and frequently also the effect of processing, has probably exercised the minds of composites workers more than most other aspects of the technology.

It would be wrong, then, to consider the fibres and the matrix as the only important constituents of a composite, even of a simple two-phased composite where no other actual component is intentionally present. The interface between two solids, especially when thermal or chemical processes have been involved in putting them together, is rarely a simple boundary between two materials of quite different character. Over a certain range, which will vary in dimensions depending upon the chemical and physical natures of the fibres and the matrix, there will be some modification of either chemical or physical characteristics, or both, resulting in a region, perhaps only of molecular dimensions, perhaps many microns, which has properties quite different from those of either of the two major components. Sometimes this occurs fortuitously, as when the two components in a ceramic-matrix composite, which has to be fired at elevated temperature to achieve compaction, react chemically together, or it may be deliberately engineered, as when, in the case of GRP materials, we apply a size coating to the fibres after drawing in order to make them adhere more readily to the resin matrix. Whatever the origin of this region, which we often refer to as an 'interphase', it is important in that its properties determine the manner in which stresses are transferred from matrix to fibres and, in consequence, many of the chemical, physical and mechanical properties of the composite itself. Control of this interphase, or interface region, is a major concern to developers and suppliers of commercial composite materials.

2.3 MANUFACTURING PROCESSES

This book is not about manufacturing, yet it is necessary to know something of the nature of the main processing methods used in order to appreciate how what happens during processing may affect the properties of the product. In this chapter, therefore, we shall take only a very general look at the major processes currently being used to produce composite artefacts. There will be no attempt to discuss the basic engineering of the individual processes, and further details of manufacturing processes may be obtained from publications listed in the *General Bibliography* for Chapter 1.

2.3.1 POLYMER-MATRIX COMPOSITES

The wide range of processes used to produce reinforced plastics is partly new, and partly derived from established methods of processing ordinary polymeric materials. The manner of combining fibres and matrix into a composite material depends very much on the particular combination in question and on the scale and geometry of the structure to be manufactured.

The commoner varieties of thermoplastic-based materials, like glass-filled Nylon and glass-filled polyacetal, are made largely by the injection moulding of granules of material in which the chopped fibres and matrix have been pre-compounded. The principal problem in such cases is that the flow of material during moulding may be non-uniform, especially in moulds of complex geometry. There may be regions in which the fibres are highly oriented and others where the degree of orientation is almost nil, as well as variations in fibre content from place to place. Preferential fibre orientation is often useful, but only if it can be adequately controlled. Some very sophisticated recent developments in multiple live-feed injection moulding by Bevis and co-workers have enabled the inventors to produce short-fibre-filled thermoplastics with very well aligned fibre arrays in different parts of a moulding with the result that the properties of these materials can compete favourably with those of many continuous-fibre composites (Allan & Bevis, 1987).

Continuous-fibre thermoset composites are produced by quite different methods. Cylindrically symmetric structures such as pressure vessels, tanks, rocket-motor casings, centrifuge cylinders, and a variety of pipes, can be made by winding fibres or tapes soaked with pre-catalysed resin onto expendable or removable mandrels. Winding patterns may be simple or complex and may be accurately calculated to resist a prescribed stress system. (*e.g.,* a given ratio of hoop stress to longitudinal stress) in service. Variations in winding pattern or in the combination of stresses subsequently applied to the structure will clearly change the extent to which the fibres are loaded purely in tension or to which shear stresses are introduced between separate layers of winding. After the resin has hardened the mandrel is removed and, if size permits, the product may be post-cured at an elevated temperature. Extremely large vessels can be made by this method, but these must usually be left to cure at ambient temperature. Since the winding procedure can be closely controlled, a high degree of uniformity is possible in the fibre distribution of filament-wound structures, but planes of weakness sometimes occur between winding layers, especially if resin-rich pockets are allowed to form. An important feature of this process is that after a structure has been designed according to well-defined principles based on stress analysis of composite materials, the design software can be extended to provide instructions to a computer or numerically-controlled winding machine to manufacture the vessel to close tolerances (Eckold, 1994; Jones *et al.*, 1996).

Large panels and relatively complex open structural shapes are easily constructed by hot-pressing sheets of pre-impregnated fibres or cloth between flat or shaped platens, or by vacuum autoclaving with the aid of atmospheric pressure to consolidate a stack of 'prepreg' sheets against a heated, shaped die. Woven reinforcements are particularly useful for constructing shapes with double curvature since they can readily be 'draped' over quite complex formers, unlike unidirectional prepregs which may wrinkle because of their anisotropy. Pressing must be carried out carefully to produce intimate association of the fibres in different layers, with expulsion of trapped air (and of excess resin unless the prepreg is of the 'zero-bleed' variety containing no excess resin), and the time/temperature cycle must be controlled so as to ensure final curing of the resin only when these conditions have been met. Higher fibre contents are obtainable with non-woven laminates than with woven-cloth composites, and the characteristics of the two types of material are quite different. The orientation of the fibres in the separate laminations is varied to suit the specific load-bearing characteristics required of the laminate or moulding. Although this is a common process and a great deal of experience has been accumulated in its use, control of the autoclave cycle is sometimes difficult to establish since there is a complex interaction between the kinetics of the resin cure, the resin curing exotherm, and the externally applied pressure and temperature. Research is currently in progress to develop the use of *in situ* fibre-optic sensors for closed-loop control of the autoclave (Roberts and Davidson, 1993, Crosby *et al.*, 1995).

For complex aeronautical and similar structures, the relationship between computer-based design procedures and robotic or computer-controlled manufacturing processes has considerably improved the production engineering of composites. In particular, the construction of the stacked laminates ready for processing or autoclaving may be carried out by a computer-controlled tape-laying machine which can lay up components rapidly, accurately, and with minimum waste of raw materials. Such a system permits of greater design flexibility than is possible in manual production and can consistently produce parts of higher strength. Most other operations, including prepreg trimming, transfer between operations, autoclaving, component trimming, and assembly can also be carried out with a high level of software control (see, for example, Hadcock and Huber, 1982).

In the pultrusion process, tightly packed bundles of fibres, soaked in catalysed resin, are pulled through a shaped die to form highly aligned, semi-continuous sections of simple or complex geometry. Curing of the resin in the die may be achieved either by heating the die itself or by the use of dielectric heating. Solid and hollow sections may be produced by this process, and because of the high fibre content (70% by volume is achievable) and the high degree of fibre alignment resulting from the tensile force used to pull the fibre bundle through the die, extremely high mechanical properties can be obtained. In more recent developments of the process, off-axis fibres may also be introduced into the structure. A detailed description of this process has been given by Spencer (1983).

Another familiar process for producing high-quality mouldings of complex shape is resin-transfer moulding or RTM. Pre-catalysed resin is pumped under pressure into a fibre preform which is contained in a closed (and often heated) mould. The preform may be made of any kind of reinforcement, but usually consists of woven cloths or fibre mats, arranged as required by the design. Thick components containing foamed polymer cores can also be produced in this way. The ease with which this can be done successfully (Harper, 1985) belies the complexity of the process which involves non-Newtonian flow of a resin (which may be curing at a significant rate during short-cycle operations) through a preform in which the fibre distribution results in flow paths of markedly different geometries in the in-plane and through-thickness directions. In high-speed versions of the process favoured by the automobile industry the cycle time may be reduced to a minute or less, and there are problems in obtaining complete wetting and uniform cure throughout the heated mould cavity. The mould cavity is often evacuated to assist wetting out. This is an important, high-volume, automatable process which, when used in association with automated fibre-preform manufacture, offers considerable advantages in the production of low-cost engineering components from GRP (Rudd *et al.*, 1990; Turner *et al.*, 1995). A wide range of low-viscosity resins — polyesters, vinyl esters, epoxies, urethane-acrylates, *etc.* — are now commercially available for use in this process.

A commercially important class of thermoset moulding compounds, the polyester dough-moulding compounds or DMC, consist of chopped glass fibres blended into a dough with a pre-catalysed resin and a substantial quantity of inert filler, like chalk. These compounds (also referred to as bulk moulding compounds or BMC) may be manufactured to final shape by injection moulding or, more commonly, by the simpler transfer-moulding process — squeezing to shape in hot dies. As in the case of the injection-moulded reinforced thermoplastics mentioned earlier, problems resulting from non-uniformity and fibre-orientation effects can also occur during the moulding of these materials.

Lower quality composites reinforced with random-fibre chopped-strand mat (CSM) or continuous-filament mat (CFM) reinforcements may be press-laminated like the higher-performance CFRP and GRP laminates but are more frequently laid up by hand methods, especially for irregularly shaped structures. Large structures, tanks, boats and pipes are often made in this way. The usual procedure is to coat a shaped former or mandrel with resin, allow the resin to gel and then build up the required shape and thickness by rolling on layer upon layer of resin-soaked cloth or mat. A final gel-coat is then applied and the finished structure may be post-cured. The distribution of fibres in such structures will usually be uneven. Resin pockets and voids are invariably present in these materials.

2.3.2 METAL-MATRIX COMPOSITES

The basic attributes of metals reinforced with hard ceramic particles or fibres are improved strength and stiffness, improved creep and fatigue resistance, and increased hardness, wear and abrasion resistance, combined with the possibility of higher operating temperatures than for the unreinforced metal (or competing reinforced plastics). These properties offer potential for exploitation in a range of pump and engine applications, including compressor bodies, vanes and rotors, piston sleeves and inserts, connecting rods, and so forth. Components of this type are still under development, but few are in commercial production (Feest, 1986) apart from one or two limited applications in Japanese automobiles. Thus, although a wide range of manufacturing methods has been used over the past twenty years on a laboratory or development scale, at this stage relatively little can be said about large-scale production processes for MMCs.

Some of the techniques that have been described in detail are

- Unidirectional solidification of eutectics or other constitutionally-appropriate alloys,
- Liquid-metal infiltration, often under vacuum, of pre-packed fibre bundles or other preforms,
- Liquid-phase infiltration during hot pressing of powder compacts containing whiskers or fibre bundles,
- Hot pressing of compacts consisting of matrix alloy sheets wrapped or interleaved with arrays of reinforcing wires,
- Hot pressing or drawing of wires pre-coated with the matrix alloy (*e.g.,* by electroplating, plasma spraying or chemical vapour deposition (CVD), and
- Co-extrusion of prepared composite billets.

The first of these processes relates to the manufacture of the so-called *in situ* composites which began with model systems, like Cu/Cr and Al/Al$_2$Cu, in most of which thermodynamic constraints impose serious restrictions on the range of materials that can be treated and in which the fibre contents are small. Better results were subsequently obtained with more complex systems such as the monovariant ternary eutectic alloy (Co, Cr)-(Cr, Co)$_7$C$_3$ which, in the directionally-solidified form, contains about 30 vol.% of the fibrous carbide phase in a Co/Cr solid solution (Thompson *et al.,*1970). Further developments were based on complex superalloy systems in which the composition of the phases can vary widely without inhibiting the formation of an aligned two-phased structure during processing. Typical of these materials are the carbide-reinforced systems exemplified by the NITAC (TaC in nickel) and COTAC (TaC in cobalt) alloys developed at ONERA in France (Bibring *et al.*, 1971). From the thermodynamic point of view, composites of this kind possess the clear advantage that they are intrinsically stable under operating conditions involving elevated temperature. There seems to be little current enthusiasm for this process.

With both liquid infiltration and solid-state bonding techniques, one of the main problems is that of ensuring the optimum degree of chemical contact (or wetting) between the fibres and the matrix. In many systems with practical potential, wetting is inhibited by oxide films or surface chemistry features of the reinforcing phase, and measures to

improve the interfacial characteristics often result in an undesirable degree of fibre/
matrix interaction and damage. Similar consequences of inherent thermodynamic
incompatibility followed when many of the early MMCs were exposed to elevated
temperatures. As a consequence of these wetting and stability problems, a considerable
proportion of the early research and development in this field was concerned with fibre
treatments both to improve adhesion and to provide diffusion barriers — two clearly
conflicting requirements — and the establishment of conditions needed to obtain optimum
interfacial bonding.

Short-fibre-reinforced MMCs are commonly made by liquid-metal infiltration (LMI)
or squeeze casting into fibre preforms, a net-shape forming technique that has much to
recommend it for localised or general reinforcement of components such as automotive
pistons, connecting rods, *etc.*, Powder processing is much less satisfactory because of
the possibility of fibre damage. A useful practical development in the liquid-state
processing of metal-matrix composites was the introduction in 1984 of a low-cost short-
fibre reinforcement consisting of δ-alumina containing a few percent of silica, called
Saffil by the manufacturers, ICI plc. These fibres, said to be some 45% stronger than
other available continuous α-alumina fibres, are easily wetted by a range of aluminium
alloys and can be fabricated into practical engineering composite components by pressure
die-casting or squeeze-casting an alloy matrix into preformed fibre arrays located in the
mould. A number of announcements of practical applications of the material followed,
but ICI's interest in composite materials proved as transitory as those of almost every
other major U.K. company which had a composites programme in the 1970s and 1980s,
and the material is no longer obtainable.

Products requiring continuous reinforcement, such as boron/aluminium composite
panels and sections that have been under consideration for aerospace applications (Renton,
1977) are preferably produced by solid-state processes such as diffusion bonding or hot
pressing in which critical pressure/time/temperature cycles are needed to obtain optimum
quality. Structures can be fabricated relatively easily (by comparison with thermoset
resin matrix composites) by brazing, welding, diffusion bonding or mechanical fastening.
There is now a wide range of continuous fibres available that are suitable for reinforcement
of MMCs, and a variety of production techniques, including liquid-metal infiltration,
squeeze casting, electroplating, diffusion bonding of plasma-sprayed tapes, and even
explosive welding, has been developed to make use of them, in addition to conventional
roll-bonding and hot-pressing. At the time of writing. however, there are few genuine
large-scale applications for composites of this kind, unlike their resin-matrix equivalents.

Of the recently renewed activity in MMCs, not all is associated with fibre-reinforced
materials, much work having been done on particulate MMCs, a class of materials that
was formerly familiar to metallurgists as 'cermets', although the classical cermet — the
cemented carbides and other metal-bonded ceramic materials mostly used for cutting
tools — contained a very low proportion of the metallic phase by comparison with modern
MMCs. The most familiar of the modern MMCs are made by the incorporation of SiC
particles (after treatment to improve particle/matrix adhesion) at volume fractions of 0.1

to 0.3 into aluminium alloys by various routes, including co-spraying, powder metallurgy, and liquid infiltration. These materials are claimed to show significant increases in specific stiffness and strength, although at the expense of ductility and toughness, with the advantage (*e.g.,* over reinforced plastics) that they can be processed by conventional metal-forming methods. A recent survey of the manufacturing processes used for MMCs has been published by Everett and Arsenault (1991).

2.3.3 CERAMIC-MATRIX COMPOSITES

Fabrication processes are complex and need to be carefully optimised because of the inevitable sensitivity of materials properties to microstructures controlled by processing conditions and interactions. Much of the recent work on CMCs in the U.S.A., Japan and Europe has largely followed relatively familiar routes in attempting to reinforce glasses (like borosilicates) and glass-ceramics (like lithium aluminosilicate or LAS and calcium aluminosilicate or CAS) with fibres such as the commercial Nicalon and Tyranno varieties of silicon carbide. Substantial improvements in mechanical properties have been achieved, by comparison with early carbon-fibre/glass composites. The fibres are usually impregnated with a slurry of fine glass powder and subsequently hot-pressed (Dawson *et al.*, 1987). If the matrix is a glass-ceramic, the final stage in manufacture is the 'ceramming' process, the final heat treatment to convert the glass to a fully dense ceramic. Companies like Corning in the U.S.A. offer such materials for commercial sale. Critical control of manufacturing conditions is needed to produce appropriate interfacial conditions for the optimum combination of strength and toughness. The advantage of the glass-ceramic route is that relatively modest processing temperatures are involved, although even so a typical residual thermal stress of the order of only 200 MPa may still be sufficient to cause cracking of the matrix in as-manufactured or lightly-loaded composites. Much of the research on monofilament-reinforced ceramics like the French SiC/SiC materials, for example (Naslain and Langlais, 1986), and on the production of carbon/carbon composites (Fitzer, 1987), has been directed towards chemical vapour infiltration (CVI) and related processes. Commercial quantities of these materials have been available for a considerable time although CVI is normally a very slow process and components produced in this way are very costly.

However, for all work involving the production of ceramic-fibre-reinforced ceramic composites by conventional pressing methods, there remains the problem of compatibility. Nicalon fibre with a starting strength as high as 3 GPa, for example, may lose half its strength when pressed into composites with a CAS or borosilicate glass matrix, and following high-temperature exposure in service this load-bearing ability may be further impaired. Current developments in the field involve the search for reliable diffusion-barrier coatings (plus ça change!).

It is perhaps also worth remarking that although the driving force for developments in the CMC field has most frequently been the needs of high-temperature applications in the aerospace industry, there are many other branches of engineering—automotive, chemical, marine, and general engineering, for example—where there is a need for reliable economic components possessing good mechanical properties and reasonable wear and corrosion resistance, in combination with adequate impact and thermal shock resistance at ordinary or only slightly raised temperatures. For many such applications there is no need for high-temperature stability, and the problems relating to fibre/matrix interactions at high temperatures are not as serious as in gas-turbine applications, for example, provided processing methods can be developed which do not require the high firing temperatures normally associated with ceramics manufacture. One such process is the sol-gel process in which low-viscosity sols are converted to solid ceramics in a manner analogous to the curing of organic resins. And by extension of the analogy, composite materials may be made by techniques of casting, filament winding and sol-transfer moulding, as described by Russell-Floyd *et al.,* (1993).

2.3.4 HYBRID COMPOSITES

Reference to hybrid composites most frequently relates to the kinds of fibre-reinforced materials, usually resin-based, in which two types of fibres are incorporated into a single matrix. The concept is a simple extension of the composites principle of combining two or more materials so as to optimise their value to the engineer, permitting the exploitation of their better qualities while lessening the effects of their less desirable properties. As such, the definition is much more restrictive than the reality. Any combination of dissimilar materials could in fact be thought of as a hybrid. A classic example is the type of structural material in which a metal or paper honeycomb or a rigid plastic foam is bonded to thin skins of some high-performance FRPs, the skins carrying the high surface tensile and compressive loads and the core providing lightweight (and cheap) structural stability. The combination of sheets of aluminium alloy with laminates of fibre-reinforced resin, as in the commercial product ARALL (aramid-reinforced aluminium, Davis, 1985) is a related variety of layered hybrid, and the mixing of fibrous and particulate fillers in a single resin or metal matrix produces another species of hybrid composite.

Some hybrids of current interest represent attempts to reduce the cost of expensive composites containing reinforcements like carbon fibre by incorporating a proportion of cheaper, lower-quality fibres such as glass without too seriously reducing the mechanical properties of the original composite. Of equal importance is the reverse principle, that of stiffening a GRP structure with a small quantity of judiciously placed carbon or aromatic polyamide fibre, without inflicting too great a cost penalty. In high-technology fields the question of cost may be insignificant by comparison with the advantages of optimising properties. In aerospace applications, a familiar purpose of using hybrids is to utilise the natural toughness of GRP or of Kevlar-fibre-reinforced plastics (KFRP) to offset a perceived brittleness of typical CFRP. From the designer's point of view the important aspect of using hybrids is that provided there is adequate understanding of the underlying mechanisms of stiffening, strengthening and toughening, they allow even closer tailoring of composite properties to suit specific requirements than can be achieved with single-fibre types of composites.

2.4 DEFECTS IN MANUFACTURED POLYMERIC COMPOSITES

All practical reinforced plastics composites are likely to contain defects of various kinds arising from the processes of manufacture. Indeed, composites are noted for the variability of mechanical properties they exhibit unless they have been produced under the most rigorously controlled conditions. The variability of material produced by hand lay-up methods is more marked than that of composites made by mechanised processes. The specific nature and severity of the defects found in any manufactured product will also be characteristic of the manufacturing process. In addition, any composite consisting of materials of widely different thermal expansion coefficients which is heated during manufacture may, on cooling, develop residual stresses sufficiently high to crack a brittle matrix. The defects that may be present in manufactured composites include:

- Incorrect state of resin cure, especially that resulting from variations in local exotherm temperatures in thick or complex sections during autoclaving,
- Incorrect overall fibre volume fraction,
- Misaligned or broken fibres,
- Non-uniform fibre distribution, with resultant matrix-rich regions,
- Gaps, overlaps or other faults in the arrangement of plies,
- Pores or voids in matrix-rich regions,
- Disbonded interlaminar regions,
- Resin cracks or transverse ply cracks resulting from thermal mismatch stresses,
- Disbonds in thermoplastic composites resulting from failure of separated flows to re-weld during moulding,
- Mechanical damage around machined holes, and
- Local bond failures in adhesively bonded composite components.

We shall discuss some of these in more detail in the context of particular manufacturing processes.

In the moulding of shaped artefacts from materials such as dough-moulding and sheet-moulding compounds (DMC and SMC) there are two main kinds of possible defects. The first results from failure of the process to control resin cure and flow, perhaps as a consequence of the use of a moulding compound that is past its shelf-life or because of incorrect control of the temperature/pressure cycle. The consequences are poor consolidation, incomplete filling of the mould, and perhaps also the failure of separated flow streams to re-amalgamate in complex mouldings. The second problem may arise either from the non-uniform filling of the mould cavity (a manual operation) prior to pressing or from poor attention to the flow patterns that will occur during pressing. In either case there may again be non-uniformity of fibre content in the finished artefact, and/or local preferred orientation of the short fibres, with resultant anisotropy in

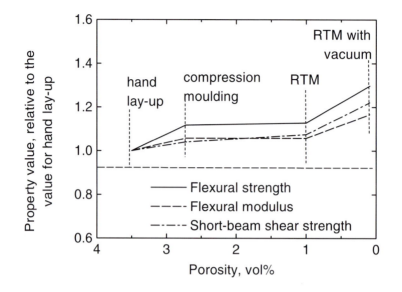

Fig. 2.1 Variation of mechanical properties of GRP materials as a consequence of variations in porosity levels introduced during manufacturing. In each case the fibre volume fraction is approximately 0.4 (Hayward and Harris, 1990).

mechanical properties. If this anisotropy is not allowed for in the design, the component may fail to meet the design specification. Similar considerations apply, of course, to the injection moulding of short-fibre-reinforced thermoplastics.

Typical defects in RTM mouldings include porosity resulting from bubbles injected with the resin, 'washing' or movement of the preform, especially near injection points or corners, failure of the resin to penetrate the preform uniformly, poor wetting-out of the fibres by the resin, distributed porosity, and premature gelation. Porosity is a particular problem, since the process relies on the flow of resin to sweep out the air contained within the mould and the fibre preform, and when filled matrix resins are being injected into a tightly packed fibre preform, there is sometimes a tendency for the filler particles to be filtered out of the resin. The quality of RTM mouldings is considerably improved, however, by slightly reducing the air pressure in the mould cavity. In the so-called vacuum-assisted RTM process (VARTM) it is sufficient to maintain the mould cavity pressure at only 0.7 bar in order to improve considerably the wetting out of the preform and reduce the level of residual porosity in the moulding to a fraction of a percent for relatively little extra cost. Figure 2.1 shows how the mechanical properties of a typical glass/polyester composite vary with the levels of porosity characteristic of different processing conditions. The average porosity levels in VARTM mouldings are comparable with those obtainable in high-quality autoclaved materials.

A schematic illustration of some of the typical faults that occur in the laying up of prepreg materials is shown in Figure 2.2. The finite dimensions of the prepreg will require, from time to time, that joins (a) will have to be made. Closely butted joins are clearly

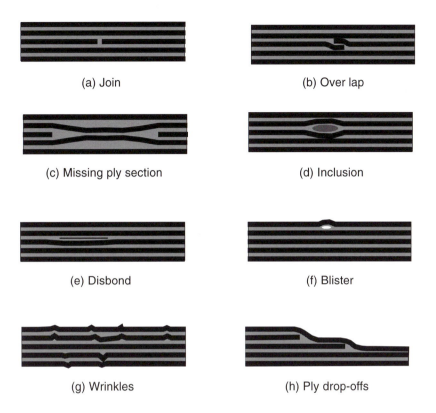

Fig. 2.2 Some typical defects in composite laminates manufactured from prepreg sheets.

more satisfactory than overlapped (b) or widely separated (c) prepreg ends since these will result in resin-rich regions, as shown. Rigorous quality-control procedures should prevent the incorporation of inclusions such as bits of rubbish (d), but a common problem with prepregs is the occurrence of small bundles of fibre debris, known as 'whorls', which are sometimes attached to the surface of the sheet and subsequently become incorporated in the laminate as small regions of random fibre alignment. These are considered to be damaging to composite strength, and an aerospace manufacturer may typically specify that such defects should not occur more than once in, say, ten metres of prepreg roll. Their effect is modest, however. In samples of $[(0/90)_2]_S$ glass/epoxy laminate hot-pressed from commercial prepreg, with deliberate concentrations of whorls to the extent of 3 defects per prepreg ply in a 150 mm x 150 mm x 1 mm plate, the tensile strength was reduced by only about 6%, and the failure strain by about 2%, the stiffness being unaffected (Ellis, 1985).

Failure to achieve a consistent level of pressure over the whole surface of a sheet during autoclaving may result in small regions where the plies are imperfectly bonded (e). These disbonds are particular sources of weakness in laminates subjected to compression or impact forces. Voids or imperfectly cured resin regions near the surface of a laminate may absorb moisture during service in wet conditions, with the result that hydrostatic pressure could develop inside the region and push out the surface layer or gel coat to create blisters (f). Poorly stored prepregs or prepregs that are draped over a former which is too sharply curved may become wrinkled (g), and this may again lead to poor surface quality or weaknesses resulting from resin-rich regions in the laminate. The last feature shown in Figure 2.2 is a deliberately introduced 'defect' rather than an adventitious manufacturing fault. It is often required to change the dimensions of a structure, and since the individual ply effectively represents a 'quantum' of laminate thickness, such changes can only be made by gradually reducing or increasing the number of plies in a cross section. This design procedure by the introduction of 'ply drop-offs' (h) nevertheless presents a problem for the stress engineer since the ply ends result in local stress concentrations the effects of which must be accounted for (see, for example, Curry *et al.*, 1985; Cui *et al.*, 1994). The prepregs themselves may also be a source of defects if the strictest quality control procedures are not employed. Broken or damaged fibre-tows, accumulations of debris (*e.g.,* the 'whorls' mentioned above), and out-of-date resin are examples of prepreg defects that can lead to poor quality in the final laminate.

Since most composites consist of materials of widely different thermal expansion coefficients they may, if heated and cooled during manufacture, develop residual stresses sufficiently high to crack a brittle matrix. Early high-performance CFRP laminates were frequently found to have suffered multiple internal cracking unless proper care was taken to control the production cycle. This type of problem has been eliminated to a large extent for polymer composites by the advent of 'modified' or 'toughened' resins with high failure strains, but such problems will inevitably be of more concern for users of such materials as CMCs.

The advantages of using hybrid laminates may be both structural and economic, as we have seen, but there is also a potential disadvantage. Over the temperature range of about 150°C from conventional curing temperatures down to ambient, the axial thermal expansion coefficient of most carbon fibres is close to zero, while that of an aramid fibre like Kevlar-49 is negative, about -3×10^{-6} K^{-1}, and that of E-glass is positive, 0.7×10^{-6} K^{-1}. Thus, in hybrid laminates containing various mixtures of these fibres, there will be a tendency to develop thermal stresses which may become great enough to cause marked undulations in the plies subject to compression — the Kevlar plies in a carbon/Kevlar hybrid, for example. A consequence of this is that the wavy plies are unable to contribute their full share of load-bearing in tension, and the compression strength of a hybrid may also be reduced because the waviness facilitates a common compression kinking mechanism (see chapter 4).

One of the virtues of polymer composites is that complex shapes can often be manufactured in single-shot operations, thereby eliminating much of the need for the joining operations that characterise most conventional manufacturing with metallic materials. Joining and machining operations may not always be completely avoidable, however, and in such cases there is always likely to be the possibility of bonding failures in imperfectly prepared adhesive joints (in thermoset composites), disbonds in welded thermoplastic composite plates, and machining damage (in sawn or drilled components, for example). But while machining damage will usually be at cut edges, and therefore visible, poor joints will rarely be detectable from the exterior.

Any of these defects can locally reduce the strength of the material below the required component design stress. They may then act as sites for the initiation of fatigue damage, or may facilitate the growth of a fatigue crack during cyclic loading. A comprehensive assessment of the quality of a composite material prior to putting it into service is therefore as important as the monitoring of the levels of damage accumulated in a composite structure during service. The use of non-destructive evaluation methods is thus of great importance in composite materials technology, and modern aerospace design procedures increasingly relate to the establishment of zero-growth thresholds for manufacturing and service defects. Many of the defects associated with hand lay-up or manual operation may be avoided, where the extra cost is justifiable, by the use of computer-controlled tape-laying machines which are able to construct a prepreg stack for autoclaving to very high standards of quality and repeatability. Similarly, errors of control in pressing can be avoided to some extent by the use of automated autoclaves with pressure-temperature cycles carefully programmed from detailed chemical knowledge of the gelation and viscosity characteristics of the resin in use. It is largely only in the aircraft industry that the high initial costs of such equipment can be borne, however.

Of particular importance to the manufacturer and user is the associated question of what are likely to be the consequences of the presence of a particular defect. At present, there is no easy way to quantify this. A few isolated spherical pores a micron or so in diameter may have no effect on any physical property, and may not therefore affect the tensile or flexural mechanical performance of a composite. But a distribution of innocuous looking pores may markedly reduce the interlaminar shear strength of a material, and by providing sites for accumulation of moisture may also ruin the electrical or dielectric performance of the material. A minor delamination between plies in a complex laminate may have no effect on the tensile strength of the material, but such defects are frequently injurious to the compression performance of laminates and, as indicated above, may grow rapidly to eventual ruin of the composite under cyclic loading conditions. The question of critical size or quantity of defects must therefore be addressed with specific reference to the particular composite and service conditions involved in a given application.

2.5 METHODS OF NON-DESTRUCTIVE EVALUATION FOR POLYMER COMPOSITES

Since the presence of defects in reinforced plastics, either *ab initio* or as a consequence of damage in service, is to some extent inevitable, it follows that manufacturers and users need sensitive techniques for the detection of these defects and damage. Much of the technology previously developed for metallic engineering materials and structures has been transferred across, with appropriate modifications, for use with fibre composites.

Some of these techniques are more useful than others, and it is often good practice to use a back-up technique where possible rather than to rely on a single method. Harris and Phillips (1983, 1990) have given summaries of some of the more common methods in current use, but at the time of writing there is a high level of activity in R & D work on fibre-optic sensors specifically for use with reinforced plastics. Brief details of some of these NDT tools are given below.

2.5.1 OPTICAL INSPECTION

In translucent GRP, inspection by transmitted light can give indications of the presence of pores, poor wetting-out, delaminations, and gross inclusions. Light transmission can also sometimes be correlated with the fibre content, V_f. Loss of transparency (stress-whitening) is associated with the development of fibre/resin debonding and resin cracking. In non-translucent composites the only feasible visual-inspection techniques relate to observation of surface damage. The detection of surface cracking may be enhanced by dye-penetrant methods. Moiré methods require photographic grids to be printed on the surface of the composite and irradiated with coherent light. The resulting interference fringes give clear indications of local stress concentrations and deformation, including those arising from sub-surface damage. Laser holographic methods and electronic speckle-pattern interferometry (ESPI) are also used for non-translucent materials.

2.5.2 RADIOGRAPHIC METHODS

Radiography is not easily applied in the field, but information about composite quality can be obtained by x-ray inspection. Contact radiography with sources of 50 kV or less yields high-contrast photographs from low-density materials like GRP because of their low inherent filtration. The linear absorption coefficient of glass is about twenty times that of most resins, and film density measurements can be correlated with V_f if the material is unpigmented. The fibre distribution, quality of weave, and the presence of large laminating defects can be easily investigated. Contact radiographs showing fibre distributions have also been analysed by optical diffraction methods to give quantitative information about the fibre distribution. The sensitivity of contact radiography can be improved by impregnating the composite with a radio-opaque material, usually an alcoholic solution of zinc iodide. By this technique, resolution of cracks of the order of a few mm long and 0.1 mm deep is feasible in GRP and CFRP, and delaminations are easily resolved.

2.5.3 THERMAL IMAGING

When a uniform heat flux is supplied to a plate, any anomalous variations in the resulting temperature distribution in the plate are indications of structural flaws in the material. Similarly, in an otherwise uniform material working under variable loads, local damage will give rise to changes in the hysteretic thermal losses in the body of the material. Thermal imaging, with limiting detection of temperature differences of about 0.2°C, is easily carried out by means of infra-red television photography. Damage has been detected in glass/epoxy laminates at low stresses and frequencies, but in materials of higher thermal conductivity detection is more difficult. Testing frequencies greater than 5 Hz are needed in order to detect damage in CFRP because of their high thermal conductivity. Thermal imaging equipment is routinely available and is widely used in the aerospace industry. The technique has the advantage over many other methods in that it is a remote monitoring technique. The use of thermography has been extended in recent years by the use of transient heat sources and rapid scanning of the induced thermal fields by TV/video-compatible infra-red imagers which forms the basis of the pulse video thermography (PVT) technique. This is a convenient and versatile extension of conventional thermography capable of rapid detection of near-surface defects. An associated thermal NDE method is that of thermoelastic stress analysis which makes use of the familiar thermoelastic effect, where a temperature change accompanies adiabatic elastic deformation of a body. This technique is said to offer spatial resolution down to 1 mm and temperature discrimination of about 0.002°C for stress changes of about 2 MPa in steel, although in composites there are potential problems of interpretation resulting from their anisotropic nature.

2.5.4 ULTRASONIC TECHNIQUES

Ultrasonic inspection techniques are perhaps the most widely-used NDE methods for composites. The velocity and attenuation of an ultrasonic pulse passing through a material provide information about the general physical properties (*i.e.*, the stiffness) and the structure (*i.e.*, the defect/damage state) in a material, provided certain basic features of the material, such as the anisotropy, are taken into account. Most composites are highly dispersive, often having regular structural patterns at several levels of scale (fibres, fibre tows, weave of cloth, laminate stacking sequence), an array of interfaces (fibre/resin interface, interlaminar interface) and an inherent distribution of defects. Ultrasonic waves in composites are therefore highly attenuated and their velocity and attenuation are dependent on frequency, this frequency-dependence being affected by the construction of the laminate. The propagation of ultrasonic waves may be affected by frequency, composite structure, structural defects, and damage induced by loading. The problem of high attenuation is overcome by the use of broad-band transducers and low wave frequencies which give shorter, more highly penetrating pulses with less pronounced near-field interference effects. As a result of the low pulse intensities used in composites testing it is generally necessary to use higher amplification than is needed for other materials, but most of the common ultrasonic techniques, including pulse-velocity measurements, echo methods, goniometry, and ultrasonic interferometry, are used with composites.

The most highly developed ultrasonic NDE method, based on measurements of through-thickness attenuation, is the C-scan technique, which is used routinely for inspecting large panels. Synchronized raster-scanning motions of the transmitter and receiver (either in a water bath or with irrigated water-jet probes) on opposite sides of the plate enable measurements of the intensity of the transmitted wave to be made as a function of position. It is common to use a focused transmitter, focusing on the back face of the plate, with the receiver as close to the front face as possible The transmitted intensity is used to modulate the brightness of a visual display, or the density of ink on a diagram, so as to build up a picture of plate quality. The resolution of the technique is limited by dispersion in the composite and by the beam dimensions. It is necessary to use something like a focused 4 MHz probe to detect defects with dimensions of the order of 1 mm, and as a consequence C-scanning is more reliable as an indicator of general quality than as a detector of specific defects. Large voids, distributions of fine porosity, areas of variation in fibre content, and delaminated regions will usually be revealed by a C-scan, but for assessment of the severity of specific defects it is necessary to make standard test plates containing flat-bottomed drilled holes of various depths and diameters for calibration purposes. Conventional C-scanning is a slow process, but new techniques for real-time operation are now on the market.

2.5.5 OPTICAL FIBRE SENSORS

Optical fibre sensors have a number of advantages over other types of sensor in that they are immune from electromagnetic interference, they can be embedded in a composite material during manufacture rather than simply being surface mounted, they can be used to monitor several parameters, including temperature, pressure, strain, and chemical characteristics, and they offer important possibilities for on-line application, both for process optimization and in-service health monitoring (Fernando, 1998).

2.5.5.1 Cure Monitoring

The mechanical properties of composites are strongly dependent upon the chemical and rheological events occurring during the cure cycle. There is considerable current interest in the possibility of using fibre-optic sensors embedded in the composite itself to carry out *in situ* measurements of cure state, cure kinetics, resin viscosity and residual stress build-up in real time during manufacture. It is expected that the signal processing procedures that are involved could, eventually, form the basis for intelligent processing of composites (Roberts & Davidson, 1993). Techniques that are being explored include refractive-index-based sensors (Crosby *et al.*, 1995), evanescent-wave spectroscopic sensors (Powell *et al.*, 1996), and transmission or reflection infra-red spectroscopy sensors (Crosby *et al.*, 1996). These sensors are able to evaluate changes in the optical properties, and hence the state of cure, of the resin surrounding a stripped length of embedded fibre, or of the resin in a gap between two lengths of fibre, or of resin at the end of a single embedded fibre.

2.5.5.2 Sensing of Deformation and Damage (Health Monitoring)

The simplest form of assessment of impact damage is based on the fact that optical fibres will fracture when the composite in which they are embedded suffers an impact event, and this reflects the related damage that will be sustained by the main load-bearing fibres. More sensitive assessment of the presence of strain or a rise in temperature of the composite can be obtained by the use of fibres which have Bragg gratings written into their cores or the somewhat cheaper Fabry-Perot sensors in which changes in the length of an air gap are measured by interferometry (Greene *et al.*, 1995). These provide a means of monitoring strain with a high degree of accuracy and resolution. They can also be used for on-line location of the position of impact events — Greene *et al.*, cite an ability to locate an impact with a resolution of 0.5 mm and an accuracy better than 5 mm. Provided the separate effects of thermal expansion and elastic strain can be isolated, sensors of this kind can also provide information about temperature changes in the material. They also therefore offer the possibility, mentioned in section 2.5.5.1, of closer monitoring and intelligent control of processing procedures, feedback methods being used to control the severity of the resin curing exotherm and therefore the quality of the finished product.

2.5.6 MICROWAVE METHODS

The dielectric constant of glass is much greater than those of resins at microwave frequencies, and any technique for measuring the average dielectric constant of a GRP composite can therefore be correlated reasonably satisfactorily with the glass content of the material. A typical instrument consists of an open-ended coaxial resonator which can be applied to a GRP surface in order to close the resonant cavity. As the instrument is moved over the surface, changes in the local average glass content are indicated by changes in the resonant frequency. Since any defect which affects the microwave penetration will also cause a resonance shift, the method is capable of monitoring the deterioration of a structure under load, once its dielectric response in the undamaged state is known.

2.5.7 DYNAMIC MECHANICAL ANALYSIS

When stresses and strains can be continuously monitored in service, we might expect that changes in elastic modulus would give non-destructive indications of deterioration of properties caused by the accumulation of damage. However, the change in stiffness of a composite as a result of cycling depends on the type of composite and the extent to which the matrix, fibres and interface are directly loaded. Measurement of the modulus itself, either by strain gauges or by ultrasonic pulse-velocity measurements, may therefore prove to be relatively insensitive to changes in the damage state in certain cases, although relative changes in resonant frequency and of the damping capacity can give more subtle indications of damage accumulation. If a structure can be made to resonate in a reproducible manner, its frequency spectrum can be rapidly analysed to establish its characteristic 'sound'. Any damage to the structure will change this characteristic spectrum in a recognisable way and this provides the basis of a simple 'coin-tap' test which can be used to monitor damage in composites.

One of the oldest-established NDE methods is the principle embodied in the Fokker debond tester manufactured by Wells-Krautkramer. This machine uses swept-frequency probes generating shear waves of constant amplitude and compares the response of the test material with that of a standard. The resonant frequency and amplitude of the transducer response are recorded in contact with the standard, and any shifts that occur when the probe is subsequently coupled to the test sample are noted. These shifts must be calibrated against some known or measurable property. The technique is said to be capable of giving more sensitive indications of the presence of flaws, voids, weak bonds, porosity, delaminations, incorrect cure conditions, poor wetting, and poor gap-filling in joints than any other ultrasonic method.

2.5.8 ACOUSTIC EMISSION METHODS

Any sudden structural change within a composite, such as resin cracking, fibre fracture, rapid debonding, or interlaminar cracking, causes dissipation of energy as elastic stress waves which spread in all directions from the source. The technique of detection of these acoustic emissions (AE) by suitable transducer/amplifier systems is now well-established and triangulation methods may be used to locate flaws in large structures and to assess their severity. There are several ways of analysis of the information obtained by AE monitoring of structures under load, some of which offer suitable quantitative procedures for proof testing or life prediction, and some of which provide deeper insight into the mechanisms of damage accumulation in composites.

2.6 THE USE OF COMPOSITES

In order to make good use of engineering materials, we have to understand their behaviour sufficiently well to be able to predict performance in both the short term and the long term. In the context of composite materials, this means developing mathematical models which represent reasonably closely the known experimental response of real composites to applied stresses and environmental conditions. If we can predict the behaviour of any given composite accurately as a result of this modelling, then we can have confidence that the material we have designed will meet the service requirements of the application concerned. One of the problems with composites is that there is such a diverse range of materials, and it is necessary to have a thorough understanding of each class: we cannot simply read across from the characterisation of one group of composites to another without carrying out extensive checks of the validity of the old models for the newer materials. Theories of the elastic behaviour of composite materials are very well developed, and predictions of elastic response are often very satisfactory. Problems arise, however, when attempts are made to predict processes of failure. The failure of composites is almost always a complex process. Damage accumulates in a widespread fashion in composites, and many individual processes occur at the microstructural level. The strength of a composite is determined by the conditions which determine the climax of this damage accumulation process, and modelling of these conditions is far from perfect at the present time. As a consequence, there is still a certain lack of confidence on the part of designers, a reluctance to use some of these new materials for which we have not, as yet, the operating experience that instils confidence, and this leads to a measure of inefficient (and often uneconomic) design for which the only cure is better understanding.

From the point of view of the designer who is using composites, there is an important constraint: composites do not behave like metals. Conventional notions of materials behaviour and design methods are therefore unlikely to be appropriate. The use of composites offers the designer a range of new solutions to engineering problems of many kinds, but it is vital to use the opportunity to rethink the approach. It is rarely a good idea simply to carry out a piecemeal substitution of composites parts for existing metallic parts. This kind of approach will certainly not liberate the designer from the old constraints, will often lead to more expensive rather than cheaper solutions, and may result in unsafe designs.

In the application of composites, as in other areas of materials engineering, it is important to consider the selection of the materials and the choice of manufacturing process as a vital and integrated part of the design process.

2.7 REFERENCES

P.S. Allan and M.J. Bevis: *Plastics Rubber & Composites: Processing & Applications*, **7**, 1987, 3-10.

H. Bibring, J.P. Trottier, M. Rabinovitch and G. Siebel: *Mém. Sci. de la Revue de Métall.*, **68**, 1971, 23-41.

P. Crosby, G. Powell, R. Spooncer and G. Fernando: *Proceedings of SPIE Conference on Smart Structures and Materials*, Society for Optical Engineering, **2444**, 1995, 386-395.

P.A. Crosby, G.R. Powell, G.F. Fernando, R.C. Spooncer, C.M: France and D.N. Waters: '*In situ Cure Monitoring in Advanced Composites Using Evanescent Wave Spectroscopy*,' *Journal of Smart Materials and Structure*, **5**, 1996, 415-428.

W. Cui, M.R. Wisnom and M. Jones: *Composites Science and Technology*, **52**, 1994, 39-46,

J.M. Curry, E.R. Johnson and J.H. Starnes: *Proceedings of 28th Conference on Structures, Structural Dynamics and Materials*, Monterey, California, (AIAA/ASME/ASCE/AHS), paper AIAA 87-0874, 1985, 737-747.

J.W. Davis: '*ARALL — from Development to a Commercial Material*,' in *Progress in*

Advanced Materials and Processing, G. Bartelds and R.J. Schlickelman, *et al.*, eds., Elsevier, Amsterdam, 1985, 41-49.

D.M. Dawson, R.F. Preston and A. Purser: *Proceedings of Ceramics Engineering Science*, **8**, 1987, 815-820.

C. Doyle and G.F. Fernando: *Condition Monitoring Engineering Materials with an Optical Fibre Vibration Sensor System,* SPIE, **3042**, 1997, 310-318.

G.C. Eckold: *Design and Manufacture of Composite Structures*, Woodhead Publishing Ltd, Abington, U.K, 1994.

K.R.J. Ellis: *'The Effect of Structure and Defects on FRP Laminate Materials,'* unpublished report, School of Materials Science, University of Bath, 1985.

R.K. Everett and R.J. Arsenault, eds.: *Metal Matrix Composites: I Processing and Interfaces*, Academic Press Boston, 1991.

E.A. Feest: *Materials Des.*, **7**, 1986, 58-64.

G. Fernando: *Materials World*, **6**(5), 1998, 340-342.

E. Fitzer: *Carbon*, **25**, 1987, 163-190.

J.A. Greene, T.A. Tran, V. Bhatia, M.F. Gunther, A. Wang, K.A. Murphy and R.O. Claus: *Smart Materials and Structure*, **5**, 1995, 93-99.

R. Hadcock and J. Huber: *Practical Considerations for Design, Fabrication and Tests for Composite Materials*, NATO (AGARD) Lecture Series 124, papers 11 and 12, AGARD, Neuilly, France, 1982.

A. Harper: *Proceedings of Conference Hands Off GRP II,* Nottingham, Plastics & Rubber Institute, London, 1985, Paper 4.

B. Harris and M.G. Phillips: *Developments in GRP Technology,* B. Harris, ed., Applied Science Publishers, London, 1983, 191-247.

B. Harris and M.G. Phillips: *Analysis and Design of Composite Materials and Structures*, Y. Surrel, A. Vautrin and G. Verchery, Editions Pluralis, Paris, Part I, 15, 1990.

J.S. Hayward and B. Harris: *Compos. Manuf.*, **1**, 1990, 161-166, also *SAMPE Journal*, **26**(3), 1990, 39-46.

D.T. Jones, I.A. Jones and V. Middleton: *Composites*, **A27**, 1996, 311-317.

J.E. Masters and F.K. Ko: *Textile Composites*, Special Issue of *Composites Science and Technology*, **56**(3), 1996, 205-386.

R. Naslain and F. Langlais: Proceedings of Conference on *Tailoring Multiphase and Composite Ceramics*, R.T. Tessler, G.L. Messing and C.G. Pantano and R.E. Newham, eds., Plenum Press, New York, 1986, 145-164.

G.R. Powell, P.A. Crosby, G.F. Fernando, C.M. France, R.C. Spooncer and D.N. Waters: *Optical Fibre Evanescent-Wave Cure Monitoring of Epoxy Resin,* SPIE, **2718**, 1996, 80-92.

W.J. Renton: *'Hybrid and Select Metal-Matrix Composites,'* AIAA, New York, S.S.J. Roberts and R. Davidson, eds., *Composites Science and Technology*, **49**, 1977, 265-276.

C.D. Rudd, M.J. Owen and V. Middleton: *Materials Science and Technology*, **6**, 1990, 656-665.

R.S. Russell-Floyd, B. Harris, R.G. Cooke, T.H. Wang, J. Laurie, F.W. Hammett and

R.W. Jones: *Journal of American Ceramic Society*, **76**(10), 2635-2643.
R.A.P. Spencer: *Developments in GRP Technology*, B. Harris, ed., Applied Science Publishers, London, 1983, 1-36.
E.R. Thompson, D.A. Koss and J.C. Chesnutt: *Metallurgical Transactions*, **1**, 1970, 2807-2813.
M.R. Turner, C.D. Rudd, A.C. Long, V. Middleton and P. McGeehin: *Advances in Composites Letters*, **4**, 1995, 121-124.

3. Elastic Properties of Fibre Composites

3.1 SIMPLE MICROMECHANICAL MODELS

The simplest method of estimating the stiffness of a composite in which all of the fibres are aligned in the direction of the applied load (a unidirectional composite) is to assume that the structure is a simple beam, as in Figure 3.1, in which the two components are perfectly bonded together so that they deform together. We shall ignore the possibility that the polymer matrix can exhibit time-dependent deformation. The elastic (Young) moduli of the matrix and reinforcement are E_m and E_f, respectively. We let the cross-sectional area of the fibre 'component' be A_f and that of the matrix component be A_m. If the length of the beam is L, then we can represent the quantities of the two components in terms of their volume fractions, V_f and V_m, which is more usual, and we know that their sum $V_f + V_m = 1$. The fibre volume fraction, V_f, is the critical material parameter for most purposes. The subscript 'c' refers to the composite.

The load on the composite, P_c, is shared between the two phases, so that $P_c = P_f + P_m$, and the strain in the two phases is the same as that in the composite, $\varepsilon_c = \varepsilon_f = \varepsilon_m$ (*i.e.* this is an 'iso-strain' condition). Since stress = load/area, we can write:

$$\sigma_c A_c = \sigma_f A_f + \sigma_m A_m$$

and from the iso-strain condition, dividing through by the relevant strains, we have:

$$\frac{\sigma_c A_c}{\varepsilon_c} = \frac{\sigma_f A_f}{\varepsilon_f} + \frac{\sigma_m A_m}{\varepsilon_m}$$

or

$$E_c = E_f V_f + E_m \left(1 - V_f\right) \tag{3.1}$$

This equation is referred to as the Voigt estimate, but is more familiarly known as *the rule of mixtures*. It makes the implicit assumption that the Poisson ratios of the two components are equal ($v_f = v_m$), thus ignoring elastic constraints caused by differential lateral contractions. More sophisticated models have been developed which allow for such effects, the most familiar being that of Hill (1964) which shows that the true stiffness of a unidirectional composite beam would be greater than the prediction of equation 3.1 by an amount which is proportional to the square of the difference in Poisson ratios, $(v_f - v_m)^2$, but for most practical purposes this difference is so small as to be negligible. For example, for high-performance reinforced plastics, $v_m \approx 0.35$ and $v_f \approx 0.25$, and for a fibre volume fraction, V_f, of about 0.6, the correction needed to account for the Poisson

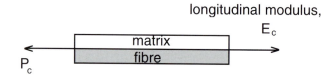

Fig. 3.1 Simplified parallel model of a unidirectional composite.

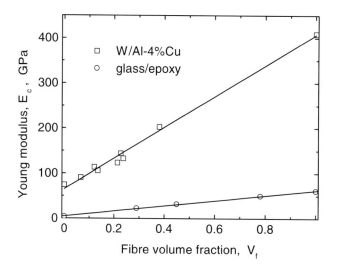

Fig 3.2 Confirmation of the rule-of-mixtures relationship for the Young moduli, E_c, of undirectional composites consisting of tungsten wires in Al-4%Cu alloy and glass rods in epoxy resin.

constraints is only about 2%. A good indication of the validity of the mixture rule for the longitudinal moduli of two composites of quite different kinds is given in Figure 3.2.

To estimate the transverse modulus, E_t, we use a similar approach with a block model such as that shown in Figure 3.3, with the same constraints as before, *i.e.* well-bonded components with similar Poisson ratios, and no visco-elastic response from the matrix. This is now an 'iso-stress' model, so that $\sigma_c = \sigma_f = \sigma_m$. The total extension of the model is the sum of the extensions of the two components:

$$\varepsilon_c L_c = \varepsilon_f L_f + \varepsilon_m L_m$$

If the cross-sections of both phases are the same, $L \equiv V$, so dividing through by the stress (and remembering that $V_f + V_m = 1$) we have:

Fig. 3.3 Simple series model of a composite.

$$\frac{\varepsilon_c}{\sigma_c} = \frac{\varepsilon_f V_f}{\sigma_f} + \frac{\varepsilon_m V_m}{\sigma_m}$$

$$\text{or} \quad \frac{1}{E_t} = \frac{V_f}{E_f} + \frac{V_m}{E_m}$$

This is referred to as the Reuss estimate, sometimes called the *inverse rule of mixtures*: the transverse modulus is therefore:

$$E_t = \frac{E_f E_m}{E_m V_f + E_f \left(1 - V_f\right)} \tag{3.2}$$

and the relationship between the Voigt and Reuss models can be seen in Figure 3.4 The Reuss estimate is sometimes modified to account for the Poisson effect in the matrix by introducing a 'constrained' matrix modulus, effectively by dividing E_m by $(1 - v^2_m)$, so that:

$$\frac{1}{E_t} = \frac{V_f}{E_f} + \frac{V_m \left(1 - v^2_m\right)}{E_m} \tag{3.3}$$

The relatively small effect of this correction is also shown in Figure 3.4.

The problem with the Reuss model for the transverse case is that the geometry shown in Figure 3.3 in no way resembles that of a fibre composite perpendicular to the fibres. And in assuming that the Poisson ratios of the phases are the same, it ignores constraints due to strain concentrations in the matrix between the fibres. As a consequence, although it appears to give the correct form of the variation of E_t with V_f, the values predicted seldom agree with experimental measurements. The model also implicitly assumes that the transverse stiffness of the fibre is the same as its longitudinal stiffness, and while this is true of isotropic fibres like glass, it is not true of reinforcements with a textile origin like carbon and Kevlar.

The need for a more refined procedure is apparent if we look at even an idealised model of the structure of a composite loaded transverse to the fibres, as shown in Figure 3.5. A model based on hexagonal (close-packing) geometry would clearly give different results from this square-packing model, and both would be different from the

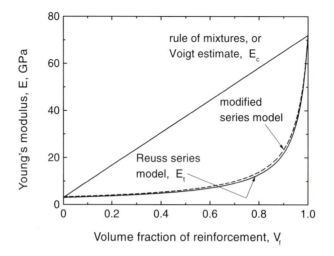

Fig. 3.4 Predicted variations of the longitudinal elastic modulus, E_c, and the transverse modulus, E_t, of composites of glass and epoxy resin connected in parallel and in series. The solid curves give the Voigt (parallel) and Reuss (series) estimates. The dashed line represents the Reuss estimate modified to include the effect of matrix Poisson constraint.

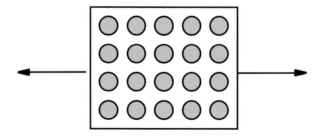

Fig. 3.5 Idealised square packing geometry in a transverse section of a unidirectional composite.

true packing geometry of any real composite (see Figure 3.6, for example) which will almost always tend to be much less regular than either of these idealised geometries. In practice the true transverse modulus lies some way above the lower bound given by the Reuss estimate.

In discussing the behaviour of anisotropic materials, it becomes inconvenient to continue using subscripts like c and t and we resort to using a numerical index system associated with orthogonal axes, x_1, x_2, and x_3. We choose the x_1 direction to coincide

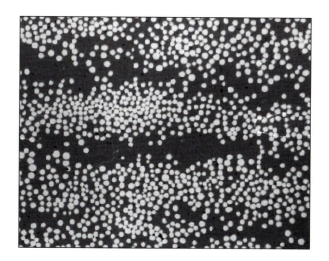

Fig. 3.6 Real packing geometry: the cross-section of a unidirectional SiC/CAS laminate of $V_f \approx 0.4$. The magnification can be judged from the fibre diameter, 15 μm.

with the fibre axis and the x_2 direction to coincide with the transverse in-plane direction (assuming that we are dealing, as is usually the case, with a thin plate of material). The longitudinal stiffness of a unidirectional composite is then referred to as E_1 and the transverse stiffness is E_2. Since in most practical reinforced plastics composites, $E_f \gg E_m$ the Reuss and Voigt estimates of stiffness given in equations 3.1 and 3.2 can be usefully approximated (in the new notation) by the relationships:

$$E_1 \approx E_f V_f \qquad (3.1a)$$

$$E_2 \approx E_m \left(1 - V_f\right)^{-1} \qquad (3.2a)$$

from which it is clear that the stiffness in the fibre direction is dominated by the fibre modulus, while that in the transverse direction is dominated by the matrix modulus.

The Poisson ratio, ν, of an isotropic material is defined as the (negative) ratio of the lateral strain, ε_2, when a stress is applied in the longitudinal (x_1) direction, divided by the longitudinal strain, ε_1, *i.e.* $\nu = -\varepsilon_2/\varepsilon_1$. Consideration of equations 3.1 and 3.2 shows that in a unidirectional composite lamina there will be two in-plane Poisson ratios, not one as in isotropic materials, and it is convenient to label these ν_{12}, called the *major* Poisson ratio (relating to the lateral strain, ε_2, when a stress is applied in the longitudinal (x_1) direction) and ν_{21}, the *minor* Poisson ratio (relating to the strain in the x_1 direction when a stress is applied in the x_2 direction). The two are not the same, since it is obvious that ν_{12} must be much larger than ν_{21}. By means of arguments similar to those above for the determination of E_1, it can be shown that for a stress applied in the x_1 direction only, the major Poisson ratio is given by:

$$\nu_{12} = \nu_f V_f + \nu_m \left(1 - V_f\right) \qquad (3.4)$$

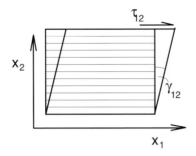

Fig. 3.7 Definition of a shear relative to the $x_1 x_2$ Cartesian axes.

i.e. it also obeys the rule of mixtures. Further thought will show that the minor Poisson ratio must be related to ν_{12} by the equation:

$$\frac{\nu_{21}}{E_2} = \frac{\nu_{12}}{E_1} \tag{3.5}$$

Values of the fibre and matrix Poisson ratios rarely differ by a great deal, so that neither matrix nor fibre characteristics dominate these two elastic constants.

When a unidirectional fibre composite is loaded by an in-plane $(x_1 x_2)$ shear force, it distorts to a parallelogram, as shown in Figure 3.7. The shear stress in the fibre direction, τ_{12}, is matched by its complementary shear stress, τ_{21}. The simplest model assumes that the fibres and matrix carry the same stress:

$$\tau_{12} = G_{12}\,\gamma_{12} = G_f\,\gamma_f = G_m\gamma_m$$

and the (apparent) in-plane shear modulus, G_{12}, is given by:

$$\frac{1}{G_{12}} = \frac{V_f}{G_f} + \frac{\left(1 - V_f\right)}{G_m} \tag{3.6}$$

By analogy with equation 3.2 for the transverse Young's modulus, it can be seen that the matrix shear stiffness again dominates the composite shear modulus unless the fibre volume fraction is very large.

3.2 THE HALPIN-TSAI EQUATIONS

The models in the last section are simple models, although, if treated with care, they can give useful approximations to the behaviour of many composites. There have been many more formal treatments, however, based on more realistic models of the transverse fibre distribution, which yield results of varying degrees of complexity. These more rigorous approaches may give predictions of elastic properties that are closer to experimentally observed values than the simple models of section 3.1, but they are seldom easy to use in practice. For design purposes it is more useful to have simple and rapid computational procedures for estimating ply properties rather than more exact but intractable solutions. Convenient interpolation procedures have been developed by Halpin and Tsai (1968; see also Halpin, 1992) who showed that many of the more rigorous mathematical models could be reduced to a group of approximate relationships of the form:

$$E_1 \approx E_f V_f + E_m (1 - V_f) \tag{3.7}$$

$$v_{12} \approx v_f V_f + v_m (1 - V_f) \tag{3.8}$$

$$E_2 = \frac{(1 + \zeta \eta V_f)}{(1 - \eta V_f)} E_m \tag{3.9}$$

$$G_{12} = \frac{(1 + \zeta \eta V_f)}{(1 - \eta V_f)} G_m \tag{3.10}$$

The first two of these equations are the same rules of mixtures that we have already discussed. In the second two equations, ζ is a factor, specific to a given material, that is determined by the shape and distribution of the reinforcement (*i.e.* whether they are fibres, plates, particles, *etc.* and what kind of packing geometry), and by the geometry of loading. The parameter η is a function of the ratio of the relevant fibre and matrix moduli (E_f/E_m in equation 3.9 and G_f/G_m in equation 3.10) and of the reinforcement factor ζ, thus:

$$\eta = \frac{\left(\dfrac{E_f}{E_m} - 1 \right)}{\left(\dfrac{E_f}{E_m} + \zeta \right)} \tag{3.11}$$

The parameter ζ is the only unknown, and values must be obtained empirically for a given composite material, although they are also sometimes derived by a circular argument involving comparison of equations 3.9 and 3.10 with one of the exact numerical solutions mentioned previously. ζ may vary from zero to infinity, and the Reuss and Voigt models are actually special cases of equation 3.9 for $\zeta = 0$ and $\zeta = \infty$, respectively. A number of analyses have been carried out to compare the predictions of equations 3.9 and 3.10 with elasticity-theory calculations, often with a great degree of success, and it is frequently quoted from the early work of Halpin and Tsai that for practical materials reasonable values of ζ are 1 for predictions of G_{12} and 2 for calculations of E_{22}. It is dangerous, however, to accept these values uncritically for any given composite. This can be illustrated by taking a specific example.

Sensitive measurements have been made of the elastic properties of a sample of unidirectional carbon-fibre-reinforced plastic (CFRP) consisting of Toray T300 carbon fibres in Ciba-Geigy 914 epoxide resin with a fibre volume fraction of 0.56 (Pierron & Vautrin, 1994). Measurements were also made of the properties of the unreinforced matrix resin. Table 3.1 gives the properties of the plain matrix and fibres.

Table 3.1 Properties of T300 Carbon Fibres and 914 Epoxy Resin

Property	Fibres	Matrix
Young's modulus, E, GPa	220	3.3
Shear modulus, G, GPa	25	1.2
Poisson ratio, ν	0.15	0.37

Table 3.2 Predictions of Composite Properties by Simple Micromechanics Models

Equation	Relationship	Predicted Values (Moduli in GPa)	Experimental Values (Moduli in GPa)
3.1	$E_1 = E_f V_f + E_m (1 - V_f)$ (RoM)	124.7	125.0
3.2	$\dfrac{1}{E_2} = \dfrac{V_f}{E_f} + \dfrac{(1 - V_f)}{E_m}$	7.4	9.1
3.6	$\dfrac{1}{G_{12}} = \dfrac{V_f}{G_f} + \dfrac{(1 - V_f)}{G_m}$	2.6	5.0
3.4	$\nu_{12} = \nu_f V_f + \nu_m (1 - V_f)$ RoM	0.25	0.34

The experimental values are first compared in Table 3.2 with the predictions of the simple micromechanical models of section 3.1. It can be seen that the fibre-dominated modulus, E_1, is well predicted by the simple rule of mixtures, while the other properties, E_{12}, G_{12} and ν_{12}, are less well predicted. It is to some extent accidental that the prediction of E_2 in Table 3.2 is as good as it is. The very rigid carbon-carbon bonds in carbon fibres are well-aligned with the fibre length, but in the transverse direction, the fibre is far less rigid, *i.e.* it is highly anisotropic, with an anisotropy ratio (the ratio of the axial to transverse moduli) of about 10. Inserting a more appropriate value of E_f (which is difficult to measure) into equation 3.2 would actually result in a value of E_2 for the composite that was even lower than that predicted in Table 3.2.

Calculations of E_2 and G_{12} by the Halpin-Tsai approximate model (equations 3.9 and 3.10) depend on knowing ζ. By solving these equations for a range of ζ we can then see

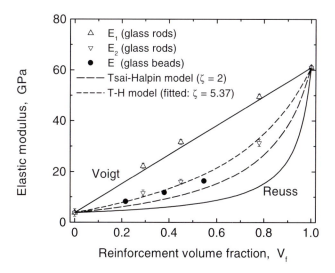

Fig. 3.8 Comparison of experimental data for the elastic moduli of model glass/resin composites with various theoretical predictions. The upper and lower bounds (rule-of-mixtures and inverse RoM) correspond to equations 3.1 and 3.2. The Halpin-Tsai curve for $\zeta = 2$ represents the commonly used function for the transverse modulus, E_2 (equation 3.9) but a value of $\zeta = 5.37$ is necessary to obtain the best fit to the data.

what values of ζ must be used in order to obtain predicted moduli that agree with the experimental values. Taking the data from Table 2, but substituting a more appropriate transverse fibre stiffness of about 22 GPa for the axial stiffness given in the table we find that the appropriate values of ζ are about 2 for E_2 and 2.5 for G_{12}, for this particular material. Use of the usually recommended values of 2 and 1, respectively, would thus be accurate for E_2 but would result in a 25% underestimate for G_{12}. This emphasises the importance of calibrating the model against experimental data rather than against predictions of other models if the method is to be valid for design purposes.

As a second example, we consider results obtained from ultrasonic pulse-velocity measurements on model composites consisting of 3 mm diameter glass rods arranged in regular hexagonal arrays in cast blocks of epoxy resin. The velocity, v, of a compressional pulse travelling through a solid is given by:

$$v = \sqrt{(C/\rho)}$$

where C is an elastic constant and ρ is the density of the material. C is not exactly equal to the conventional engineering Young modulus, E, although it has been shown that for ultrasonic rod waves C and E are practically identical for a wide range of GRP materials (Harris and Phillips, 1972). Figure 3.8 shows some experimental values for the longitudinal and transverse moduli of the model composites, determined by this method, as a function of volume fraction of the glass rods (Shiel, 1995).

The E_1 values are well represented by the rule-of-mixtures line joining the experimental moduli of the plain resin and the glass rods, but the E_2 values lie well above the line representing the Reuss estimate (equation 3.2) and significantly above the predictions of the Halpin-Tsai equation (3.9) with the recommended value of ζ of 2. In order to obtain a good fit to the data points, a value of ζ of 5.37 is required. It is interesting to note that the stiffnesses of model 'particulate' composites consisting of 3mm diameter glass beads randomly dispersed in the same resin (Birch, 1995) are almost indistinguishable from the transverse stiffnesses of the rod-reinforced resin composites, as might be expected. Nielsen (1978) has discussed at length the applicability of the Halpin-Tsai equations to particulate composites.

3.3 THE COMPRESSION MODULUS

We might expect that the compression stiffness of a unidirectional composite would be similar to its tensile stiffness, until we remember that the fine-scale structures of certain fibres — particularly those derived from textiles — are known often to result in unpredictable behaviour of these fibres under shear and bending conditions. We might take as an analogy the behaviour of a piece of rope: tensile forces draw the fibres into tighter bundles, but any attempt to squeeze a length of rope axially will force the fibres apart. Piggott and Harris (1980) illustrated the problem by measuring the compression moduli, E_c, of polyester-based composites reinforced with glass, Kevlar-49 and two varieties of carbon fibres. Their results, illustrated in Figure 3.9, show that the variation with V_f is linear up to only about 50 vol% of reinforcement, and that beyond this the rate of increase of stiffness falls or even becomes negative, depending on the fibre. The change in behaviour is evidently related to the diminishing ability of the matrix to constrain the fibre bundles to deform axially as V_m falls. As the inset figure plotting the initial slopes of these curves against the fibre *tensile* modulus shows, the rule of mixtures is not properly obeyed except by the homogeneous and isotropic glass fibre. The slopes of these curves, dE_c/dV_f, for the two carbon-fibre composites fall slightly below the 1:1 rule-of-mixtures values, while that for the aramid fibre composite falls well below the RoM. These results confirm what we know of the structures of these reinforcing fibres. Some recent elegant experiments by Young *et al.* (1996), who used laser Raman spectroscopy to study the strains in single filaments embedded in transparent resin, clearly showed the markedly non-linear compression behaviour of carbon and aramid fibres, in keeping with the behaviour of the composites illustrated in Figure 3.9.

3.4 THE ANISOTROPIC PROPERTIES OF A THIN LAMINA

Having considered the properties of a lamina parallel with and perpendicular to the fibres, it is clear that the elastic properties of a piece of unidirectional composite must

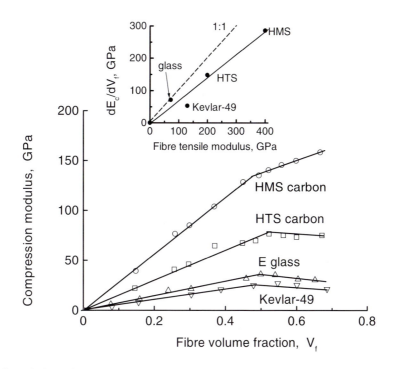

Fig. 3.9 Variation of the compression moduli of polyester-matrix composites with volume fraction of four species of reinforcing fibre.

vary continuously with orientation between these two extremes, and that in order to be able to design with such anisotropic materials it is necessary to have a means of predicting this angular variation.

Most common engineering materials — metals, plastics or ceramics — are usually considered to be isotropic, which is to say that they have the same properties in whichever direction the property is measured. This is not strictly true: wood is an excellent example of an anisotropic material in which the stiffness in the axial direction (parallel to the trunk centre) is greater than the stiffness in any other direction. Cold-worked metals also often have different properties parallel with and perpendicular to the direction of working, although designers usually treat them as isotropic. An isotropic material has four characteristic elastic constants, the Young modulus, E, the shear or rigidity modulus, G, the bulk modulus, K, and the Poisson ratio, ν. In isotropic bodies, however, only two of these are independent properties. If, for example, we know E and ν, we can calculate the other two constants from the familiar relationships (see, for example, Dugdale, 1968):

$$G = \frac{E}{2(1+\nu)}, \quad K = \frac{E}{3(1-2\nu)}$$

In an anisotropic fibre composite material these simple relationships cannot be valid. But in order to discuss anisotropy, we first need a frame of reference, and we choose for

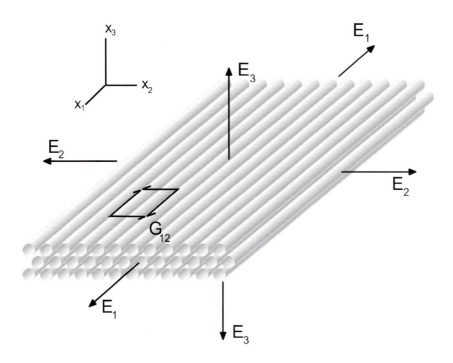

Fig. 3.10 Definition of elastic constants for an anisotropic thin composite lamina.

convenience to use an orthogonal set of axes, (x_1, x_2, x_3), in which the fibres are aligned in the x_1 direction and transverse to the x_2 direction, as shown in Figure 3.10.

For this thin plate of composite, we might guess that there must be three values of Young's modulus, E_1, E_2, and E_3, three shear moduli, G_{12}, G_{13}, and G_{23}, and six values of the Poisson ratio, v_{12} and v_{21}, v_{13} and v_{31}, and v_{23} and v_{32}, (not all of these are shown in Figure 3.10, for the sake of clarity) instead of the two independent constants that we need for an isotropic material. However, consideration of the geometry of the laminate suggests that E_3 is not likely to be very different from E_2 and G_{13} is similar to G_{12}, although G_{23} will probably be different from the other two shear moduli. Similarly, v_{12} will be close in value to v_{13}. By contrast with the shear moduli, however, for which symmetry implies that $G_{12} = G_{21}$, it is clear, from the definition of Poisson's ratio given earlier, that $v_{12} \gg v_{21}$, and so forth. The response of a unidirectional lamina, such as that shown in Figure 3.10, to applied stresses must therefore be more complex than that of an isotropic material.

The general state of stress on a body can be written in the form of a group of six independent components, $\sigma_1, \sigma_2, \sigma_3, \tau_{12}, \tau_{23}, \tau_{31}$, where the single suffices represent normal components (tension or compression) and the double suffices represent shear components (see Dugdale, 1968). Likewise, the strains form a similar series of six components, $\varepsilon_1, \varepsilon_2, \varepsilon_3, \gamma_{12}, \gamma_{23}, \gamma_{31}$. When a general stress system consisting of the six stress components acts on an isotropic body, the familiar Hooke's-law relationship between stress and strain is usually written in the form of a family of equations for the strain components:

$$\varepsilon_1 = \frac{1}{E}\left[\sigma_1 - v(\sigma_2 + \sigma_3)\right]$$

$$\varepsilon_2 = \frac{1}{E}\left[\sigma_2 - v(\sigma_3 + \sigma_1)\right]$$

$$\varepsilon_3 = \frac{1}{E}\left[\sigma_3 - v(\sigma_1 + \sigma_2)\right] \tag{3.12}$$

$$\gamma_{12} = \frac{1}{G}\tau_{12} \quad \gamma_{23} = \frac{1}{G}\tau_{23} \quad \gamma_{31} = \frac{1}{G}\tau_{31}$$

These equations can also be written in matrix format:

$$
\begin{bmatrix} \varepsilon_1 \\ \varepsilon_2 \\ \varepsilon_3 \\ \gamma_{23} \\ \gamma_{31} \\ \gamma_{12} \end{bmatrix}
=
\begin{bmatrix}
1/E & -v/E & -v/E & & & \\
-v/E & 1/E & -v/E & & & \\
-v/E & -v/E & 1/E & & & \\
& & & 1/G & & \\
& & & & 1/G & \\
& & & & & 1/G
\end{bmatrix}
\times
\begin{bmatrix} \sigma_1 \\ \sigma_2 \\ \sigma_3 \\ \tau_{23} \\ \tau_{31} \\ \tau_{12} \end{bmatrix}
\tag{3.13}
$$

The order of the stress and strain components is arbitrary, and the sequences above are those conventionally used. In this form, the individual terms in the matrix of constants relating strain and stress are referred to as *compliances* and, as we have already said, there are only two independent elastic constants in the matrix. The blanks in the matrix, which are really zeros, show that the normal stress components do not cause shear deformations, and the shear stresses do not cause tensile or compressive deformations. For a thin rod of isotropic material under a simple tension stress, σ_1, equations 3.12 and 3.13 reduce to the familiar form of Hooke's law given by:

$$\sigma_1 = E\,\varepsilon_1$$

and for a thin plate subject to a plane-stress state only, since no stress components containing a suffix 3 are then of any importance ($\sigma_3 = \tau_{13} = \tau_{23} = 0$), the Hooke's law relationship of equation 3.13 reduces to:

$$
\begin{bmatrix} \varepsilon_1 \\ \varepsilon_2 \\ \gamma_{12} \end{bmatrix}
=
\begin{bmatrix}
1/E & -v/E & 0 \\
-v/E & 1/E & 0 \\
0 & 0 & 1/G
\end{bmatrix}
\begin{bmatrix} \sigma_1 \\ \sigma_2 \\ \tau_{12} \end{bmatrix}
\tag{3.14}
$$

By contrast, in an anisotropic solid like a fibre composite, *each* of the strain components has to be related to *all* of the stress components, and it is possible that all of the elements in the matrix of equations 3.13 could have real values — a total of 36 elastic constants. In reality, however, the symmetry of a unidirectional material makes it possible to reduce the number of constants and since the most demanding of current applications for composites call for the use of thin plates, the plane-stress state allows further simplification.

Since we shall want to make some limited use of the formal tensor theory of anisotropic elastic behaviour, we must first introduce some new definitions. First, a unidirectional lamina, as discussed with reference to Figure 3.10, is regarded as transversely isotropic, that is to say that it has one major axis of anisotropy — the fibre direction — and perpendicular to this the properties are independent of orientation (*i.e.* $E_2 = E_3$, *etc.*). By convention we take the fibre axis to be parallel to x_1 and the lamina is then referred to as *orthotropic*. Second, in tensor notation the definitions of shear strains are slightly different from the engineering definitions that we have been using so far. Although the engineering shear stress component identified as τ_{12} is the same as the tensor stress σ_{12}, the tensor strain ε_{12} is only half the engineering strain γ_{12}. For the orthotropic lamina loaded in plane stress, as we have seen, no stress component with a suffix 3 is present, and the Hooke's law tensor relationship, modifying equations 3.13, is then written:

$$
\begin{bmatrix} \varepsilon_1 \\ \varepsilon_2 \\ \varepsilon_{12} \end{bmatrix} = \begin{bmatrix} S_{11} & S_{12} & 0 \\ S_{12} & S_{22} & 0 \\ 0 & 0 & S_{66} \end{bmatrix} \times \begin{bmatrix} \sigma_1 \\ \sigma_2 \\ \sigma_{12} \end{bmatrix}
\tag{3.15}
$$

or

$$
\varepsilon_{ij} = S_{rs}\,\sigma_{kl}
\tag{3.16}
$$

where the S_{rs} terms are referred to as the *elastic compliances*. A similar relationship gives stress as a function of strain:

$$
\sigma_{ij} = C_{rs}\,\varepsilon_{kl}
\tag{3.17}
$$

where the C_{rs} terms are referred to as *elastic constants* or *elastic coefficients*, and it should be borne in mind that in general C is not simply the reciprocal of S.

Note: although the suffices for stress and strain, as in ε_1, ε_{12}, *etc.* are directly related to the orthogonal axis designations, x_1, x_2, those used for the S_{rs} terms are not: they simply reflect the rows and columns in the compliance tensor — hence, S_{66} simply corresponds to the last of the compliance components in equations 3.13.

Again we see that the elasticity tensor contains some zero values: shear stresses do not cause normal strains and normal stresses do not cause shear strains, and we describe this formally by saying that there is no *tension/shear coupling*. The significance of this will become apparent later. The compliances, S_{rs}, in equations 3.15 are related to our normal engineering elastic constants for the composite lamina:

$$
S_{11} = 1/E_1 \qquad S_{22} = 1/E_2 \qquad S_{66} = 1/2G_{12}
$$

$$
S_{12} = -\nu_{12}/E_1 = -\nu_{21}/E_2
$$

3.5 ORIENTATION-DEPENDENCE OF THE ELASTIC PROPERTIES OF A UNIDIRECTIONAL LAMINA

Composites are rarely used in the form of unidirectional laminates, since one of their great merits is that the fibres can be arranged so as to give specific properties in any desired direction. Thus, in any given structural laminate, predetermined proportions of the unidirectional plies will be arranged at some specific angle, θ, to the stress direction. In order to calculate the properties of such a multi-ply laminate, it is first necessary to know how the elastic response of a single unidirectional lamina, such as that which we have been considering so far, will vary as the angle to the stress direction is changed. This is done by transforming the $x_1 x_2$ axes through some arbitrary angle, θ, a procedure that will be familiar to anyone who has studied the derivation of the Mohr's circle construction. The derivation in given in Appendix 1, and for our present purpose we need only the result of this analysis to illustrate the full anisotropic behaviour of a single unidirectional lamina. If both the stresses and the strains are transformed through an angle θ from the x_1, x_2 co-ordinates to an arbitrary pair of axes (x,y), we find that the appropriate form of Hooke's Law is now:

$$
\begin{bmatrix} \varepsilon_x \\ \varepsilon_y \\ \varepsilon_{xy} \end{bmatrix} = \begin{bmatrix} \overline{S}_{11} & \overline{S}_{12} & \overline{S}_{16} \\ \overline{S}_{12} & \overline{S}_{22} & \overline{S}_{26} \\ \overline{S}_{16} & \overline{S}_{26} & \overline{S}_{66} \end{bmatrix} \times \begin{bmatrix} \sigma_x \\ \sigma_y \\ \sigma_{xy} \end{bmatrix}
\tag{3.18}
$$

Except for the case when the (x,y) axes coincide with the composite orthotropic axes (the fibre direction, x_1, and the transverse direction, x_2) the compliance matrix, $\overline{S}_{rs}$, now contains six elastic constants instead of the four in the matrix of equations 3.15. Since there are now no zeros in the compliance matrix each component of stress will contribute to all of the components of strain. A shear stress will therefore induce normal strains and normal stresses will induce shear strains. This *shear-tension coupling* has important consequences for designers since a thin plate of a unidirectional composite will show complex out-of-plane distortion if loaded off-axis. This is illustrated in Figure 3.11 for a plate of a unidirectional lamina with fibres at 30° to the loading axis which is subjected to a central point bending load. The out-of-plane distortion is calculated by the commercial laminate analysis programme LAP (Anaglyph Ltd, U.K.).

The compliance terms in the matrix in equations 3.18 are barred to remind us that these are no longer material elastic constants like the S_{rs} in equations 3.15, but are functions of stress, as shown by the presence of the matrix functions T and T^{-1} in their definitions (Appendix 1). It can be seen that if only a tensile stress, σ_x, is applied, the strain measured in the direction of the stress, ε_x, is related to σ_x by the familiar Hooke's law, *i.e.*

$$\varepsilon_x = \overline{S}_{11}\sigma_x$$

and the compliance $\overline{S}_{11}$ therefore represents the reciprocal Young's modulus for this deformation:

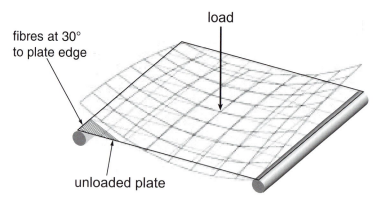

Fig. 3.11 Out-of-plane distortion when an off-axis unidirectional laminate is loaded in bending.

$$\bar{S}_{11} = \frac{1}{E_x}$$

But, clearly, $\bar{S}_{11}$, and therefore also E_x, are functions of the orientation, θ. The matrix algebra needed to evaluate the barred compliances in equations 3.18 can be carried out painlessly with Mathcad, and it can be shown that the full expression for $\bar{S}_{11}$ is:

$$\bar{S}_{11} = S_{11}\cos^4\theta + S_{22}\sin^4\theta + 2(S_{12} + S_{66})\cos^2\theta\sin^2\theta \qquad (3.19)$$

where the compliances S_{11}, S_{22}, S_{12}, and S_{66} are the four elastic constants for the orthotropic lamina relative to the (1,2) axes, as defined in terms of the normal engineering constants at the end of section 3.3. If those definitions are substituted in equation 3.19, we have the orientation dependence of the elastic modulus, which we shall call $E(\theta)$:

$$\frac{1}{E(\theta)} = \frac{\cos^4\theta}{E_1} + \frac{\sin^4\theta}{E_2} + \left(\frac{1}{G_{12}} - \frac{2\nu_{12}}{E_{11}}\right)\cos^2\theta\sin^2\theta \qquad (3.20)$$

Similarly, the shear compliance component, $\bar{S}_{66}$, can be evaluated to obtain the orientation dependence of the shear modulus, $G(\theta)$:

$$\frac{1}{G(\theta)} = 4\left(\frac{1+2\nu_{12}}{E_1} + \frac{1}{E_2} - \frac{1}{2G_{12}}\right)\cos^2\theta\sin^2\theta + \frac{1}{G_{12}}\left(\cos^4\theta + \sin^4\theta\right) \qquad (3.21)$$

Equations 3.20 and 3.21 are plotted to show the angular dependence (and the high degree of anisotropy) of two typical unidirectional laminates in Figure 3.12. It can be seen that the stiffness of the unidirectional composite falls rapidly as the loading direction shifts away from the fibre axis, the effect being more marked the higher the stiffness of the reinforcing fibres. By contrast, the shear stiffness of the composite is less sensitive to

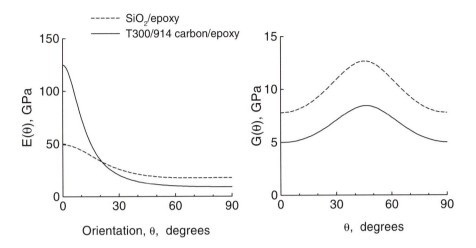

Fig. 3.12 Dependence of Young's modulus, $E(\theta)$, and the shear modulus, $G(\theta)$, on the angle, θ, between the fibres and the stress axis for a carbon-fibre composite and a silica/epoxy composite.

the fibre characteristics, as would be expected, being largely controlled by the matrix rigidity. The shear modulus is a maximum, however, in a direction at 45° to the fibre axis. The validity of equations 3.20 and 3.21 for predicting the off-axis elastic properties of laminates is usually very good, as can be judged from Figure 3.13 which compares the predicted values of $E(\theta)$ for the silica/epoxy composite in Figure 3.12 with experimental data for this material (Pabiot, 1971).

For most reinforced plastics composites, where the modulus ratio E_f/E_m is large and the axial lamina properties are dominated by the fibre stiffness, the influence of the values of in-plane shear modulus and Poisson ratio on the form of the orientation dependence defined by equation 3.20 is not significant. But in certain materials, like ceramic-matrix composites where the stiffnesses of the fibres and matrix are often similar, variations in the value of Poisson's ratio, caused by changes in processing conditions, for example, may modify the form of $E(\theta)$ substantially, as illustrated by the surface plot of $E(\theta)$ for a range of values of v_{12} for a SiC-reinforced calcium aluminosilicate CMC manufactured by Corning (as shown in Figure 3.14).

3.6 MULTI-PLY LAMINATES

Unidirectional composites are rarely used in practice because of the high level of anisotropy demonstrated by Figure 3.12. In practice it is more sensible to make use of the versatility of composite materials by arranging the fibres in different directions to

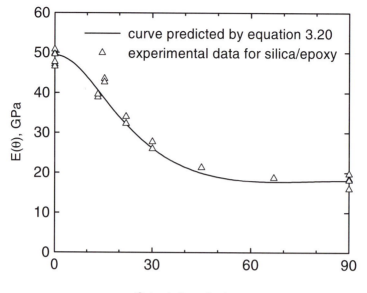

Fig. 3.13 Comparison of predicted and experimental results for the angular variation of Young's modulus of a unidirectional silica/epoxy composite (data of Pabiot, 1971).

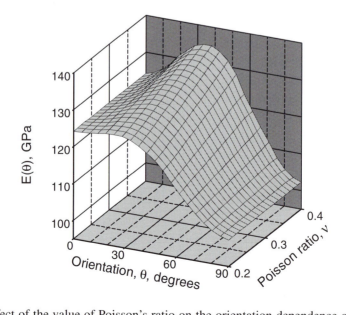

Fig. 3.14 Effect of the value of Poisson's ratio on the orientation dependence of Young's modulus for SiC/CAS ceramic-matrix composites.

suit the design requirements. There are various ways in which this is done, depending on the level of sophistication of the required product. It is useful at this stage, however, to refer to the most common procedure for producing high-performance laminates, which is to stack groups of single-ply laminae cut at various angles from the continuous sheet of 'prepreg' — a continuous layer of fibres preimpregnated with resin which is then dried (but not cured) — and then hot-press them between the heated platens, flat or shaped, of a press. The stacking sequence of such a laminate is described in short-hand form by formulae of the following kind:

$(0)_{12}$ — a unidirectional composite with 12 plies

$(0,90)_{2s}$ — a cross-plied laminate with four pairs of 0/90 laminae arranged symmetrically (indicated by the 's') (0,90,0,90,90,0,90,0)

$[(\pm45,0_2,90_2)_2]_s$ — a quasi-isotropic laminate with 24 plies arranged symmetrically

(+45,-45,0,0,90,90,+45,-45,0,0,90,90,90,90,0,0,-45,+45,90,90,0,0,-45,+45), the external ± 45 plies providing protection against impact damage.

The choice of laminating sequence may vary widely, depending upon the particular design requirements, but an indication of the potential for modifying the lay-up to achieve a desired result can be seen by comparing the orientation dependence of the stiffness of a unidirectional laminate with the stiffness of a series of composites of lay-up ± θ, where in each case θ varies from 0 to π/2, as shown in Figure 3.15. Cross-laminating is a compromise solution which results in improved torsional and transverse rigidity at the expense of some loss in longitudinal stiffness. It can be seen, for example, that by using a ± 20° laminate instead of a unidirectional composite, for only a 34% loss in axial stiffness, the torsional rigidity may be doubled. The extent of the axial stiffness reduction caused by cross-laminating is greater the stiffer the reinforcing filaments. Nielsen and Chen (1968) demonstrated the significance of this by showing the effect of randomising the fibre distribution in the plane of the laminate. The modulus, E(θ), given by equation 3.20, is randomised over all θ:

$$\langle E \rangle = \frac{\int_0^{\pi/2} E(\theta)\, d\theta}{\int_0^{\pi/2} d\theta} \tag{3.22}$$

The solution to this equation, also obtained in Mathcad, for a pair of composites with different modulus ratios, E_f/E_m is shown in Figure 3.16. For most practical purposes (Hull, 1981) the solution to equation 3.22 can be approximated reasonably closely by:

$$\langle E \rangle = \tfrac{3}{8} E_1 + \tfrac{5}{8} E_2 \tag{3.23}$$

where E_1 and E_2 are the known or calculated longitudinal and transverse moduli for a unidirectional lamina. An analogous approximate expression can be obtained for the shear modulus of a random continuous-fibre composite:

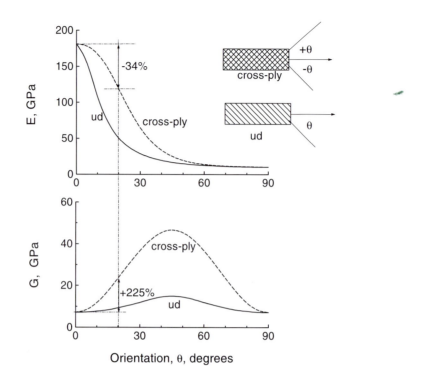

Fig. 3.15 The effect of cross-laminating on Young's modulus and the shear modulus of a T300/5208 carbon-fibre/epoxy composite.

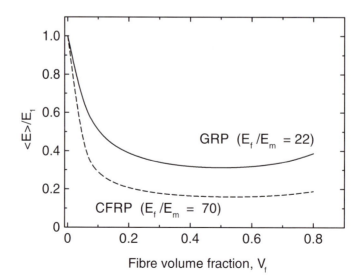

Fig. 3.16 Effect of randomising the fibre orientations on the stiffnesses of CFRP and GRP laminates.

$$\langle G \rangle = \tfrac{1}{8} E_1 + \tfrac{1}{4} E_2 \tag{3.24}$$

Multiply laminates usually, but not always, consist of regular arrays of plies of the same material and of the same thickness. When a laminate contains equal numbers of plies in each orientation, it is said to be *balanced*. A $(0,90)_3$ laminate is thus a balanced laminate, but such a laminate would have 0° and 90° plies on opposite faces and may behave in unexpected ways if loaded in other than the 0° and 90° directions. Laminates that are arranged to have the plies on either side of a mid-plane matching in both material and orientation are described as *symmetric:* such laminates have the advantage that they remain flat after curing and during deformation. Symmetric laminates are therefore preferred for most purposes, but careful tailoring of a laminate lay-up can be used deliberately to produce a material that will change shape in some desired manner under stress.

For design purposes, it is necessary to have accurate methods of calculating the elastic properties of a composite with a given lay-up as part of the process of tailoring a material to suit a given set of requirements. For a simple 0/90 lay-up it is a simple matter to treat the laminate as a parallel combination of materials with two different elastic moduli and apply the mixture rule as given by equation 3.1, provided accurate values of the two moduli E_1 and E_2 are known. Thus, from dynamic measurements on an 11-ply unidirectional GRP composite of $V_f = 0.64$, values of E_1 and E_2 of 46.1 GPa and 19.6 GPa were measured (Harris *et al.*, 1978: see Table 3.3). For an 11-ply 0/90 laminate from the same manufacturer with the same raw material specifications, we may rewrite the rule of mixtures:

$$E_0 = \tfrac{6}{11} E_1 + \tfrac{5}{11} E_2$$

$$E_{90} = \tfrac{5}{11} E + \tfrac{6}{11} E_2$$

to give predicted values of E_0 and E_{90} in the two orthogonal directions. As Table 3.3 shows, the estimates would be adequate for most practical purposes.

A preliminary estimate of the elastic response of a general multi-ply laminate can often be obtained by a netting analysis (Cox, 1952; Krenchel, 1964) which sums the contributions from each group of fibres lying at a specific angle, θ, to the applied stress. If we assume that the fibres are continuous and that there are no elastic Poisson constraints ($v_f = v_m$), each group of fibres (a proportion a_n of the total fibre content) can be considered to have a reinforcing efficiency $a_n \cos^4\theta$, and an overall composite efficiency factor, η_θ, can be defined as:

$$\eta_\theta = \sum a_n \cos^4\theta \tag{3.25}$$

The approximate composite modulus can then be obtained from a modified rule of mixtures as:

$$E_c = \eta_\theta E_f V_f + E_m \left(1 - V_f\right) \tag{3.26}$$

The fourth-power geometrical term in the efficiency factor clearly indicates its relationship with equation 3.19. Shear and transverse behaviour are ignored, however, and the limitation of the model is therefore that it allows no stiffness contribution from

Table 3.3. Comparison of Some Predicted and Experimental Elastic Modulus Values for Typical GRP Laminates

Material	V_f	Density 10^3 kg.m^{-3}	Orientation	Dynamic Modulus, GPa	Predicted Modulus, GPa	Model
Pultruded Rod	0.70	2.09	0 90 90	54.4 24.7 24.7	51.3 9.1 24.6	RoM Inverse RoM Halpin-Tsai ($\zeta = 4.6$)
UD 11-Ply Laminate (Permali, U.K.)	0.64	2.04	0 90 90 45	46.1 19.6 19.6 20.4	47.2 7.8 19.6 12.6	RoM Inverse RoM Halpin-Tsai ($\zeta = 4$) Krenchel
0/90 11-Ply Laminate (Permali, U.K.)	0.63	2.03	0 (6 plies) 90 (5 plies) 45	35.3 30.5 21.6	34.0 31.6 12.6	RoM RoM Krenchel
0.90 Woven 35-ply Laminate (Permali, U.K.)	0.47	1.85	0 90 45	26.5 27.5 16.7		

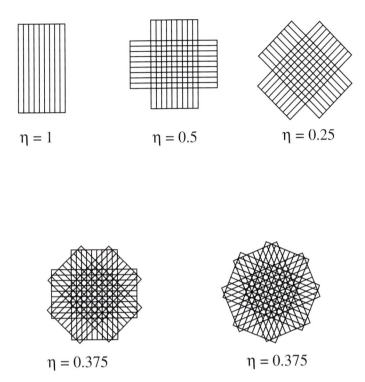

$\eta = 1$ $\eta = 0.5$ $\eta = 0.25$

$\eta = 0.375$ $\eta = 0.375$

Fig. 3.17 Values of the Krenchel efficiency factor for various fibre groupings.

transverse fibres and ignores interlayer constraints. For composites containing high volume fractions of high-stiffness fibres, none of which are at 90° to the stress axis, it can give reasonable estimates of the laminate stiffness, but it is not acceptable for detailed design purposes. For the GRP materials referred to in Table 3.3, for example, the predictions of the stiffnesses E_{45} are poor. Values of the efficiency factor, η_θ, for a range of fibre distributions are shown schematically in Figure 3.17. For a 3-D random array η_θ is about 1/6. It can be seen from Figure 3.16 that although the Krenchel factor is approximately in agreement with the more general randomisation model for GRP, it gives a poor prediction for CFRP.

In real laminates the individual plies which are stacked together to produce the multi-ply structure each have their own characteristic anisotropy. If a ply is at an angle to the load axis the simplified form of Hooke's law for the orthotropic case (equation 3.15) is no longer applicable and the more complex form of equation 3.18 must be used. As we have seen, the zeros in equation 3.15 are replaced by non-zero values and there is then tension/shear coupling — *i.e.* the application of a tensile force will induce shear strains and the application of shear forces will induce normal strains. This coupling can lead to unexpected distortions of composites under load and unless these are carefully balanced in constructing the laminate interlaminar shear forces may be generated that are high

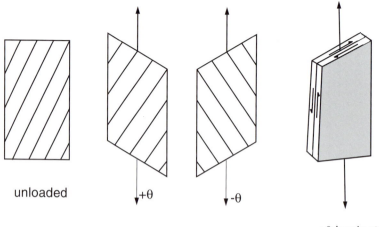

unloaded   +θ   -θ

±θ laminate

Fig. 3.18 Out-of-plane distortion in an unsymmetric laminate consisting of a single pair of off-axis (θ) plies. The anisotropic shape changes induced in each separate ply can only be accommodated in the laminate by the distortion shown, accompanied by the development of interlaminar shear stresses.

enough to cause delamination (separation of the individual plies). A schematic illustration of the effect of shear/tension coupling in causing out-of-plane distortion in a simple ± θ laminate is shown in Figure 3.18 (Soden and Eckold, 1983).

A full treatment of the elastic behaviour of thin composite laminates is beyond the scope of this book (readers are referred to the standard texts listed in the bibliography to Chapter 1) but an indication of the approach used in the classical theory may be given. Assuming symmetry about the centre line and in-plane loads only, as shown in Figure 3.19, the elastic response of the kth ply will, in the general case, be described by equations A6 and A6a of Appendix 1, *i.e.*

$$[\varepsilon]_k = [\overline{S}]_k [\sigma]_k$$
$$[\sigma]_k = [\overline{Q}]_k [\varepsilon]_k$$

in simplified form. The second of these relates each of the stress components to all of the strain components through a matrix of elastic coefficients:

$$\begin{bmatrix} \sigma_x \\ \sigma_y \\ \sigma_{xy} \end{bmatrix} = [A_{ij}] \begin{bmatrix} \varepsilon_x \\ \varepsilon_y \\ \varepsilon_{xy} \end{bmatrix} \tag{3.27}$$

where $[A_{ij}]$ is a stiffness matrix whose elements (the $\overline{Q}$ coefficients in equation A6a) are functions of the unidirectional lamina elastic constants E_1, E_2, G_{12}, ν_{12}, and the

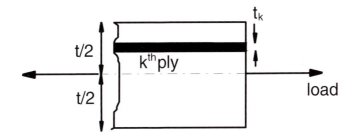

Fig. 3.19 In-plane loading of a laminate of thickness t, and the location of the kth ply relative to the mid-plane.

transformation angle θ. If the total composite is composed of a large number of evenly stacked layers, its properties can be determined by summing the stiffness matrices, $[A_{ij}]$, of each layer proportional to the number of plies lying in each direction:

$$\left[A_{ij}\right]_{composite} = \frac{1}{t}\sum_{k=1}^{k=n}\left(A_{ij}\right)_k . t_k \tag{3.28}$$

where n is the number of plies, t_k is the thickness of the kth ply, and t is the total laminate thickness.

On the other hand, if the laminate contains relatively few plies, the effect of stacking sequence becomes important. Unsymmetric arrangements of plies, for example, may cause out-of-plane deformations (twisting or bending) such as that illustrated in Figure 3.18. These effects are mostly undesirable, but on occasion they may be used to advantage, as in the aeroelastic tailoring of aircraft structures, designing the lay-up so that it changes shape (and therefore its aerodynamic performance) in a predictable way in response to aerodynamic forces. In the full classical theory of thin laminates, therefore, it is necessary to obtain the resultant forces and moments acting on the laminate by integrating the stresses in each ply through the laminate thickness. In addition to the A matrix of equation 3.27 which relates in-plane forces to in-plane deformations, the full constitutional equation for laminate deformation then includes relationships between in-plane forces, N, moments, M, in-plane deformations, ε, and curvatures, x, which are conveniently represented in the form of a partitioned matrix:

$$\begin{bmatrix} N \\ M \end{bmatrix} = \begin{bmatrix} A & B \\ \hline B & D \end{bmatrix} \begin{bmatrix} \varepsilon \\ x \end{bmatrix} \tag{3.29}$$

where A is the extensional stiffness matrix that we have already met, D represents the bending stiffnesses, and B represents the stiffness matrix for coupling between these two.

The calculations involved in analysing a complex laminate fully are tedious but straightforward, and are ideally carried out on one of the many pieces of computer software

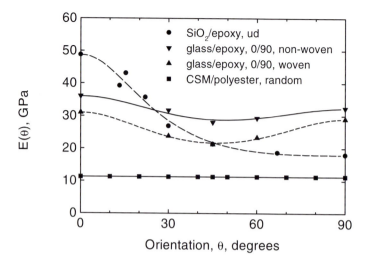

Fig. 3.20 Orientation-dependence of Young's modulus for four types of glass-fibre-reinforced plastic. The curves fitted to the upper three data sets correspond to equation 3.20.

currently available. The LAP programme by Anaglyph, U.K., previously mentioned is one such programme which operates under Microsoft Windows, but there are many others, of different degrees of sophistication, offering a variety of calculation facilities and graphics.

As an illustration of the general effects of structure and orientation on the elastic stiffness of glass-fibre laminates, Figure 3.20 shows the variation with orientation of Young's modulus for three types of laminate, the unidirectional silica/epoxy material of Figure 3.13, the 0/90 non-woven and woven glass/epoxy laminates referred to in Table 3.3, and an isotropic GRP laminate made with randomly arranged chopped-strand mat (CSM) reinforcement.

3.7 SHORT-FIBRE COMPOSITES

High-performance composites are generally made from continuous fibres, but there are many applications for which the requirements are less demanding, or for which the appropriate manufacturing route cannot handle long fibres: it is then natural to consider using short fibres. Some reinforcing fibres are also available only in the form of short filaments.

The ends of a fully-embedded short fibre cannot be fully loaded by shear at the fibre/matrix interface. This can be seen by embedding different lengths of a fibre in a resin block and attempting to pull the free end out. The pull-out force is resisted by

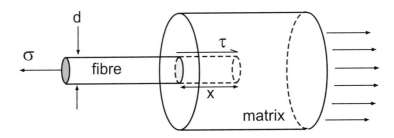

Fig. 3.21 Forces on a fibre being pulled from a block of matrix material.

interfacial friction or bonding, and if the embedded length, x, is greater than some critical length, l_c the tensile stress in the fibre will break the free fibre before the embedded end can be pulled out. This critical length can be determined by considering the force required to pull an embedded length, x, of fibre of diameter, d, from a block of matrix (Figure 3.21). A force P applied to the fibre develops a tensile stress $\sigma = 4P/\pi d^2$ in the unembedded length and this is balanced by a mean shear force at the interface equal to $\tau \pi dx$. Thus, the ratio of embedded length to fibre diameter, $x/d = \sigma/4\tau$. If the fibre can just be withdrawn from the matrix by a tensile load only marginally smaller than the fibre breaking load, x must be equal to half the critical length, $l_c/2$, since only one fibre end is concerned. Thus, the critical aspect ratio is:

$$\frac{\ell_c}{d} = \frac{\sigma_f}{2\tau_i} \tag{3.30}$$

where σ_f is the breaking stress of the fibre and τ_i is the interfacial shear strength. This is a very simplistic model which assumes that all deformations are entirely elastic and that the shear stress is independent of embedded length, but it gives useful insights into composite behaviour.

We imagine an aligned array of short fibres of identical length embedded in a composite which is loaded parallel with the fibres, as shown in Figure 3.22. Since the load on the composite must be transferred into the fibre by shear at the interface, it can be seen that no transfer can occur at the extremities of the fibre, and that the tensile stress in the fibre will build up from zero at the ends. At the same time, the interface shear stress will be high at the fibre ends, and will fall to zero when the full extent of load transfer has been achieved, as shown in Figure 3.23.

The calculations on which this figure is based are the result of what is referred to as a 'shear-lag' analysis (Cox, 1952) which involves determining the conditions for equilibrium of the shear and tensile forces in a simple twin-cylinder model such as that shown in Figure 3.24. For the results shown in Figure 3.23, Cox's equations for the fibre tensile stress and the interfacial shear stress have been used to simulate the situation for

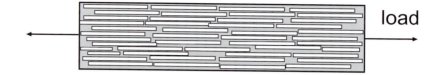

Fig. 3.22 Idealized short-fibre composite containing discontinuous but aligned fibres.

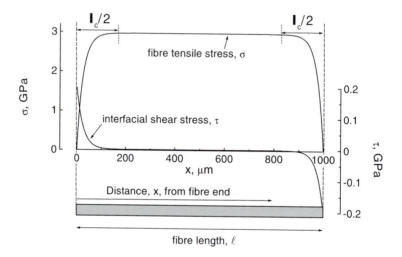

Fig. 3.23 Variation of the tensile stress, σ, in the fibre and the shear stress, τ, at the interface along the length of a short fibre embedded in a matrix. The calculations, based on the shear-lag model of Cox (1952), relate to a 1000 μm length of carbon fibre in an epoxy resin matrix.

a 1 mm length of carbon fibre (Young's modulus 300 GPa) embedded in an epoxy resin matrix. Other solutions for the stress distributions also exist, but the general validity of the functions which give rise to Figure 3.23 have been demonstrated both by photoelastic experiments and, in recent years, by means of laser Raman spectroscopy which is able directly to determine, at the molecular scale, the strain at any point along a single fibre embedded in a resin block under external load (see, for example, Galiotis, 1991).

In the central portion of a long fibre, the Cox stress functions vary only very slowly towards the fibre centre, so that we can usually make the assumption that there is a plateau, as illustrated in Figure 3.23. The ends of a short fibre thus constitute a notional 'ineffective' length which reduces the fibre reinforcement efficiency. The contribution

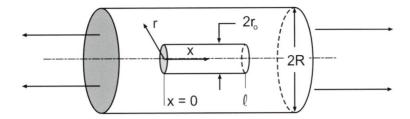

Fig. 3.24 Concentric cylinder model used to determine the stress distributions shown in Figure 3.23.

of these short, aligned fibres to the stiffness of the composite will therefore be less than the rule-of-mixtures contribution, $E_f V_f$, by an amount that will depend on the efficiency of the stress transfer, which is determined by the relative stiffnesses of the fibre and the matrix.

Thus, the mixture-rule equation needs to be modified to allow for this fibre-end effect by introducing another efficiency factor, usually designated η_l, so that:

$$E_c = \eta_l E_f V_f + E_m (1 - V_f) \tag{3.31}$$

The Cox shear-lag model leads to a value for the efficiency factor, η_l,

$$\eta_l = 1 - \frac{\tanh\left[\dfrac{\beta\ell}{2}\right]}{\dfrac{\beta\ell}{2}} \tag{3.32}$$

where β, which determines the rate at which the tensile stress builds up at the fibre ends, is a function of the fibre and matrix stiffnesses and the composite geometry, *viz.*

$$\beta = \sqrt{\frac{2G_m / E_f}{r_o^2 \log_e (R/r_o)}} \tag{3.33}$$

In a regular fibre arrangement, the fibre radius, r_o, and the matrix cylinder radius, R, are related to the fibre volume fraction. For example, if we assume a regular array, as in Figure 3.25, it can be seen that the triangle defines a unit cell which, if duplicated many times, would result in hexagonal symmetry. The area of the triangle is $\frac{1}{2}(2R)^2 \sin 60$. This area contains three fibre sectors, each of area $(\pi r_o^2)/6$, and the total area of fibre is $\frac{1}{2}\pi r_o^2$. From the definition of fibre volume fraction (and assuming $A_f \approx V_f$), it can be seen that:

$$V_f = \frac{\frac{1}{2}\pi r_o^2}{\frac{1}{2}(2R)^2 \sin 60} = \frac{\pi r_o^2}{2\sqrt{3}R^2} \tag{3.34}$$

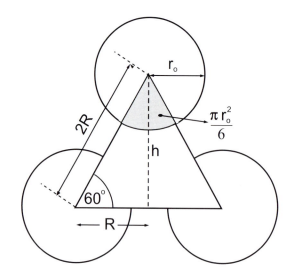

Fig. 3.25 Fibre packing in a simple hexagonal arrangement.

and the value of R in equation 3.33 can therefore be written in terms of the fibre distribution. If equations 3.30 to 3.34 are now used to determine values of the stiffnesses of typical GRP and CFRP composites reinforced with short fibres, the results can be plotted as functions of the fibre length or the fibre aspect ratio, l/d, as in Figure 3.26, to show these stiffnesses as percentages of the moduli of unidirectional laminates of the same composites but reinforced with continuous fibres, as given by the normal rule of mixtures, equation 3.1. We see that η_l tends to unity as $l \rightarrow \infty$, and the efficiency of stress transfer is marginally higher in the GRP than in the CFRP. For aspect ratios of the order of 100 (*i.e.* for fibre lengths of the order of only 1 mm) some 95% of the full potential stiffening effect of fine reinforcing fibres like glass and carbon can be achieved in composites reinforced with chopped, rather than continuous, fibres.

In deriving Figure 3.26, we have ignored certain problems in applying the Cox model. It has been pointed out (Galiotis and Paipetis, 1998) that in reality the matrix shear stiffness in equation 3.33 is affected by interface chemistry, fibre surface treatments, elastic constraints due to fibre volume fraction, *etc.*, and R is much more difficult to define for a real composite than is implied by equation 3.34. An analytic derivation of β is therefore questionable. They propose instead that β should be treated as an empirical index of the efficiency of stress transfer which can be considered as an 'inverse length'. For values of $\beta l \geq 10$, the Cox hyperbolic equation for the fibre tensile stress, $\sigma(x)$, which is plotted in Figure 3.23 can be simplified to give:

$$\sigma(x) = \sigma_\infty \left(1 - \exp(-\beta x)\right) \tag{3.35}$$

Direct estimates of β can be made by the laser-Raman spectroscopy technique mentioned earlier.

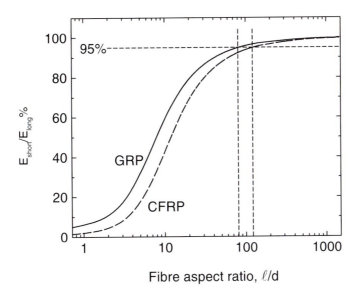

Fig. 3.26 Variation with fibre aspect ratio, l/d, of the stiffnesses of aligned short-fibre composites of GRP and CFRP for a fibre volume fraction of 0.5. The results are given as percentages of the normal continuous-fibre composite stiffnesses.

If the short-fibre composite consists of layers of material with the fibres arranged at different angles to the stress direction, equation 3.31 can be further modified by the inclusion of the appropriate value of the Krenchel factor, η_θ, thus:

$$E_c = \eta_\ell \eta_\theta E_f V_f + E_m \left(1 - V_f\right) \tag{3.36}$$

to give an estimate of the composite stiffness.

The short-fibre composite model described above is clearly based on an idealised structure consisting of short fibres, all of identical length, arranged in a unidirectional array. Although there are manufacturing processes which can achieve such structures (see chapter 1, section 1.3.6), they are rather specialised and composites reinforced with chopped or short fibres are, for the most part, likely to be of the type which are produced by the injection moulding of chopped-fibre reinforced thermoplastics or by the hot-press-moulding of dough moulding compounds (*e.g.* glass-fibre-filled polyester resins). During such moulding processes, there may well be some fibre alignment as a result of the high shear forces acting in the injection die or in the mould, but the fibres will be imperfectly oriented, so that the mean angle between the fibres and the material flow direction will be greater than zero and the fibre lengths will be distributed about some mean value as a result of damage leading to fibre breakage during processing. Thus, in order to estimate the elastic modulus of a short-fibre composite, the spreads of both the fibre orientations and the fibre lengths have to be taken into account.

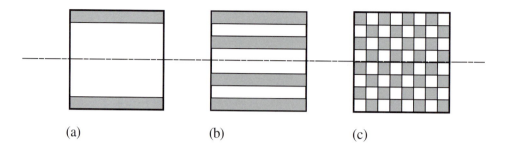

Fig. 3.27 Schematic illustration of three possible types of hybrid laminate structure: (a) skin-core, (b) ply-by-ply and (c) tow-by-tow.

3.8 HYBRID COMPOSITES

Fibres of different types are sometimes mixed in a single matrix to produce *hybrid* composites. The mixing may occur at several structural levels, as illustrated in Figure 3.27, from 'skin-core' structures, through ply-by-ply laminates to mixed-tow laminates. Random, intimately mixed fibre hybrids are rarely considered. The purpose of hybridisation is to extend the concept of 'tailoring' the material's properties to suit particular design requirements, and to offset the disadvantages of one component by the addition of another. For example, GRP are cheaper and tougher than CFRP, but they are significantly less rigid, as we have seen. In principle, then, the addition of a proportion of glass fibres to a CFRP composite offers the possibility of making the material cheaper while improving its toughness, ideally without paying too great a penalty in terms of loss of stiffness. Reinforcing cloths and braids are available which are woven as hybrid reinforcements with two or more species of fibre appropriately placed.

Skin-core structures are common in structural engineering and behave, in a sense, like I-beams, the core playing the same rôle as the web of the I-beam and therefore being required to carry only shear forces, while the skin layers carry the major tensile and compression forces. In composite structures of this kind, the core may be a structural polymer foam or a paper or metal honeycomb. The bending stiffness, EI, of a skin-core beam can be found in the usual way:

$$\left(EI\right)_{beam} = \sum_{i=1}^{n} E_i l_i \tag{3.37}$$

provided the core is an isotropic material and the skin materials are symmetric and balanced laminates (Stavsky and Hoff, 1969; Marshall, 1982). For example, for a skin-core sandwich beam of width b consisting of thin skins of a fibre composite of thickness t and a foamed polymer core of thickness c, the second moment of area of the core about the neutral plane is:

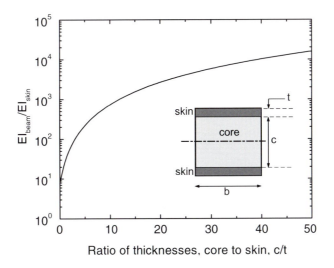

Fig. 3.28 The bending stiffness of a sandwich beam of width 10cm, normalised with respect to the bending stiffness of the skin panels, as a function of the ratio of the thicknesses of the core and skin. The core is a polyurethane foam and the skin panels are 1 mm thick. The normalised bending stiffness of the beam is independent of the stiffness of the skin material.

$$I_c = \frac{bc^3}{12}$$

and that of each of the skins about the neutral plane is:

$$I_s = \frac{bt^3}{12} + btz^2$$

where z is the distance from the neutral plane to the mid-plane of the skin. The bending stiffness of the sandwich beam is thus:

$$(EI)_{beam} = 2E_s I_s + E_c I_c$$

When the skin thickness is small relative to that of the core, $t^2 \ll z^2$, and since $E_c \ll E_s$ the expression can be approximated to:

$$(EI)_{beam} = 2E_s btz^2$$

which is sufficiently accurate for most purposes. Significant benefits are obtained by separating two skins of a fibre composite with a low stiffness core, as illustrated by the graph in Figure 3.28 which shows the variation of beam bending stiffness, given by equation 3.37, as a function of the ratio of the thicknesses of the core and the skin, c/t. The illustration is for a skin thickness, t, of 1mm, a beam width, b of 10 cm, and a polyurethane foam core of stiffness 20 MPa. The beam bending stiffness, $(EI)_{beam}$, is normalised with respect to the bending stiffness, $(EI)_s$, of the skin material, and the graph

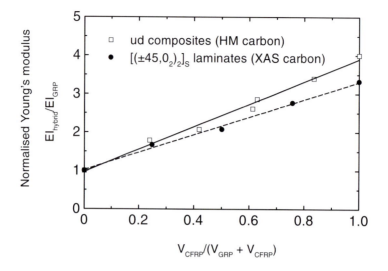

Fig. 3.29 Variation of the Young's moduli of two families of unidirectional carbon/E-glass hybrid composites with composition. The rule of mixtures fits both sets of data reasonably well.

is independent of the actual skin stiffness. For a 1mm skin and a 25 mm core, the beam is over 4000 times as stiff as the skin materials. Since the overall density of the beam falls as the ratio c/t increases, there is an optimum skin thickness for a beam of given weight, as discussed by Farebrother and Raymond (1977).

For simple hybrids consisting of unidirectional plies or mixed, aligned fibre tows, the tensile moduli obey the familiar rule of mixtures. With appropriate change of symbols, equation 3.1 becomes:

$$E_{hybrid} = E_A V_A + E_B V_B + E_C V_C + ...$$

(3.38)

where E_A, E_B, E_C ...... are the moduli of the individual component unidirectional *composites* (*e.g.* CFRP and GRP in a carbon/glass/epoxy hybrid) and V_A, V_B, V_C are the volume fractions of the components A, B, C, *etc.* As in the case of a simple composite, $V_A + V_B + V_C = 1$. An illustration of the validity of the mixtures rule for two varieties of unidirectional CFRP/GRP hybrid composites is given in Figure 3.29. Two quite different composite structures are represented, one being an early pultruded polyester rod material with intimately mixed E-glass and HM carbon fibres (Harris and Bunsell, 1975), and the other a laminate with alternate plies of E-glass/epoxy and XAS-carbon/epoxy (Dickson *et al.*, 1989). When a mixed-ply laminate is tested in bending, the resulting flexural stiffness will clearly be affected by the distributions of the different species of fibre. This was studied by Wagner *et al.* (1982a and b) who found apparent deviations from the mixtures rule in CFRP/GRP hybrid laminates. These deviations were, however, predictable by their theoretical analysis based on the Hoff composite-beam model of equation 3.37. Chamis and Lark (1977) have shown that fair agreement can be obtained between measured values of the flexural modulus of hybrid beams and those calculated by the formula:

$$E_{hybrid} = \frac{1}{3t} \sum_{i=1}^{k} \left[z_{i+1}^3 - z_i^3 \right] E_i \qquad (3.39)$$

where t is the laminate thickness, z_i is the distance to the bottom of the ith ply and z_{i+1} is the distance to the top of the ith ply.

The tensile and flexural stiffnesses of most laminate hybrids are predicted satisfactorily by the methods of the classical thin-laminate theory, and most computer packages allow the construction of laminates from plies of any given material for which the properties of the basic unidirectional ply are known. A detailed discussion of models of hybrid laminates is given by Chou (1992).

3.9 RESIDUAL STRAINS

Residual thermal strains occur in newly manufactured composites as a result of differential thermal contraction during cooling from the moulding, post-curing, or final heat-treatment temperature (*e.g.* the 'ceramming' temperature for a CMC with a glass-ceramic matrix). The state of thermal stress will also change as a consequence of any temperature changes during service. Since the axial thermal expansion coefficient of many modern reinforcing fibres is small (slightly negative in the case of carbon) and that of a resin matrix relatively large, the fibres in a composite laminate will be put into compression and the matrix into tension during cooling from a typical curing temperature of 140°C. A rough estimate of the level of the axial stress in a unidirectional composite can be obtained by application of simple compound beam mechanics. The difference in the thermal strains in the two components, $\Delta\varepsilon$, is given by:

$$(\varepsilon_m - \varepsilon_f) = (\alpha_m - \alpha_f)\Delta T$$

or

$$\frac{\sigma_m}{E_m} - \frac{\sigma_f}{E_f} = (\alpha_m - \alpha_f)\Delta T$$

where α_m and α_f are the coefficients of thermal expansion (CTEs) of the matrix and fibres, respectively, and ΔT is the temperature change. With the condition for equilibrium that no external load, P, is applied, the forces in the fibres and matrix will balance:

$$P = P_f + P_m = \sigma_f A_f + \sigma_m A_m = 0$$

where A_f and A_m are the cross-sectional areas of the two components, and equal to the respective volume fractions. Solving these two equations for the fibre and matrix residual stresses, σ_{fr} and σ_{mr}, we have:

Table 3.4 Residual Thermal Stresses in Some Typical Unidirectional Composites

Matrix	Fibre	V_f	Temperature Range, ΔT, K	Fibre Residual Stress, MPa	Matrix Residual Stress, MPa
Epoxy (high T cure)	T300 carbon	0.65	120	-19	36
Epoxy (low T cure)	E glass	0.65	100	-15	28
Epoxy (low T cure)	Kevlar-49	0.65	100	-16	30
Borosilicate glass	T300 carbon	0.50	520	-93	93
CAS glass ceramic	Nicalon SiC	0.40	1000	-186	124

$$\sigma_{fr} = \frac{E_f E_m V_m (\alpha_f - \alpha_m) \Delta T}{E_c}$$

$$\sigma_{mr} = \frac{E_f E_m V_f (\alpha_m - \alpha_f) \Delta T}{E_c} \tag{3.40}$$

Values of these residual stresses for a group of polymer- and ceramic-matrix composites are given in Table 3.4. It can be seen that in each case the fibre is in compression and the matrix in tension, as expected. The fibre stresses are low compared with their tensile strengths, although it has been shown that in very low-V_f resin-based composites fibres may be buckled or broken as a result of the residual strains (Ferran and Harris, 1971). The matrix strains in the resin-based composites are also low, but it can be seen that in the CMCs the residual matrix tensile stresses are of the order of 100 MPa, which is close to the expected matrix tensile strengths of these materials. As a consequence, matrix microcracking occurs very readily in CMCs under very small applied tensile loads (Harris et al., 1993), a matter of some concern for users of these materials.

In addition to the axial residual stresses, differential thermal contraction also results in residual radial stresses which may make an important contribution to the mechanical behaviour of the composite since they provide one of the mechanisms whereby the matrix grips the fibre and therefore allows the transfer of shear stress from matrix to fibre, as we saw in section 3.6. If there is no mechanical gripping of this kind, the composite must rely exclusively on a chemical fibre/matrix bond. A rough estimate for the radial residual stress, $E_m \alpha_m \Delta T$, when a resin matrix of $E_m = 3$ GPa and CTE $100 \times 10^{-6} K^{-1}$ contracts onto a fibre of infinite rigidity as it cools through a temperature range, ΔT, of $100°C$, is about 30 MPa. If the environment surrounding the fibre were to be a homogeneous one having the transverse properties of a composite instead of pure matrix, the effective stiffness of the matrix would be higher and its CTE would be lower. The net effect on the radial residual pressure would therefore depend on relative changes in E_2 and α_2 with increasing V_f. Harris (1978) used a simple micro-mechanical model to show that the radial pressure was of the order of 25 MPa in glass/resin systems, but a reworking of his calculation to incorporate more realistic models for the effects of V_f on the transverse stiffness (Halpin-Tsai) and CTE (Schapery, 1968, see below) suggests values as high as 40 MPa for practical levels of V_f. These values are considerably higher than those commonly quoted from other models, but are in accord with experiments involving the pulling out of glass fibres from blocks of resin (Harris *et al.*, 1975).

One of the difficulties in carrying out such calculations is that of knowing precisely what value of ΔT to use. Although it is common practice to select the glass-transition temperature, T_g, or the post-cure temperature for a thermoset-matrix composite, it is known that some stress relaxation can occur in resins like epoxies some way below these temperatures. The actual residual stress levels in manufactured composites are therefore likely to be considerably lower than those predicted from simple models.

The thermal expansion coefficients in composite laminae will, like other elastic properties, be anisotropic. Schapery (1968) derived the following expressions for the CTEs parallel with and transverse to the fibres in a unidirectional lamina:

$$\alpha_1 = \frac{E_f \alpha_f V_f + E_m \alpha_m V_m}{E_c} \tag{3.41}$$

$$\alpha_2 = (1 + v_m) \alpha_m V_m + (1 + v_f) \alpha_f V_f - \alpha_1 (v_f V_f + v_m V_m) \tag{3.42}$$

α_1 and α_2 being, respectively, the axial and transverse CTEs. In polymer-matrix composites α_1 is strongly fibre-dominated and falls much more rapidly than a mixtures-rule prediction as V_f increases, whereas α_2 initially rises slightly before following an approximate mixtures rule between the fibre and matrix values. The maximum difference between the CTEs occurs at a volume fraction of about 0.2 for a typical laminate of E-glass and epoxy resin. A consequence of the expansion anisotropy is that laminates containing plies at different angles will distort in unexpected ways unless care is taken to ensure that the constraints are always matched. A single pair of unidirectional CFRP plies bonded together at $0°$ and $90°$ will behave like a bimetallic strip during unconstrained cooling from the curing temperature — a phenomenon that can be used to determine the residual stress levels in laminates by direct measurements of the curvature that occurs.

Equations 3.40 may also be used to calculate the residual stresses in 0/90 lay-ups or in mixed-ply hybrid laminates which can develop significant levels of thermal stress if the expansion coefficients of the two species are significantly different, as in a CFRP/

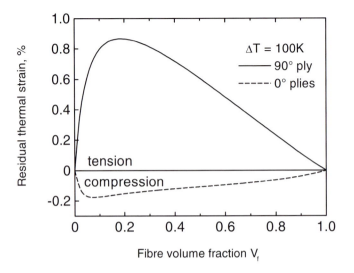

Fig. 3.30 Residual thermal strains in the 0° and 90° plies in a 0/90/0 glass/epoxy laminate.

GRP hybrid, for example. As an illustration, Figure 3.30 shows the predicted levels of thermal strain, as functions of the ply fibre content, in the plies of a 0/90/0 GRP laminate. The values of the CTEs were obtained from equations 3.41 and 3.42 and the value of E_2 required to determine the transverse strain in the 90° ply was obtained from the Halpin-Tsai equation (equation 3.9) with $\zeta = 4$. It can be seen that the 0° plies are in a modest state of compression, while for a practical V_f of 0.6 the strain level in the transverse ply is of the order of 0.5%. Similar results were obtained experimentally for model glass/resin laminates by Jones and Mulheron (1983). The development of residual thermal stresses in hybrid laminates has sometimes been observed to result in a waviness of the fibre tows in the laminae that are under compression as a consequence of in-plane buckling of the tows before the resin ceases to be capable of flow. Such a defect would be likely to have serious consequences for the compression strength of the material (see Chapter 4).

The thermal strains arising during either post-cure cooling or subsequent service at raised temperatures in both single-fibre and hybrid laminates can be determined satisfactorily by means of most commercial software applications, and the theoretical basis for the calculations can be found in standard texts, such as that of Jones (1975). It should be born in mind, however, that most matrix resins are visco-elastic to some extent, and their ability to undergo time-dependent stress relaxation is accentuated by even slightly elevated temperatures or exposure to moisture which plasticizes some thermosets. Residual stresses arising during initial manufacture may therefore fall to insignificant levels during storage under ambient conditions. An illustration of this is shown in Figure 3.31 where the stresses determined from the curvature of two-ply strips of a unidirectional glass/carbon/epoxy laminate are observed to be dissipated as a result of hygrothermal treatment.

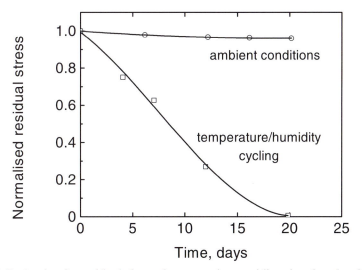

Fig. 3.31 Reduction in residual thermal stresses in a unidirectional carbon/glass/epoxy hybrid laminate under different hygrothermal conditions.

3.10 TEXTILE COMPOSITES

Although many high-performance composites are manufactured from prepreg sheet consisting of thin layers of aligned fibres preimpregnated with semi-cured resin, there are many practical advantages in using woven fabrics as the reinforcing medium. Textiles can be produced in a wide variety of forms, starting from simple plain-weave cloth, with equal numbers of tows of fibre in the warp and weft directions, *via* tubular braided structures, to cloths with fibres in three or five directions (sometimes confusingly called 3D and 5D weaves) and genuinely three-dimensional fabrics with a proportion of fibres running through the thickness. The characteristics of these materials as reinforcements may be of great value to composites designers, particularly since woven fabrics drape much more satisfactorily over complex curved surfaces than unidirectional prepregs can. A detailed mathematical treatment of textile composites manufactured from woven products is given by Chou (1992), and an important recent collection of papers on the subject has been edited by Masters and Ko (1996).

Since, in woven cloths, varying proportions of the reinforcing filaments will be lying out of the plane of the laminate into which they are incorporated, the elastic reinforcing efficiency of the fabric will be somewhat lower than that of a ply derived from a prepreg sheet. This can be illustrated by way of a simple example for composites containing a 0/90 textile. The general appearance of a typical balanced plain-weave glass fabric is shown in Figure 3.32. The fibre tows in this fabric are flattened, and the extent to which they deviate from a single plane depends both on the degree of flattening and the tightness of the weave. In satin-weave fabrics, the ratio of the numbers of tows (or ends) lying in the 0° and 90° directions may be greater than unity, and the use of a satin-weave

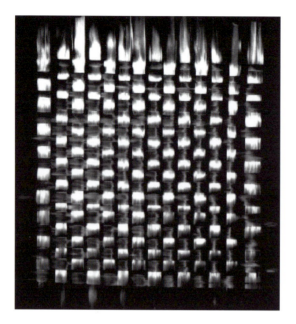

Fig. 3.32 Photograph of a piece of commercial plain-weave glass cloth.

reinforcement therefore allows the designer greater flexibility in meeting service requirements. A simple model to represent the structure of such a weave is shown in Figure 3.33 which depicts an idealised weave described as a six-harness satin weave. For a harness number of two, the model is equivalent to a plain weave. Many models of woven-fabric composites treat the undulations of the warp fibres as they pass over and under the weft tows as a sine wave, but for a rough estimate we can reasonably adopt the simpler linear approach shown in Figure 3.33.

The 'representative volume element' (RVE) shown in this figure can be modelled as two series-connected elements — the kinked section, component 1, and the unkinked section, component 2. The first of these is itself a parallel-connected trio of elements, one at the out-of-plane angle θ to the main composite axis, the second at 90°, and the third a plain resin element. Component 2 is equivalent to a simple non-woven 0/90 laminate, also parallel connected. Thus, the mechanical analogue of the weave is illustrated in Figure 3.33(b).

Component 2 is modelled as in section 3.5 of this chapter, the required values of E_1 and E_2 required to calculate the component stiffness, $E(2)$, being determined from the Halpin-Tsai equations with $\zeta = 5$ for a notional V_f of 0.65. Simple geometrical considerations allow us to determine the volumes of the elements in component 1 and their effective fibre volume fractions. The kinked component is then analysed by a modified Krenchel approach:

$$E(1) = \cos^4(\theta)\, E_1 V_f(\theta) + E_2 V_f(90) + E_m V_m$$

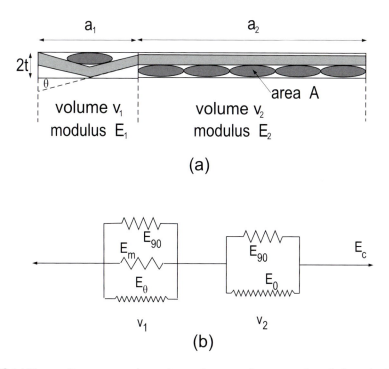

Fig. 3.33 (a) Two-unit representative volume element of a composite reinforced with a six-harness satin-weave cloth and (b) Mechanical model of the RVE.

where $V_f(\theta)$ and $V_f(90)$ are the proportions of fibre at the two specified angles. Unlike the conventional Krenchel estimate, this therefore includes a contribution from the $90°$ fibre tow. The overall modulus of the RVE is now obtained as a Reuss (series) estimate for the stiffness contributions of the appropriate proportions of components 1 and 2. The harness number will clearly affect these proportions and the degree of flattening of the tows determines the angle θ and the overall thickness of the RVE. Figure 3.34 shows the results of these calculations, plotting the composite Young modulus as a function of harness number for various tow-flattening ratios, w/t (ratio of major to minor axes of the elliptical tow shape). Crude though this model is, the agreement with the stiffnesses of the plain-weave and 0/90 non-woven GRP laminates referred to in Table 3.3 is adequate for many design purposes.

3.11 PLIANT COMPOSITES

Pliant composites are, generally speaking, textile fabrics coated with waterproof polymers which are used in various applications requiring flexibility, toughness, gas-tightness, *etc.* Typical applications include air-house membranes, dirigible air-ship

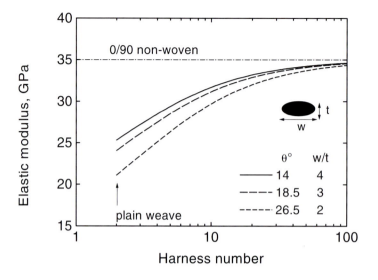

Fig. 3.34 Predicted moduli for satin-weave GRP composites as a function of harness number. The angle θ is the mean angle of deviation from the laminate long axis of the fibre tows in the weave cross-over sections. The flatter the fibre tows, the higher the aspect ratio, w/t, and the lower the misalignment angle, θ.

envelopes, and rubber boats. Depending upon requirements, the reinforcement may be fabrics woven from either glass fibres or from conventional textiles, such as polyesters (*e.g.* Terylene/Dacron) or polyamides (Nylons). For special requirements, aramids like the lower-performance Kevlar fibres may be appropriate. The range of possible fibre stiffnesses thus covers an order of magnitude or so, from about 5 GPa for Nylon to 73 GPa for glass. The coatings may be low-cost materials like PVC, polyurethanes and rubbers, or more expensive fluoroelastomers like Hypalon. The more costly and more rigid fluoropolymers like PTFE have excellent long-term weathering and fire-resistance properties.

The elastic behaviour of pliant composites is more difficult to characterise than that of rigid composites. When a typical uncoated, woven fabric is stretched the effective stiffness will include apparent strain contributions from several mechanisms other than the true elastic extension of the fibres themselves. These mechanisms, which do not occur to the same extent in rigid composites, may be identified as follows.

i.   Yarn shear or rotation may occur if the principle stresses are not aligned with the warp and weft of the fabric.

ii. Crimp interchange. If a highly crimped weft yarn is directly loaded, it will not begin to extend elastically until it has lost the crimp which will in some measure be transferred to the originally straight warp yarns. A similar type of deformation occurs, although to a lesser extent, in woven cloth-reinforced resins.

iii. Yarn flattening, leading to a reduction in effective yarn diameter.

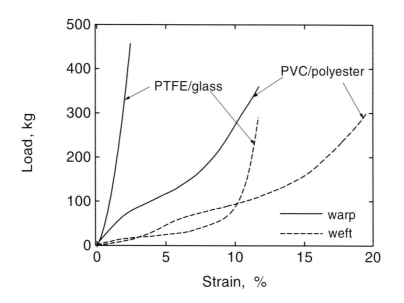

Fig. 3.35 Tensile load/strain curves for two types of architectural coated fabrics. Test strips were 5 cm wide.

iv. Yarn compaction, resulting in an increase in packing density.

v.   Fibre straightening and/or rotation with the yarn structure.

The first two occur at low stresses and depend on the principal stress ratios (not on their magnitudes), whereas the other mechanisms become active only at high stresses. In normal fabric behaviour true elastic extension of the individual filaments probably never occurs, and since this is the only mechanism considered in conventional netting analyses like that of Cox (1952) such analyses are not useful for predicting the behaviour of fabrics or pliant composites. Shanahan *et al.* (1978) have shown that in some respects fabrics may be treated as sheets of an elastic continuum and have applied the basic framework of laminate theory. For membrane strains the tensile modulus, $E(\theta)$, from equation 3.18 gives a reasonable approximation to measured properties of some types of fabric, but coupling between membrane modes of deformation, bending and twisting, and large non-linear elastic strains cannot be dealt with by the continuum model.

The application of an extensible or rubber coating to a woven fabric is considered, for design purposes, not to change the cloth characteristics. For small-strain extensional behaviour in the warp or weft directions this may be an acceptable assumption, but even a non-rigid matrix will significantly raise the in-plane shear resistance of a coated fabric. The shear stiffness may also be affected by a stress-system which increases the inter-yarn frictional forces. The wall of an internally pressurised tube, for example, appears to become stiffer as the internal pressure is increased and rotation of the yarns is inhibited by the increased inter-yarn friction (Topping, 1961). An illustration of the comparative stress/strain behaviour of two quite different types of coated fabric is shown in Figure 3.35 (Ansell and Harris, 1980): the effect of the matrix rigidity is clear.

The generalised two-dimensional form of Hooke's Law for a textile or coated fabric (Alley and Faison, 1972) contains nine stiffness coefficients. Since the straining of such materials is not necessarily conservative (*i.e.* the values of the coefficients may be dependent on the state of strain in the material and its loading history) all nine coefficients are independent, necessary, and multi-valued. Furthermore, in a coated fabric some of the coefficients are likely to be time dependent. In a non-linear anisotropic membrane, then, it is not surprising that the setting up of practicable constitutive relationships should be an unwieldy process and of little practical use for design purposes. Useful surveys of the problems of designing with coated fabrics are given in two published conference proceedings on air-supported structures (Evans *et al.*, 1980, and Happold *et al.*, 1984).

3.12 REFERENCES

V.L. Alley and R.W. Faison: *Journal of Aircraft*, **9**, 1972, 55-60 and 211-216.
M.P. Ansell and B. Harris: *Air-Supported Structures: The State of the Art*, Institution of Structural Engineers, London, 1980, 121-130.
S.J. Birch: Unpublished data, School of Materials Science, University of Bath, 1995.
C.C. Chamis and R.F. Lark: *Hybrid and Select Metal-Matrix Composites,* W.J. Renton, AIAA, New York, 1977, 13-51.
T.W. Chou: *Microstructural Design of Fibre Composites*, Cambridge University Press, 1992.
H.L. Cox: *British Journal of Applied Physics*, **3**, 1952, 72-79.
E.M. de Ferran and B. Harris: *Journal of Materials Science*, **6**, 1971, 238-251.
R.F. Dickson, G. Fernando, T. Adam, H. Reiter and B. Harris: *Journal of Materials Science*, **24**, 1989, 227-233.
D.S. Dugdale: *Elements of Elasticity*, Pergamon, Oxford, 1986.
C.J. Evans, S.C.C. Bate, E. Happold and G.H.A. Piall, eds.: *Proceedings of Symposium on Air-Supported Structures: The State of the Art*, Institution of Structural Engineers, London, 1980.
T.H. Farebrother and J.A. Raymond: *Polymer Engineering Composites,* M.O.W. Richardson, ed., Applied Science Publishers, London, 1977, 198-235.
C. Galiotis: *Composites Science and Technology*, **42**, 1991, 125-150.
C. Galiotis and A. Paipetis: *Journal of Materials Science*, **33**, 1998, 1137-1143.
J.C. Halpin and S.W. Tsai: *Effects of Environmental Factors on Composite Materials*, Report AFML-TR 67-423, U.S.A. Air Force Materials Laboratory, Dayton, Ohio, 1969.
J.C. Halpin: *Primer on Composite Materials Analysis,* 2nd edition, Technomic Publishing, Lancaster, Pennsylvania, U.S.A., 1992.
E. Happold, M.G.T. Dickson, D.H. Halstead, S.B. Tietz and C.J.K. Williams, eds.: *Proceedings of Symposium on the Design of Air-Supported Structures*, Institution of

Structural Engineers, London, 1984.

B. Harris and A.R. Bunsell: *Composites*, **6**, 1975, 197-199.

B. Harris and M.G. Phillips: *Developments in GRP Technology,* B. Harris, ed., Applied Science Publishers, London, 1983, 191-247.

B. Harris: *Journal of Materials Science*, **13**, 1978, 173-177.

B. Harris F.J. Guild and C.R. Brown: *Journal of Physics D: Applied Physics*, **12**, 1979, 1385-1407.

B. Harris F.A. Habib and R.G. Cooke: *Proceedings of Royal Society*, London, **A437**, 1992, 109-131.

B. Harris, J. Morley and D.C. Phillips: *Journal of Materials Science*, **10**, 1975, 2050-2061.

R. Hill: *Journal of Mechanics and Physics of Solids*, **12**, 1964, 199-212 and 213-218.

D. Hull: *An Introduction to Composite Materials*, Cambridge University Press, 1981.

F.R. Jones and M. Mulheron: *Composites*, **14**, 1983, 281-287.

R.M. Jones: *Mechanics of Composite Materials*, Scripta Book Co, Washington, and McGraw-Hill, 1975.

H. Krenchel: *Fibre Reinforcement*, Akademisk Forlag, Copenhagen, Denmark, 1964.

A. Marshall: *Handbook of Composites,* G. Lubin, ed., Van Nostrand Reinhold New York, 1982, 557-560.

J.E. Masters and F. Ko, eds.: *Composites Science and Technology, Textile Composites Special Issue*, Elsevier Science, Oxford, **56**, 1996, 205-386.

L.E. Nielsen and P.E. Chen: *Journal of Materials*, **3**, 1968, 352-358.

L.E. Nielsen: *Predicting the Properties of Mixtures*, Marcel Dekker, New York, 1978.

J. Pabiot: *Composite Materials,* Conference proceedings number 63, AGARD/NATO, Neuilly, Paris, Paper 6, 1971.

F. Pierron and A. Vautrin: *Composites Science and Technology*, **52**, 1994, 61-72.

M.R. Piggott and B. Harris: *Journal of Materials Science*, **15**, 1980, 2523-2538.

J. Schapery: *Journal of Composites Materials*, **2**, 1968, 380-404.

W.J. Shanahan, D.W. Lloyd and J.W.S. Hearle: *Textile Research Journal*, **48**, 1978, 495-505.

D. Shiel: Unpublished Data, School of Materials Science, University of Bath, 1995.

P.D. Soden and G.C. Eckold: *Developments in GRP Technology,* B. Harris, ed., Applied Science Publishers, London, 1983, 91-159.

Y. Stavsky and N.J. Hoff: *Engineering Laminates*, A.G.N. Dietz, ed., MIT Press, Cambridge, Massachusetts, 1969, 5-59.

A.D. Topping: *Aerospace Engineering*, **20**, 1961, 53-58.

H.D. Wagner, I. Roman and G. Marom: *Fibre Science and Technology*, **16**, 1982a, 295-308.

H.D. Wagner, I. Roman and G. Marom: *Journal of Materials Science*, **17**, 1982b, 1359-1363.

R.J. Young, R.B. Yallee and M.C. Andrews: *Proceedings of 7th European Conference on Composite Materials*, ECCM7, London, Woodhead Publishing, Abington, U.K., **2**, 1996, 383-388.

4. The Strength of Fibre Composites

Strength is more difficult to predict than elastic properties because it depends on the mechanisms of damage accumulation and failure as well as on the properties of the constituents, and the failure behaviour of fibre composites is often complex. The manner in which damage occurs in the material, and the way in which it accumulates to reach some critical level which precipitates final failure depends on many aspects of the composite construction, including the fibre type and distribution, the fibre aspect ratio, ℓ/d, and the quality of the interfacial adhesive bond between the fibres and the matrix. Much of the software currently available for predicting the strengths of fibre composites must be treated with circumspection on account of this uncertainty relating to the modelling of failure. Because of the complex nature of failure in many types of composite, considerations of strength and toughness are closely interrelated. And although in this book we have opted, for the sake of clarity, to treat them in separate chapters, we should bear in mind when designing with composites that the strength and toughness may not be independent.

4.1 TENSILE STRENGTH

4.1.1 UNIDIRECTIONAL CONTINUOUS-FIBRE COMPOSITES

Consider a simple unidirectional composite lamina reinforced with continuous fibres which are initially well-bonded to the matrix so that under load fibres and matrix deform together (Figure 4.1). The load is shared between them and, following the argument at the beginning of chapter 3, we can write the stress on the composite, $(\sigma)_c$, as:

$$(\sigma)_c = (\sigma)_f V_f + (\sigma)_m (1 - V_f)$$

where $(\sigma)_f$ and $(\sigma)_m$ are the *stress levels* in fibres and matrix (as opposed to being *properties* of fibre and matrix), and are equal to $E_f \varepsilon$ and $E_m \varepsilon$, respectively, where ε is the composite strain. What happens as loading continues depends on the nature of the individual components. There are two common situations which may be taken as examples, as illustrated in Figure 4.2. In most practical composites, the high-performance reinforcing fibres can usually be regarded as being brittle, *i.e.* they deform elastically to failure, showing little or no non-linear deformation. In metal- and polymer-matrix composites, the unreinforced matrix is usually capable of some irreversible plastic deformation and in such materials the matrix failure strain is usually much greater than that of the fibres. By contrast, in ceramic-matrix composites the matrix is also brittle, and although much

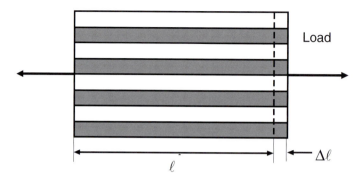

Fig. 4.1 A simple unidirectional composite lamina under tensile load.

weaker than the reinforcing fibres it is often almost as rigid as the fibres. The matrix failure strain is therefore usually less than that of the fibres.

As the diagram in Figure 4.2 shows, when the stress in the ductile matrix reaches the matrix yield stress, σ_{my}, the matrix continues to bear load, although the slope of the stress/strain curve falls somewhat. If the fibres are carrying most of the load, when the stress on the fibres reaches σ_{fu} (the fibre tensile strength), failure will occur and the stress on the composite at this point, $(\sigma)_c$, defines the composite *strength*, σ_c, thus:

$$\sigma_c = \sigma_{fu} V_f + \sigma'_m (1 - V_f) \qquad (4.1)$$

where σ'_m is the stress in the matrix at the fibre failure strain. With the minor modification, then, that the matrix stress, σ'_m, is not a normal *property* of the matrix, equation 4.1 is another rule of mixtures, although it is only valid above a certain critical value of V_f. Failure in this kind of composite may be gradual if the fibre content is not too high but will be rapid at high V_f levels.

In all-brittle systems, such as most CMCs, the matrix reaches its own failure stress, σ_{mu}, at relatively low composite stress levels, and will begin to crack throughout the body of the material (Harris *et al.*, 1994). During this stage, the composite may deform without increase in stress, as shown in Figure 4.2b, and only when the matrix is completely permeated by an array of closely spaced cracks does the composite stress begin to increase again. Beyond this point, only the fibres are carrying load, and failure occurs when the fibres reach their failure stress. In such materials, the problem for the designer is whether or not the development of an array of cracks during service can be tolerated.

Considering now the brittle-ductile system, equation 4.1 can be represented on a diagram showing composite strength as a function of fibre content, V_f (Figure 4.3). When there are very few fibres present (*i.e.* near the left-hand axis, for $V_f < V_{min}$), the stress on the composite may be high enough to break the fibres. But even if the fibres rupture and cease to carry load, because of its work-hardening ability the matrix is still able to support the load on the composite until its tensile strength, σ_{mu}, is reached. In this simplistic approach, we consider the effect of the broken fibres to be similar to that of an

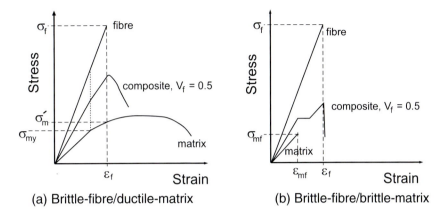

Fig. 4.2 Schematic illustrations of the stress/strain curves of two types of composite, (a) with a ductile matrix and (b) with a brittle matrix, as derived from the stress/strain behaviour of the constituents.

array of aligned holes, so that, to begin with, as the number of fibres increases, the strength of the composite actually falls, following the equation:

$$\sigma_c = \sigma_{mu} (1 - V_f) \qquad (4.2)$$

in Figure 4.3. Up to V_{min} it is therefore the matrix that exerts control of failure behaviour. When V_{min} is reached, there are sufficient fibres to bear some of the load when the matrix reaches its failure stress, and the rule-of-mixtures relationship defined by equation 4.1 then begins to operate. An increase in strength beyond the matrix tensile strength only occurs above the critical V_f level, V_{crit}, which can be found from the geometry of Figure 4.3 to be:

$$V_{crit} = \frac{\sigma_{mu} - \sigma'_m}{\sigma_f - \sigma'_m} \qquad (4.3)$$

In practical composites where $\sigma_{fu} \approx \sigma_{mu}$, V_{crit} is very low — of the order of only 0.01 — and the relationship between strength and composition therefore approximates to a rule of mixtures.

4.1.2 THE PROBLEM OF THE STRENGTH OF BRITTLE FIBRES

The simple models that we have been considering make the assumption that the reinforcing filaments have a unique failure stress, σ_{fu}, and that they all fracture simultaneously when the load on the composite reaches this critical level. In reality, since almost all of the common reinforcing filaments exhibit brittle behaviour, there is a statistical distribution of fibre strengths rather than a unique value of σ_{fu}. The mean

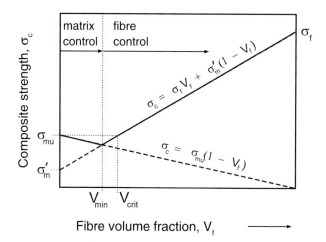

Fig. 4.3 Schematic illustration of the variation of the strength of a unidirectional composite with fibre content.

filament strength, $\overline{\sigma}_{fu}$, and the spread of values is determined by the statistical distribution of defects or flaws in the fibre or on its surface. In general, the spread of strengths in any commercial tow of fine fibres is quite large, as shown in Figure 4.4 by some typical results for commercial HTA carbon fibres manufactured by Enka.

Although we often speak of the 'mean strength' and the 'standard deviation' of the strengths of the fibres, the distributions are seldom Gaussian (or Normal) distributions. We usually use an exponential distribution described by Weibull (1951) which gives the probability of failure at a stress level, σ, of any sample from a given batch of fibres of identical length as:

$$P(\sigma)=1-\exp\left[-\left(\frac{\sigma}{b}\right)^{m}\right]$$

(4.4)

The parameters m and b are known as the shape and scale parameters, respectively.[*] The shape factor, m, sometimes called the Weibull modulus, relates to the uniformity of the distribution of flaws in a brittle material: a high value of m implies a highly uniform distribution of defect sizes and therefore a low level of variability of fibre strengths. Conversely, a low value of m implies highly variable flaw sizes and a large spread of measured strengths, as in Figure 4.4. When the data are ranked and plotted as described in Appendix 2 (see Figure 4.5), it can be seen that the two-parameter Weibull model of equation 4.4 provides a reasonable fit to the data, despite the fact that for carbon fibres one would expect to find a minimum stress level (the location parameter) below which no failures would occur. The parameters of the distribution shown are m = 6.3

---

[*]A brief introduction to the statistical background to this method of analysis is given in Appendix 2.

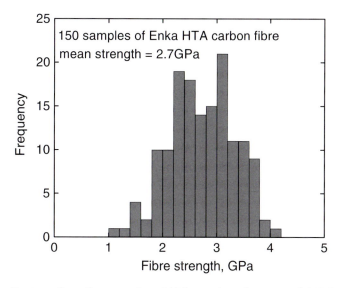

Fig. 4.4 Distribution of tensile strengths of 150 samples of commercial Enka HTA carbon fibre. The sample test length was 3 cm (Harris *et al.*, 1992a).

and b = 2.96 GPa (*cf.* the arithmetic mean of 2.76 GPa). A value of 6.3 is very low when compared with the behaviour of normal engineering materials (m ≈ 30 to 100), but it is not untypical of ceramics.

For composites users, there are two important consequences of the weakest-link model upon which the statistical analysis is based. First, the breaking load of a large bundle of brittle fibres (tows of carbon fibres typically contain between 3,000 and 10,000 individual filaments) is lower than the arithmetic mean strength of all the filaments in the bundle. This can be shown by the following simplified argument.

We assume that all of the filaments in a large bundle have the same length and the same elastic modulus and are equally loaded. As a bundle is loaded, the weaker fibres in the bundle will fail at low loads and the applied load must then be redistributed among the remaining fibres. At some point, the failure of one more filament will mean that the load carried by those remaining is too great for them to support, and the bundle fails. Suppose that at a load, P_F, on the bundle, n fibres out of a total of N have already failed. The stress on the remaining filaments is then:

$$\sigma_F = P_F / A_{N-n} \qquad (4.5)$$

A_{N-n} being the cross-sectional area of the remaining unbroken fibres. The ratio of unbroken fibre area to total area, A_{N-n}/A_N, is 1 - n/N, and since this describes the same failure probability as that given by the Weibull function, $P(\sigma)$, in equation 4.4, we can rewrite the expression for P_F, the load on the remaining fibres in the bundle, in equation 4.5, as:

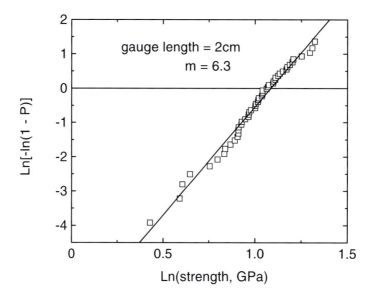

Fig. 4.5 Two-parameter Weibull plot of strengths of untreated ENKA carbon fibres: gauge length = 2 cm.

$$P_F = \sigma_F A_N \left(1 - \frac{n}{N}\right) = \sigma_F A_N \exp\left[-\left(\frac{\sigma_F}{b}\right)^m\right]$$ (4.6)

To find the critical load to fracture the bundle, we maximize P_F by setting the derivative $dP_F/d\sigma_F = 0$, with the result, after some arithmetic, that:

$$\sigma_F = b(m)^{-1/m}$$

Inserting this into equation 4.6, we obtain the maximum bundle load:

$$P_F(max) = b\,A_N\,(m)^{-1/m}\,e^{-1/m}$$

and the strength of the bundle, σ_B, which is given by $P_F(max)/A_N$, is therefore:

$$\sigma_B = b e^{-1/m}\,m^{-1/m}$$ (4.7)

In Appendix 2 we see that the mean (or first moment) of a Weibull distribution is given by the function $b\Gamma(1 + 1/m)$, Γ being the Gamma function. The ratio of the bundle strength to the mean filament strength, $\overline{\sigma}_{fu}$, can therefore be found:

$$\frac{\sigma_B}{\sigma_{fu}} = \frac{b e^{-\frac{1}{m}} m^{-\frac{1}{m}}}{b\Gamma\left(1 + \dfrac{1}{m}\right)} = \frac{(m.e)^{-\frac{1}{m}}}{\Gamma\left(1 + \dfrac{1}{m}\right)}$$ (4.8)

and this ratio is plotted as a function of m in Figure 4.6. As we see from the graph, the values of m for brittle fibres like carbon and glass are typically less than 10, so

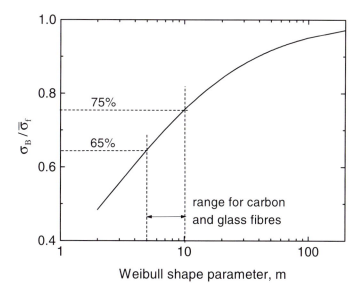

Fig. 4.6 Dependence of the strength of a bundle of fibres on the Weibull shape parameter, m.

that the strength of a bundle of such filaments may be only 70% of the mean fibre strength.

Another consequence of the statistical model of failure of brittle solids is that their strengths depend on the size of the piece being tested and for samples of different sizes equation 4.4 contains a volume term, as shown in Appendix 2. For a batch of samples of fibres of reasonably constant cross-section, this size-dependence manifests itself as a variation of strength with sample length. This can be shown by restating equation 4.4 to include a term for the volume, V:

$$P(\sigma) = 1 - \exp\left[-V\left(\frac{\sigma}{b}\right)^m\right] \tag{4.9}$$

Assuming some fixed failure probability for two groups of fibres of lengths ℓ_1 and ℓ_2, it is then easy to see that the ratio of the strengths of these two groups must be:

$$\frac{\sigma(\ell_1)}{\sigma(\ell_2)} = \left[\frac{\ell_1}{\ell_2}\right]^{-\frac{1}{m}} \tag{4.10}$$

The value of m alone thus determines the magnitude of the strength/length effect. This effect is a feature common to all brittle filaments, and is illustrated for some commercial carbon fibres in Figure 4.7. It has always been a problem for designers of composite structures that when dealing with data from fibre manufacturers the fibre strength should only be quoted in relation to the sample test length.

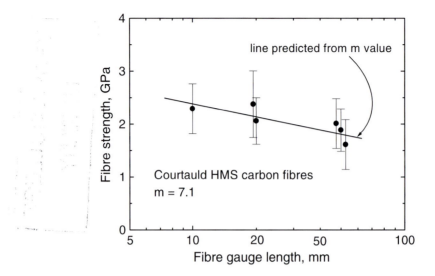

Fig. 4.7 Variation of fibre strength with test length for samples of Courtauld HMS carbon fibres (Dickson, 1980). The straight line is predicted from the value of m of 7.1 obtained from individual fibre tests.

We see now that when trying to calculate the composite strength the choice of the value of σ_{fu} that should be used in equation 4.1 is a matter of some concern. It is clearly *not* the mean fibre strength, and although it may seem that the fibre bundle strength, σ_B, is the appropriate value to use, this may also prove to be an incorrect assumption. To try to answer this question, we have to consider in more detail the mechanism by which a composite breaks down under load.

4.1.3 PROGRESSIVE DAMAGE AND FAILURE IN UNIDIRECTIONAL COMPOSITES

A fibre bundle embedded in a matrix does not behave like a free fibre bundle. When the local load level reaches the failure stress at a weak point in a given fibre it breaks and the load it carried is transferred back into the neighbouring matrix regions. But away from the broken ends the fibre carries its full share of the load, by contrast with what happens in an unbonded bundle. The stress carried by neighbouring fibres in the vicinity of the break will be perturbed, as shown schematically in Figure 4.8, but the stress concentration may not be great enough to break a neighbouring fibre (or fibres). As the load on the composite increases other fibre breaks will occur, but each fibre may break many times without seriously damaging the overall load-bearing ability of the composite since the tensile load supported by the broken fibre within a short distance, δ, known as the *ineffective length*, from each broken end, will rapidly build up again to its original level.

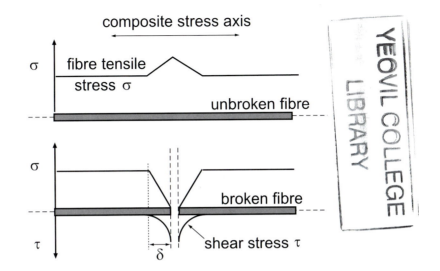

Fig. 4.8 Schematic illustration of the perturbation of the stresses in the neighbourhood of a broken fibre.

If the scatter in filament strengths is small (large value of m) and the fibres are closely spaced, the extra load carried across to a neighbouring fibre may be sufficient to break that one too. The local stress concentration is then even higher. If the process is repeated a number of times, the cross section where the breaks are occurring will rapidly become too weak to support the applied load and catastrophic brittle failure will follow. In reinforced plastics containing brittle fibres like glass or carbon, however, this mode of fracture rarely occurs. The second fibre seldom breaks at the same cross section as the first because the stress concentration adjacent to the first break is unlikely to coincide with a weak point in the second fibre. Instead, fibre breaks accumulate randomly throughout the whole sample or structure, and final failure may then occur when the number of breaks in any one cross section has effectively reduced the local V_f below that required to support the applied load at that instant.

This process of damage accumulation may be illustrated in a simple fashion by considering the model composite shown in Figure 4.9. It represents a single 'ply' consisting of ten parallel fibres. In order to model the statistical distribution of flaws in the fibres, each fibre 'chain' is divided into 'links', each of which is of length ℓ_c ($= 2\delta$): as we shall see later, this is the smallest length of fibre that can just be broken by a tensile force. In a realistic statistical model, such as the Monte Carlo treatment of Fukuda and Kawata (1977), each fibre section would be allocated a 'strength' value from a Weibull distribution, and a notional 'load' acting on the composite would be increased in small steps. The links would fail as the local stress reached the Weibull 'strength', and this process would be allowed to continue until some preset failure criterion was reached. A similar result can be demonstrated by simply using a random-number generator to cause random failures throughout the model, as displayed by the crosses in Figure 4.9.

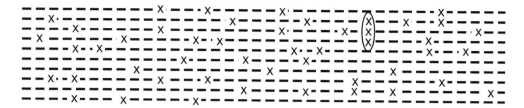

Fig. 4.9 Model composite for studying the random failure of fibres leading to composite failure. Three adjacent fibre failures is assumed to be the failure condition (x = fibre failure).

A suitable failure condition can be deduced from the work of Jamison (1985) who deconstructed samples of composite material that had been loaded to failure, or just below, and searched for groups of adjacent fibre fractures throughout the material. He concluded that since the largest groups that could be found were triplets, a group of four must constitute a critical-sized nucleus that led immediately to failure. If the computer running the simple stochastic failure model is programmed to recognise an appropriate group of adjacent failures, then the time to reach that condition (which may equally well represent an applied load on a composite being loaded at a constant rate) is the composite life (or failure stress). The result of repeating such a simulation 100 times is shown in Figure 4.10: for this simple model, the failure condition was three adjacent breaks, as illustrated in Figure 4.9 and, to accommodate the effective-length concept, in-line chain links immediately adjacent to an existing failure were not allowed to break. The total number of fibre breaks that may occur prior to failure varies widely, from 19 to 147 in the case of this series of simulated tests, and the times to failure (or failure stresses) are also very widely distributed as the figure shows. Somewhat surprisingly, the data fit a two-parameter Weibull distribution. This is not a very realistic model: we might ask, for example, whether three breaks in a row would be more damaging than two in a row and one slightly oblique, and such matters have to be addressed in serious calculations. However, we can easily appreciate the progressive nature of damage from such a model and the reasons for the variability in composite strengths and fatigue lives (see chapter 6).

4.1.4 CALCULATION OF THE TENSILE STRENGTH OF A UNIDIRECTIONAL COMPOSITE

As we see, damage builds up progressively in the composite until sufficient fibre breaks occur in one local region to constitute a crack which can then propagate catastrophically and cause fracture. This progressive failure has been investigated by a variety of methods, including acoustic emission studies and by microscopy or x-ray techniques. If we ignore the (low) matrix contribution in the strength equation (4.1) we can see that the mean filament strength and the fibre bundle strength might well define the upper and lower bounds on the strength of a unidirectional fibre composite:

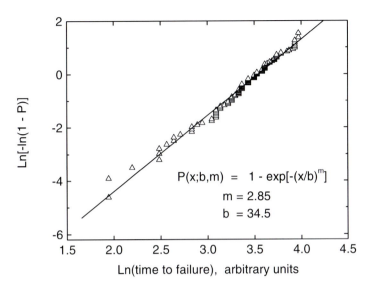

Fig. 4.10 Two-parameter Weibull plot of times to failure generated by a random-break computer program for a model fibre composite (100 'tests').

$$\sigma_B V_f < \sigma_c < \overline{\sigma}_{fu} V_f \qquad (4.11)$$

But there still remains the problem of knowing what is the relevant 'test length' for which we should choose the value of $\overline{\sigma}_{fu}$. The whole length of a fibre embedded in a composite matrix is never stressed uniformly, and it is usually assumed (Rosen, 1965) that the effective 'test length' of fibres breaking within a matrix is of the order of the critical length, ℓ_c, or 2δ (see Figure 4.8). Rosen's cumulative weakening-model assumes that fibre flaws are distributed randomly throughout the composite and it extends the fibre-bundle model to the situation inside the composite. As in the simple composite simulation described above, fibre breaks occur singly and randomly throughout the material. The load near the breaks in the fibres must be redistributed because of the ineffective end sections, and this redistribution will be controlled by the strength of the fibre/matrix interface and the relative stiffnesses of the fibre and matrix, as we have already seen.

In the expression for the bundle strength (equation 4.7) the characteristic strength b relates to whatever fibre test gauge length has been used. However, for a given fibre population with shape factor m, the value of b appropriate to a 'gauge length' ℓ_c can be obtained by applying equation 4.10. Combining equations 4.7 and 4.10, therefore, we obtain an expression for the strength, σ_c^*, of the bundle *in the composite*:

$$\sigma_c^* = b\left(\frac{\ell_c}{L}\,me\right)^{-\frac{1}{m}} \qquad (4.12)$$

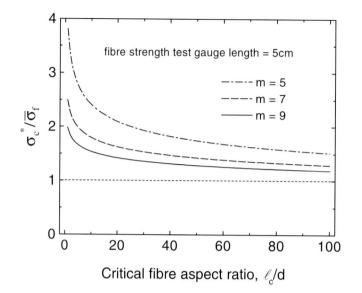

Fig. 4.11 Variation of predicted composite strength, $\sigma_c^* V_f$, normalized with respect to the value predicted from the mean fibre strength, $\bar{\sigma}_f\, V_f$, as a function of the critical aspect ratio, ℓ_c/d.

where the strength of the reinforcing filaments was determined on samples of gauge length L. The upper bound on composite strength is then equal to $\bar{\sigma}_{fu}\, V_f$. Dividing this estimate of composite strength with that predicted from the mean filament strength (ignoring any contribution from the weak matrix) we obtain the ratio:

$$\frac{\sigma_c^*}{\bar{\sigma}_{fu}} = \frac{\left[\dfrac{\ell_c}{L}.m.e\right]^{-\frac{1}{m}}}{\Gamma\left(1+\dfrac{1}{m}\right)} \tag{4.13}$$

As we saw in chapter 3, the fibre critical length, ℓ_c, is determined by the elastic characteristics of the fibre and matrix and by the strength of the interfacial bond between the two. The fundamental reinforcing potential of the fibres is therefore affected by a number of factors other than their nominal tensile strength. In Figure 4.11 the strength ratio of equation 4.13 is plotted as a function of the critical aspect ratio, ℓ_c/d, for different values of the Weibull shape parameter and assuming a fixed fibre test length (L) of 5 cm.

For many practical composites the values of m and ℓ_c/d fall within the ranges in this graph, and it is not surprising, therefore, that the strengths of some unidirectional composites may rise above a mixture-rule prediction based on the single-fibre strength. An illustration of this coated-bundle effect is shown by some results of Dickson (1980) in Figure 4.12. In this figure, the failure strains of the bundles are plotted instead of

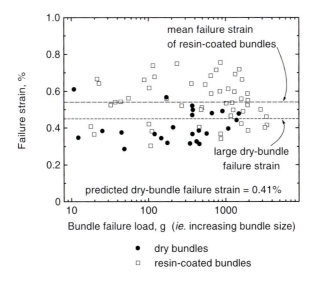

Fig. 4.12 The breaking strains of fibre bundles containing different numbers of Courtauld HMS carbon fibres. The filled symbols are for dry bundles and the open symbols are for resin-coated bundles. The breaking load of the bundle is used as a rough indicator of the number of fibres in the bundle (approx. 10 g per fibre).

stresses, and the bundle failure *load* is used as an indicator of bundle size in order to avoid the tedious business of counting large numbers of fine filaments. It can be seen that the failure of dry bundles (filled symbols) is subject to substantial scatter, but that the degree of variability falls as the bundle size increases. This is what would be expected of the bundle-failure model: the dry-bundle failure strain of 0.41% is predicted from the single-filament data of Figure 4.7. By contrast, the variability of the resin-coated bundles, which resemble real composites, is much greater and not markedly dependent on bundle size. The mean failure strain of the resin-coated bundles is 0.55%, 20% greater than that of the dry bundles and over 95% of the dry bundles failed at strains below the mean coated-bundle failure strain. The mean filament failure strain, however, is 0.62% and it is therefore clear that although many of the resin-coated bundles do indeed fail above the line represented by $\sigma_c^*/\overline{\sigma}_{fu} = 1$ in Figure 4.11, on average they do not do so.

In practice experimental composite strengths are never as high as the Rosen model would lead us to expect. And although the model was reasonably successful in a number of cases, it is limited by the fact that it takes no account of the stress concentrations that result from single and multiple fibre breaks, *i.e.* of the *interactions* between individual damage events which depend on the surface characteristics of the fibre, the interfacial adhesion, and the resin flexibility, as well as on the fibre strength statistics. It has also been suggested that the notional 'test length' of ℓ_c is far too small and that the Weibull model cannot be extrapolated to such small test lengths. Hughes *et al.* (1980) and Harris *et al.* (1992a) have shown, nevertheless, that for some types of carbon-fibre composites the extrapolated bundle strength gives good agreement with measured tensile strength values.

There are perhaps two extreme situations. If there is a great deal of scatter in the strengths of individual fibres, failures occur randomly throughout the composite and are non-interacting. On the other hand, if the fibre failure strength is very uniform, there will be a high degree of interaction between individual fibre failures in one cross section of the composite. In general, most composites will behave in a complex fashion between these two extremes and it will usually be difficult to predict the composite strength accurately. Bader (1980) has developed a model for the tensile failure of unidirectional composites which suggests that the critical size of the group of fibres which initiates failure is of the order of 9 fibres (*i.e.* about three in a row in any given direction (*cf.* the results of Jamison described earlier)) but that the final failure sequence is determined by the relative resistance of the laminate to splitting parallel to the fibres which occurs as a consequence of the shear stress developed around the groups of failed fibres. Manders and Bader (1981) have also described a statistical model to predict the strengths of mixed-fibre (hybrid) composites in which they suggest that the critical number of fibre fractures is of the order of three, and Curtis (1986) published details of a computer model for tensile failure of unidirectional composites which showed the sequential development of fibre failures and suggested that four adjacent broken fibres is the maximum stable group that could be tolerated in standard high-strength-fibre composites.

4.1.5 Transverse Strength

The strengths of unidirectional composites are highly anisotropic. Perpendicular to the reinforcing fibres most composites are weak and failure is controlled by rupture or plastic flow of the matrix, or by fibre/matrix decohesion: the precise mechanism will depend on the capacity of the matrix for plastic deformation and the strength of the fibre/matrix bond. If the matrix is a metal or polymer, its intrinsic yield strength will govern the transverse behaviour of a composite containing only a few fibres. If this yielding is restricted to the ligaments between the fibres, the transverse strength of the composite might be estimated from a crude rule-of-mixtures model in which the deforming matrix cross-section is obtained from a knowledge of V_f and the reinforcement geometry. For the square array shown in Figure 4.13, for example, the effective volume fraction of the deforming matrix normal to the fibres is:

$$V_m(\text{eff}) = 1 - \sqrt{\frac{4V_f}{\pi}}$$

Assuming no contribution from the fibres, the transverse strength of the composite is then:

$$\sigma_t = \sigma_{my}\left(1 - \sqrt{\frac{4V_f}{\pi}}\right) \qquad (4.14)$$

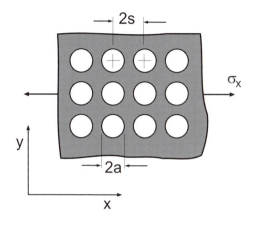

Fig. 4.13 Kies's model (1962) for strain magnification during transverse loading of a unidirectional composite.

As the volume fraction increases, however, the relatively undeformable fibres impose an increasing degree of constraint on the matrix. Contraction in the fibre direction (the width) is prevented, and extension under stress can only be accompanied by deformation in the thickness direction, *i.e.* the deformation is under plane-strain conditions. A consequence of this is that the effective yield strength of the matrix rises from the 'free' value of σ_{my} to the plane-strain yield stress, 1.15 σ_{my} (Cottrell, 1964). Despite this, the transverse strengths of practical composites fall rapidly with increasing fibre content while their axial strengths rise linearly, and even if corrections are made in equation 4.14 (*vide* Cooper and Kelly, 1969) to allow for the effect of a positive adhesive bond between the fibres and matrix, the degree of anisotropy of a unidirectional composite rises dramatically with increasing V_f, as illustrated in Figure 4.14. Two materials are represented in this figure, a typical MMC (mullite-fibre-reinforced aluminium) and a common epoxy-based CFRP, for which the ratios of fibre strength to matrix strength are different by an order of magnitude (about 7 for the MMC and 88 for the CFRP). A crude correction has been made to allow for the plastic constraint just mentioned (the *effective* matrix yield strength increases in proportion to $\sqrt(V_f)$) but the effect is insignificant by comparison with the geometrical effect of equation 4.14. Figure 4.14 illustrates an important advantage of typical MMCs over reinforced plastics which derives from the much higher inherent matrix strengths of the former. Typical ratios of transverse to longitudinal strengths of GRP, for example, are only of the order of 0.05 for glass/epoxy (Harris *et al.*, 1979) and 0.04 for glass/polyester (Guild *et al.*, 1982). Such low levels of transverse strength mean that unidirectional reinforced plastics are incapable of bearing transverse loads, and cross-plied laminates may even develop cracks during processing as a result of thermal stresses normal to the fibres if careful control is not exercised.

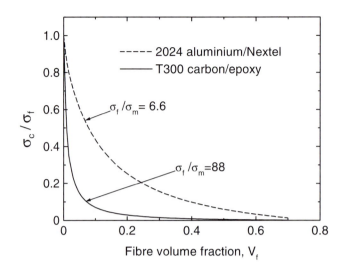

Fig. 4.14 Variation of predicted ratio of composite transverse strength to axial strength with fibre content for two types of composite with different ratios of fibre-to-matrix strengths. The matrix strength varies from σ_{my} at $V_f = 0$ to 1.15 σ_{my} at $V_f = 0.7$.

Strain concentrations in the matrix may also lead to brittle failure at low strains, even in composites with otherwise ductile matrix materials. The magnitude of this strain concentration may be obtained from an approximate elastic model by Kies (1962) which defines a strain magnification factor, SMF, due to composite strains in a direction transverse to the fibres (Figure 4.13). For this simple rectangular array, the increase in local matrix strain in the x direction, ε_m, as a fraction of the mean composite strain, ε_c, is given by the SMF:

$$SMF = \frac{\varepsilon_m}{\varepsilon_c} = \left[\left(\frac{a}{s}\right)\left(\frac{E_m}{E_f}\right) + \left(1 - \frac{a}{s}\right)\right]^{-1} \tag{4.15}$$

The geometrical ratio, s/a, defined in Figure 4.13, depends on the fibre diameter and volume fraction, and on the geometrical arrangement of the fibres. In typical GRP materials with 65 vol.% of 10 μm diameter fibres the strain magnification factor given by Kies's model would be about 5 for a hexagonal fibre array and 7.5 for a square array. Thus, even in resin matrices with failure strains of the order of 4%, transverse resin cracking perpendicular to the fibres could occur at applied composite strains as low as 0.5% at a very small percentage of the full composite potential strength. Indeed, in the Permali 0/90 GRP which we have already used as an example in discussing elastic properties, transverse cracking was shown to occur in the 90° plies at strains as low as 0.2%, only a tenth of the composite failure strain (Harris *et al.*, 1979): such cracking is shown in the photomicrograph in Figure 4.15. And while this transverse cracking may

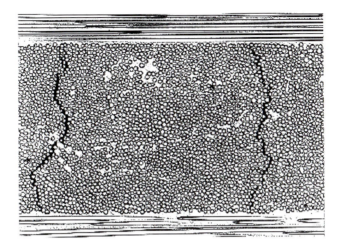

Fig. 4.15 Photomicrograph of transverse-ply cracking in a 0/90 GRP laminate.

not impair the load-bearing ability of the material in a major fibre direction, it could have serious consequences, if, for example, the material were required to contain liquid or gas under pressure.

There is some doubt, however, whether failure in composites of high V_f ever occurs in the manner described by these simple models for, as Bailey *et al.* (1979) have shown, it is the fibre/matrix bond that is usually the weakest element and transverse cracking occurs by the linking up of local fibre/resin decohesion failures, which is how we also expect a ceramic- or glass-matrix composite to fail. The transverse strength of brittle-fibre/brittle-matrix laminates is probably best treated by a modified fracture mechanics approach where the critical flaw size is related to the transverse dimension of the fibres.

4.1.6 ORIENTATION-DEPENDENCE OF COMPOSITE STRENGTH

When the fibres in a loaded unidirectional composite lie at an angle to the stress axis the effective strength of the composite is reduced. In chapter 3 we examined the variations of stress and strain as functions of orientation, and it is easy to see how some kind of failure criterion could be combined with the stress or strain functions to provide indications of the variation of strength with orientation. The strength of a composite is highly dependent on the local damage mechanisms, however, and these must change as the active stress system affects different components of the composite. In our discussion of the axial tensile strength, we emphasised the rôle of the fibres to the exclusion of all other effects. But when the level of shear stress builds up in a composite this can lead to failure modes of other kinds, including fibre/matrix interfacial shear failure and interlaminar shear. And we have already discussed the possibility of tensile failure of

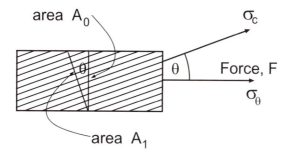

Fig. 4.16 Resolution of the force, F, acting on a composite of arbitrary orientation. The strength of the composite in the fibre direction is σ_c.

the interface or of the matrix between the fibres during transverse loading. It is vital in using composite strength models to ensure that the appropriate failure mode is being considered. This is shown by studying one of the simplest of the models, the maximum-stress theory of Stowell and Liu (1961).

The central assumption is that failure will occur when the stress in a given direction reaches some critical value (*i.e.* the critical stress for the operative failure mode). If the measured in-plane strength properties of a unidirectional composite lamina are the normal axial tensile strength, σ_c, the transverse tensile strength, σ_t, and the in-plane shear strength, τ_c, the maximum-stress model is developed as follows.

Consider, first, the case where the angle between the stress and the fibres is relatively small, as shown in Figure 4.16, so that composite failure is still going to be dominated by fracture of the fibres. The fibre tensile stresses reach the fibre fracture stress before any other failure event can occur. Resolving the applied force, F, and the area over which it acts, we have:

$$\sigma_c = F\cos\theta/A_1 = F\cos\theta.\cos\theta/A_0 = \sigma_\theta\cos^2\theta$$

and hence the failure stress, σ_θ, of the composite at the angle θ to the fibre direction is given by:

$$\sigma_\theta = \sigma_c \sec^2\theta \tag{4.16}$$

For larger values of θ, the shear component of stress acting at the fibre/matrix interface will gradually increase until shear failure occurs. Resolution again yields the angular-dependence of the composite strength appropriate to an in-plane shear mode of failure:

$$\sigma_\theta = \tau_c / \sin\theta\cos\theta = 2\tau_c.\mathrm{cosec}\,2\theta \tag{4.17}$$

At very large angles, the level of applied tensile stress resolved normal to the fibre/matrix interface results in a failure passing partly through that interface and partly through the matrix between the fibres, which occurs at relatively low stresses, like interface shear failure. Further resolution leads to the result:

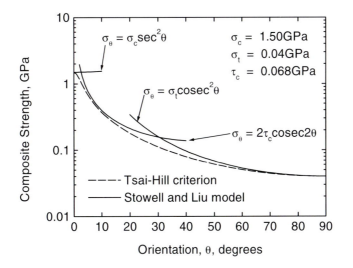

Fig. 4.17 Comparison of the maximum-stress and Tsai-Hill failure criteria for a T300/5208 unidirectional carbon/epoxy composite.

$$\sigma_\theta = \sigma_t \cos ec^2\theta \qquad (4.18)$$

The three equations 4.16, 4.17 and 4.18 define the three régimes of failure of the maximum-stress criterion and their predictions are in reasonably good agreement with experimental results for many unidirectional composites. One difficulty with this model is that it is appropriate for unidirectional fibre composites only and cannot allow for constraints resulting from the presence of fibres at other angles (*i.e.* in a real composite). It is therefore likely to underestimate real composite properties.

There are many other models of composite failure, and many arguments concerning the validity of the underlying reasoning. One example is that derived from the von Mises maximum-distortion-energy criterion for the failure of isotropic metals which was modified by Hill (1964) for anisotropic materials. The Hill argument was adapted by Azzi and Tsai (1965) for use with transversely isotropic laminates under plane stress, and in one of its forms it predicts the orientation dependence of composite strength as:

$$\frac{1}{\sigma_\theta^2} = \frac{\cos^4\theta}{\sigma_c^2} + \frac{\sin^4\theta}{\sigma_t^2} + \cos^2\theta\sin^2\theta\left(\frac{1}{\tau_c^2} - \frac{1}{\sigma_c^2}\right) \qquad (4.19)$$

where the stress terms are as already defined. This model, which is almost universally known as the Tsai-Hill criterion, has the advantage over the maximum-stress criterion that it is a mathematically continuous function, and the two models are almost indistinguishable under the simple tensile conditions which we have been discussing, as shown in Figure 4.17. In this graph, experimental results for σ_c, σ_t, and τ_c for a standard CFRP material, T300/5208, have been used to generate the maximum-stress

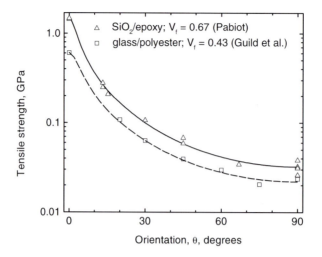

Fig. 4.18 Comparison between experimental data and the predictions of the Tsai-Hill criterion(full and dashed lines) for the orientation-dependence of the strength of a ud SiO_2/ epoxy composite (Pabiot, 1971) and a ud E-glass/epoxy composite (Guild *et al.*, 1982).

and the Tsai-Hill failure criteria, and it can be seen that they agree well except in the range $20° < \theta < 40°$ and for this material even this difference is relatively small. It is interesting to note that for slight changes of θ away from $0°$ the maximum-stress model predicts a slight increase in strength as a consequence of the resolution of forces. However, it is very difficult to manufacture composites with perfect fibre alignment and equally difficult to test them at a precise angle to the fibres, so that experimental errors will make it almost impossible to distinguish between the two models at small θ. The Tsai-Hill criterion is a common feature of commercial computer software programs for strength prediction (see section 4.4).

The predictions of the Tsai-Hill criterion are compared in Figure 4.18 with strength measurements for two types of unidirectional composite, SiO_2/epoxy (Pabiot, 1972) and E-glass/polyester (Guild *et al.*, 1982). The mean measured values of σ_c, σ_t, and τ_c for the SiO_2/epoxy are similar to those for the T300/5208 CFRP material just referred to, namely 1.46 GPa, 0.32 GPa, and 0.68 GPa, respectively, while those for the lower V_f glass/ polyester are proportionately lower. The comparison between theory and experiment for both kinds of material is seen to be excellent.

4.1.7 THE STRENGTHS OF MULTI-PLY LAMINATES

When a 0/90 (non-woven) laminate is stressed parallel to one set of fibres the transverse plies will begin to crack at low strain levels, as we have seen. This cracking will continue over a range of composite strains determined by the statistical spread of

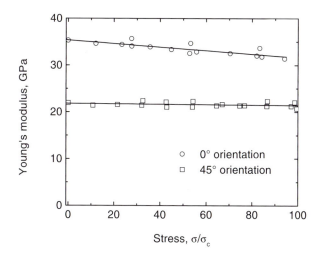

Fig. 4.19 Elastic modulus of Permali XE6 non-woven glass/epoxy laminate as a function of pre-stress. The stress level is normalized with respect to the composite failure stress, σ_c (results of Harris *et al.*, 1979).

transverse strengths of the individual laminae, and each transverse lamina may crack many times, the crack spacing gradually being reduced until each block of transverse ply is too short to be cracked any further, as illustrated by Figure 4.15. Aveston *et al.* (1980) have proposed that for 0/90 glass/resin laminates the spacing, c, of transverse ply cracks as a function of applied stress, σ_a is given by the relationship:

$$c = -\sqrt{\frac{E_L E_T b d^2}{G_T E_C (b+d)}} \, ln\left[1 - \frac{\varepsilon_{Tu} E_C}{\sigma_a}\right] \tag{4.20}$$

where E_C, E_L, and E_T are the moduli of the composite and of the longitudinal and transverse plies, with respect to the stress direction, G_T is the shear modulus, and ε_{Tu} is the failure strain of the transverse plies. The transverse plies are of thickness d while the longitudinal plies are of thickness b. Cracking of the transverse plies could be supposed to start when the composite strain reaches the normal failure strain, ε_{Tu}, of the transverse plies.

Transverse-ply cracking releases some of the inter-ply coupling constraint and residual thermal stresses, and impairs the load-bearing ability of the laminate. It is often assumed that the fully cracked transverse plies no longer contribute to the stiffness of the laminate. On this basis, the 0° axial modulus, E_0, of the Permaglass XE6 laminate referred to earlier should fall by about 45% after preloading whereas, as Figure 4.19 shows, the reduction in 0° stiffness after loading to 95% of the failure load is only about 11% and there is no reduction at all in the 45° stiffness. The fully-cracked transverse plies therefore continue to contribute to the stiffness of the damaged composite, both through direct stress transfer from the longitudinal plies and because of the lateral (Poisson) constraint they exert on the extensional strain of

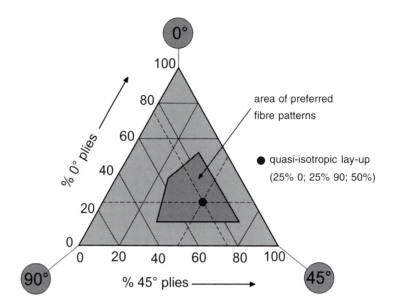

Fig. 4.20 Preferred fibre patterns for aerospace composite laminates (adapted from Hart-Smith, 1993).

the longitudinal plies. The models of Bailey *et al.* (1979) and Aveston *et al.* (1980) predict that the critical strain for first cracking of the transverse plies depends on the ply thicknesses, and that the thinner the plies the higher the strains to which cracking is deferred by the elastic constraint.

Approximate estimates of the strengths of multiply laminates may be obtained by the use of the Krenchel model discussed in chapter 3, with the same values of the efficiency factor, η_θ, defined by equation 3.25, as were used to calculate elastic response. This factor is then used in a mixture-rule calculation:

$$\sigma_c = \eta_\theta \sigma_{fu} V_f + \sigma'_m (1 - V_f) \tag{4.21}$$

The approach may again be criticized because it ignores contributions of fibres at 90° to the stress axis and elastic interactions between the individual anisotropic laminae, but these are less important in calculating the strength of a laminate than in determining its elastic properties. The choice of an appropriate value for σ'_m is also of no great significance in a fibre-dominated laminate.

Another useful approximate approach is that known as the 'Ten-Percent Rule'. Hart-Smith (1993) points out that while, in principle, fibre patterns could vary widely throughout any given structure, depending on requirements, practical aerospace laminates have for many years been designed within a relatively limited range of options. They are usually balanced laminates, with equal numbers of +45° and –45° plies, and the ratio of 0° and 90° plies varies to accommodate local loads. Preferred fibre arrangements usually fall within the shaded area surrounding the quasi-isotropic lay-up shown in the ternary

Table 4.1 Prediction of Laminate Efficiencies by Laminate Theory and Approximate Methods

Laminate Lay-Up	Comparison of Laminate Efficiencies Predicted by Laminate Theory and Approximate Methods		
	Laminate Theory Software	Krenchel Factor	Hart-Smith 10% Rule
UD	1	1	1
$(0,90)_S$ (cross-ply)	0.530	0.500	0.550
$(\pm45)_S$	0.139	0.250	0.100
$[(\pm45,0_2)_2]_S$	0.575	0.625	0.550
$[(\pm45)_2,0_3,90)]_S$	0.489	0.500	0.438
$(0,\pm45,90)_S$ (quasi-isotropic)	0.385	0.375	0.325

'phase diagram' in Figure 4.20. In the ten-percent rule, each 45° or 90° ply is considered to contribute one tenth of the strength or stiffness of an axial (0°) ply to the overall performance of the laminate. Justification for this lies in the fact that there is a rapid fall-off in both the strength and stiffness of a unidirectional laminate with orientation, as illustrated by Figure 3.12 (stiffness) and Figure 4.17 (strength). The effective properties at 45° and 90° are therefore indeed very similar, although in reality the appropriate value for the anisotropy ratio may be nearer to 4% than 10% for many unidirectional composites.

The ten-percent rule offers a rapid, and reasonably accurate, means of estimating the strength and stiffness of any practical laminate consisting only of 0, 45 and 90° plies, without the need for elaborate calculations or extensive experimental evaluation of properties. For example, for a lay-up consisting of 37.5% fibres in the 0° direction, 50% of ±45 fibres, and 12.5% of 90° fibres, a simple mixture-rule sum gives the laminate strength and stiffness in the 0° direction as $(0.375 \times 1 + 0.5 \times 0.1 + 0.125 \times 0.1) = 0.438$ times the value of the relevant property measured in axial tension on the unidirectional laminate. Table 4.1 gives a comparison of the predictions of the Krenchel and Hart-Smith methods for the efficiencies of some representative laminates, together with values obtained from analysis by laminate-theory software for T300/5208 carbon/epoxy composites. Except for the $(\pm45)_S$ laminate, the agreement is very good, despite the assumptions on which the approximate models are based.

We saw in chapter 3 how the analytical procedures of the classical thin-laminate theory may be used to determine the levels of stress and strain in a multiply laminate. If the level of stress in a ply at some point in the laminate exceeds the load-bearing ability of that ply, local failure will occur. Because of the way in which the microstructure controls the accumulation of damage in fibre composites, however, the composite rarely fails when the first ply breaks. The classical thin-laminate theory provides a means of estimating the laminate strength. Since all stresses in a loaded laminate can be calculated by the theory, a single stress analysis serves to determine the stress field that causes failure in any one of the constituent laminae by comparing the individual lamina stresses, σ_x, σ_y, τ_{xy}, with some suitable failure criterion, such as the Tsai-Hill model discussed in section 4.16. The thermal contraction stresses following cure can also be determined and included in the overall level of stress, mechanical plus thermal. If no laminae are found to have failed following cure the computer program can adjust the external stress system until the first lamina fails according to the chosen failure criterion. This lamina is then eliminated, the stiffness matrices are recalculated to accommodate the redistributed loads and the cycle is repeated. At the point when the failure of one more lamina results in violation of the failure criterion for the remaining laminations, the composite is considered to have reached its macroscopic failure stress. The procedure is thus somewhat similar to that of calculating the strength of a fibre bundle. Hart-Smith (1996) has criticized this process of ply-by-ply decomposition, followed by assessment of each ply independently from the others. He argues that such a process cannot be valid when matrix cracking accompanies the deformation process, and he also considers that values of the transverse strength obtained by measurements made on a single ply are not relevant for the failure of a 90° ply embedded in a laminate.

The ply-discounting procedure is probably conservative, however — *i.e.* it predicts lower strengths than actual values because even after a single lamina has failed one or more times it is still capable of bearing load, the stresses in the neighbourhood of the break being diffused away into adjacent laminae in the same manner as those surrounding a break in a single fibre.

A central feature of analytical models and software for the prediction of the strength of laminates is that they must depend on the use of an appropriate failure criterion. We have already discussed two of these, the maximum-stress and the Tsai-Hill (maximum-strain-energy) criteria, but there is currently much disagreement about the validity of many of the commonly used models. It is necessary to make further reference to this problem, and we shall return to it in section 4.5.

4.1.8 SHORT-FIBRE COMPOSITES

We saw in section 3.7 of chapter 3 that if the fibres are short, stress must be transferred into them by shear and their ends are not fully loaded. Only if a fibre is longer than the critical length, ℓ_c, can it be broken by loading the composite and its full reinforcing potential realized. To illustrate this further, we consider three situations, where the fibres in the composite are assumed to be aligned and of uniform lengths which are less than, equal to, or greater than the critical length, as shown in Figure 4.21. We also make the simplifying assumption that the tensile stress in the fibre builds up from the ends in a linear fashion, rather than as given by equation 3.33, but this does not greatly affect the argument.

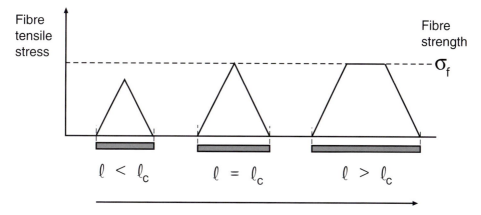

Fig. 4.21 Simplified illustration of the variation of tensile stress in a short fibres as a function of fibre length. σ_f is the fibre breaking stress and ℓ_c is the fibre critical length.

In the case where ℓ is equal to ℓ_c, the tensile breaking stress in the middle of the fibre can just be reached, and the fibre can therefore be broken, but the load-bearing ability of the whole composite must be less than that of a continuous-fibre composite containing an identical type of fibre. If the fibre strength is σ_{fu}, ignoring statistical aspects of the problem, the average stress in each fibre of length ℓ_c will be only $\frac{1}{2}\sigma_{fu}$ (middle sketch in Figure 4.21). Thus, in a longer fibre, such as that in the right-hand sketch of this figure, there will be two inefficiently loaded end sections of total length ℓ_c in which the average stress is $\frac{1}{2}\sigma_{fu}$ and a central section of length $(\ell - \ell_c)$ in which the stress is σ_{fu}. By a rule-of-mixtures argument based on these lengths, then, the overall level of stress, $\overline{\sigma}_f$, in a fibre of length $\ell > \ell_c$ will be:

$$[\sigma_{fu}(\ell - \ell_c) + \tfrac{1}{2}\sigma_{fu}\,\ell_c]/\ell$$

i.e.

$$(\overline{\sigma})_f = \sigma_{fu}\left(1 - \frac{\ell_c}{2\ell}\right) \qquad (4.22)$$

for the linear stress variation model illustrated above, where $(1 - \ell_c/2\ell)$ represents an efficiency factor for length. The mixture rule may then be modified to give the strength of the composite:

$$\sigma_c = \sigma_{fu}V_f\left(1 - \frac{\ell_c}{2\ell}\right) + \sigma'_m\left(1 - V_f\right) \qquad [\ell \geq \ell_c] \qquad (4.23)$$

This equation can be used to calculate the strengthening efficiency (the ratio σ_c/σ_{fu}) as a function of fibre length (ℓ/ℓ_c), as shown in Figure 4.22 for a typical composite. It can

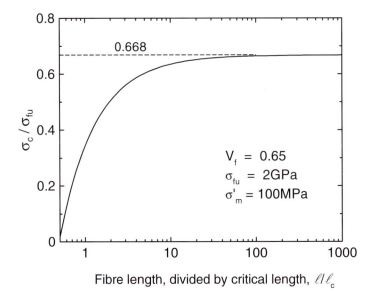

Fig. 4.22 Efficiency of strengthening in short-fibre composites, as given by equation 4.23.

be seen that provided the fibre length is about 10 times its critical length, some 95% of the strength obtainable in a continuous-fibre system can be achieved with short, aligned fibres. The validity of this conclusion is not affected by the simplistic stress-transfer model assumed above. The problem with discontinuous-fibre reinforcement, however, as we have suggested, is that it is very difficult in conventional manufacturing processes to keep the fibres well aligned.

If the fibres are shorter than ℓ_c they cannot be loaded to their failure stress and the strength of the composite is then determined by the strength of the fibre/matrix bond, not by the fibre strength. This is sometimes the case in GRPs manufactured by injection moulding, especially if the lengths of the fibres are reduced by damage during the manufacturing process. The average stress in the fibres can be estimated from the balance of forces illustrated schematically in the diagram of Figure 3.21. If the stress in the fibre is identified as $(\sigma)_f$, the maximum fibre tensile force, $(\sigma)_f \cdot \pi d^2/4$, is balanced by an internal frictional force which must exceed the fibre/resin bond strength, τ_i, or the matrix shear yield stress, τ_{my}, if failure by a mechanism other than fibre failure is to occur. If we let this limiting shear stress be equal to τ_i (the more common failure mode) the force balance at failure will be:

$$\frac{\pi d\ell\tau_i}{2} = \frac{\pi d^2 (\sigma)_f}{4} \qquad \left[(\sigma)_f < \overline{\sigma}_{fu} \right]$$

and the average tensile stress in the fibre, $(\overline{\sigma})_f$, is therefore:

$$(\overline{\sigma})_f = \tfrac{1}{2}(\sigma)_f = \frac{\tau_i \ell}{d} \qquad\qquad (4.24)$$

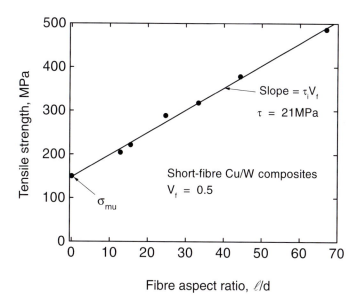

Fig. 4.23 Relationship between the tensile strength and fibre aspect ratio in tungsten-wire/ copper composites containing short, aligned fibres (Harris & Ramani, 1975).

The composite strength, from the mixtures rule, is therefore:

$$\sigma_c = \left(\frac{\tau_i \ell}{d}\right) V_f + \sigma_{mu}\left(1 - V_f\right) \qquad \left[\ell < \ell_c\right] \tag{4.25}$$

where σ_{mu} is the matrix tensile strength. This relationship emphasises the importance of good fibre/matrix bonding if reasonable strength levels are to be obtained in short-fibre composites. If the bond is poor, a soft matrix like polyethylene or a pure metal like aluminium will deform readily around the more rigid fibres and the tensile strength falls as the fibre length is reduced. This is illustrated in Figure 4.23 by some tensile strength measurements on MMCs consisting of aligned short tungsten fibres in a matrix of pure copper (Harris and Ramani, 1975). The strength of these short-fibre composites only approaches that of a continuous-fibre composite at an aspect ratio of about 100. Since this is only about twice the critical aspect ratio, ℓ_c/d however, it appears that the stress transfer at the fibre ends is rather more efficient than is assumed by Figure 4.21.

So far we have assumed the short fibres to have been arranged in a single direction, and comparisons with the properties of unidirectional, continuous-fibre composites are therefore valid. The shorter the fibres, however, the more difficult it is to obtain good alignment, and without good alignment close packing cannot be achieved. The strengths of short-fibre composites are therefore limited both by fibre misorientation and by the lower fibre contents than those which can be obtained in continuous-fibre composites in addition to the fibre-length effects already discussed. Methods of producing composites reinforced with almost perfectly-aligned chopped fibres have been described by Parratt *et al.* (1971) but short-fibre composites are used most frequently as moulding materials. The polyester dough-moulding compounds (DMCs, sometimes referred to as bulk moulding compounds) and sheet moulding compounds (SMC) are typical of thermoset moulding materials of this kind, while for injection moulding a wide range of glass-fibre-reinforced thermoplastic composites is available.

The strengths of materials of this kind are hard to predict on the basis of equations 4.23 and 4.25 because of the combination of variables, the uncertainty about parameters such as τ_i and ℓ_c (which are difficult to measure for conditions appropriate to the real composite) and lack of knowledge of the effects of processing on the strength of the reinforcing filaments. On the basis of assumptions relating to probable values of ℓ_c and τ_i and with the application of a Krenchel factor of $\eta_\theta = 0.375$ for randomly-oriented composites, Harris and Cawthorne (1974) estimated strengths for typical DMC and SMC that were well within the experimental scatter for their measured strength values (37 MPa and 100 MPa, respectively, for DMC with 15 wt.% of 6.4 mm long fibres and SMC with 25 wt.% of 50 mm fibres in chopped-strand mat). The strength of the DMC, with fibres shorter than the probable critical length, is in fact no greater than that of the resin matrix, but its advantage lies in its much higher toughness, as we shall see in chapter 5.

One of the problems encountered in the manufacturing of short-fibre composites is that the fibres may sustain damage as a result of the mechanical processes involved. Very often the lengths of the filaments are reduced, either as a consequence of the blending of the constituents in a thermoset moulding compound, or of the injection moulding of thermoplastic compounds through narrow die openings. And although the starting lengths of the fibre stock may initially be well above the critical length, their lengths in the finished mouldings may be well below ℓ_c. In DMCs, the final fibre lengths are also affected by the nature of the sizing compound used on the fibre tows. Figure 4.24 shows examples of normalized length distributions for fibres extracted from polyester moulding compounds after processing (Andrews, 1998). Three sets of data are included: two of them for fibres initially 6 mm and 13 mm long coated with a soluble size, and the third for 13 mm fibres with an insoluble size. It can be seen that when the size dissolves in the resin during processing, the fibre bundles become easily separated and they are reduced in length so as to have roughly the same modal value of length, only about 1 mm, whatever their initial length. If an insoluble size is used, the bundles do not filamentize to quite the same extent, but the modal length is only marginally greater. And although 1 mm may be equal to or slightly greater than ℓ_c, a large fraction of the fibres are much shorter than this. In typical fibre-length distributions from injection moulded Nylon 66 with 15 vol.% of glass fibres, Bowyer and Bader (1972) showed that some 80% of the fibres were below the critical length, and they emphasised the need to control moulding procedures to minimize this processing damage. Some materials manufacturers have developed commercial 'long-fibre' glass/thermoplastic moulding compounds to improve on the properties of conventional mouldings: ICI's Verton, containing 10 mm long fibres, is an example of such a material.

Bowyer and Bader (1972) attempted to account for the effect of damage to fibres during the injection moulding of reinforced thermoplastics by studying the fibre length

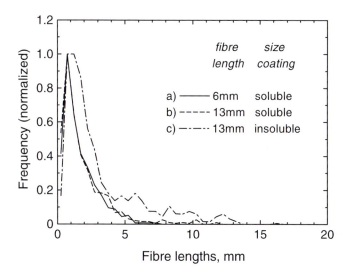

Fig. 4.24 Normalized distributions of the lengths of fibres extracted from moulded DMCs. The mean lengths of the fibres after processing are: (a) 1.63 mm, (b) 1.86 mm and (c) 3.10 mm (Andrews, 1998).

distributions in mouldings. They used a model in which contributions to composite strength from fibres of sub-critical and super-critical lengths were summed separately over the effective range of lengths:

$$\sigma_c = \eta_\theta \left[\sum_i \frac{\tau L_i}{d} V_i + \sum_j E_f \varepsilon_c \left(1 - \frac{E_f \varepsilon_c d}{4 L_j \tau} V_j \right) \right] + E_m \varepsilon_c (1 - V_f) \qquad (4.26)$$

The first summation represents the sub-critical fibre distribution, sub-fraction V_i with fibre length spectrum $L_i(\ell < \ell_c)$, and corresponds to the first term in equation 4.25. The second summation gives the super-critical contribution, from sub-fraction V_j with fibre-length spectrum $L_j(\ell > \ell_c)$, and the third term is the usual matrix contribution. η_θ is the Krenchel factor, 0.375 for a planar-random moulding and 0.167 for 3-dimensional material. In this equation, τ represents the fibre/matrix interfacial shear strength that we have previously called τ_i, the subscript having been temporarily dropped to avoid confusion with the summation indices, i and j. This model gave reasonable agreement with their experimental results on glass-filled Nylon and polypropylene.

The strength of short-fibre composites is strongly affected by the fibre/matrix bond strength, and since the reinforcing efficiency of short fibres is low, it is essential to ensure that they are as strongly bonded as possible to the matrix. The strengths of some typical short-fibre composites, commercial and experimental, are shown in Figure 4.25. Improvements in strength by a factor of two or three are obtainable in practical moulding

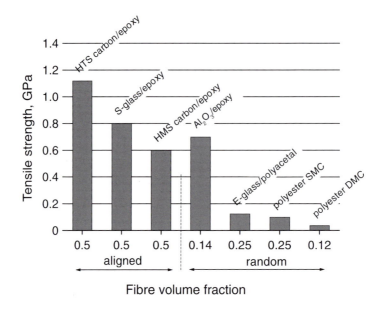

Fig. 4.25 Tensile strengths of some short-fibre composites, aligned and random.

systems, but the strengths of these materials are very low by comparison with those of continuous-fibre composites. By contrast, the well-aligned, short-fibre composites produced by Parratt and Potter (1980) can be seen from Figure 4.25 to have 60 to 70% of the strength of comparable continuous-fibre composites. The outstanding reinforcing potential of whiskers indicated in the table is not used in practice because of the cost of this kind of fibre.

The effects of fibre characteristics and distribution on the expected strengths of aligned fibre composites are summarized in the schematic diagram of Figure 4.26. Increasing the intrinsic fibre strength naturally increases the expected composite strength in the ways that we have discussed in this chapter. But if the fibres are discontinuous, the reinforcing efficiency is reduced for two reasons: first, as a result of the aspect-ratio effect described by equation 4.23 and, second, because the shorter the fibres the more likely they are to be misoriented, to a greater or lesser extent. This not only results in a further efficiency loss from the orientation effect (Krenchel factor), but it also limits the maximum fibre content that can be incorporated into the composite since the closeness of packing is reduced.

4.1.9 HYBRID COMPOSITES

As we saw in chapter 3 the elastic properties of hybrid composites can be calculated to within limits acceptable for most design purposes by the application of well-established

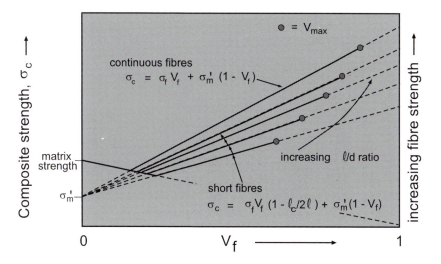

Fig. 4.26 Schematic illustration summarising the effects of choice and arrangement of fibres on the strengths of fibre-reinforced composites.

principles of composite mechanics and classical thin-laminate theory. But although elasticity calculations are largely independent of the nature of fibre distributions, predictions of properties such as strength, toughness, and fatigue behaviour are more difficult because they depend on micromechanisms of damage accumulation which, in turn, are determined by the construction of the laminate and the scale of dispersion of the mixed fibres. At the grossest level of hybridization, strips of GRP incorporated into a CFRP laminate can act as efficient crack arresters (Bunsell and Harris, 1976; Sun and Luo, 1985), provided the width of the strips is sufficient to dissipate the energy of a crack moving rapidly in the CFRP by localized debonding and splitting.

The commonest type of hybrid, however, and that which is most easily achieved in practice, is made by laminating some balanced sequence of plies within each of which there is only a single species of fibres. The composition-dependence of the strength of a unidirectional composite of this kind is obtained by analogy with the response of a single-fibre composite by considering the failure strains of the two separate constituents (Aveston and Kelly, 1979). When a small quantity of fibres with low failure strain, ε_{le}, is added to a composite consisting of fibres of high-failure-strain, ε_{he}, the low-elongation (*le*) fibres will experience multiple failure before the failure strain of the high-elongation (*he*) fibres is reached. By analogy with the case of a flexible resin containing only one species of brittle fibre, the strength of the hybrid composite, σ_H, will be given at first (Figure 4.27) by the residual load-bearing capacity of the *he* composite, assuming the *le* component makes no contribution at all, *i.e.*

$$\sigma_H = \sigma_{hu} V_{he} \tag{4.27}$$

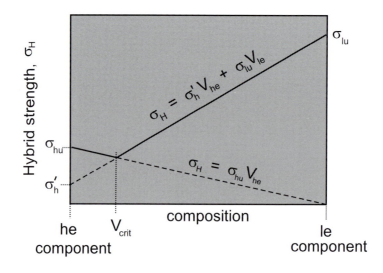

Fig. 4.27 Schematic representation of the variation of the strengths of a family of unidirectional, two-component hybrid composites with proportions of the two constituents, referred to as the high-extensibility (*he*) and low-extensibility (*le*) materials.

where V_{he} is the volume fraction of the *he* component, $V_{he} + V_{le} = 1$, and σ_{hu} is the strength of the pure *he* fibres.

At the other extreme, if a small amount of *he* fibre is added to an *le* composite, the strength is again reduced because the *he* fibres cannot carry the load when the *le* component fails. Failure of the *le* fibres therefore leads to single, catastrophic fracture of the composite. The hybrid strength in this mode is then:

$$\sigma_H = \sigma'_h V_{he} + \sigma_{lu} V_{le} \qquad (4.28)$$

where σ_{lu} is the strength of the pure *le* fibres, and the intercept of this line on the *he* axis ($V_{he} = 1$) is the stress σ'_h in the *he* component at the failure strain of the *le* component, equal to $\varepsilon_{lu} E_h$ (ε_{lu} and ε_{hu} are the failure strains of pure *le* fibres and the pure *he* fibres, respectively). The critical composition, V_{crit}, for the transition from single to multiple fracture is found by solving the simultaneous equations 4.27 and 4.28, whence:

$$V_{crit} = \frac{\sigma_{lu}}{\sigma_{lu} + \sigma_{hu} - \varepsilon_{lu} E_h} \qquad (4.29)$$

and this transition point also defines the hybrid composition of minimum strength. Figure 4.28 illustrates that this simple model is adequate for two unidirectional hybrid combinations, CFRP/GRP and CFRP/KFRP. It also shows that in the case of the KFRP/CFRP combination, for which the failure strains, ε_{lu} and ε_{hu}, are markedly different, the hybrid composite failure strain varies non-linearly with composition and falls below a

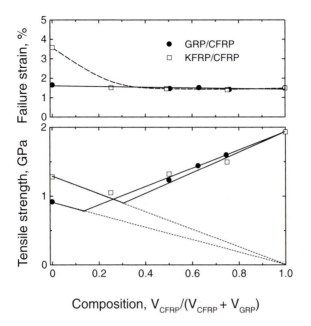

Fig. 4.28 Variation of tensile strength and failure strain with composition in ud hybrid laminates of CFRP with GRP and with KFRP. The full lines in the lower figure represent equations 4.27 and 4.28 for the two sets of strength data.

straight line joining ε_{lu} and ε_{hu}. This type of variation is characteristic and has given rise to the notion of 'failure-strain enhancement', also known as the 'hybrid effect'. This idea stems from the fact that since there is a continuous variation of composite failure strain between the two end points, some hybrid compositions must inevitably fail at strains greater than the failure strain of the lower extensibility component, ε_{lu}. An analysis by Zweben (1977) shows that the failure process is affected by the statistical spread of failure strains of the two fibre species and predicts that the addition of *he* fibre to an *le* composite raises the strain level needed to propagate fibre breaks because the *he* fibres behave like crack arresters at a microstructural level. The situation becomes more complicated when the fibres are intimately mixed because micro-mechanisms of failure change as the pattern of nearest neighbours around a given fibre alters and the strength is governed by failure either of single fibres or of small bundles of fibres (Parratt and Potter, 1980).

4.2 COMPRESSION STRENGTH

Studies of the compression behaviour of composites have been complicated by the fact that when testing small, free blocks of material, failure modes occur which are not

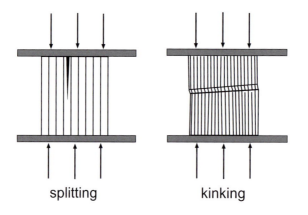

splitting   kinking

Fig. 4.29 Schematic illustration of compression failure modes in unidirectional composites.

necessarily determined by fibre behaviour, and the compression strength of a composite often appears to be lower than its tensile strength. As in the case of wood compressed along the grain, longitudinal splitting and/or linking of the sample often occur, as illustrated schematically in Figure 4.29, at stresses well below the equivalent tensile load-bearing ability of the composite.

Measurements to obtain the true compression strength are difficult to carry out. A range of complex jigs has been developed for the purpose of preventing macroscopic (Euler) buckling of the sample during testing, but these often constrain the sample in an unnatural fashion, and it is uncertain whether the measured value of the compression strength is valid. Even after many precautions have been taken, measured compression strengths are often below the tensile strength of the same material, as shown by some experimental results for a variety of composites in Table 4.2. By contrast with polymer- and metal-matrix composites, however, the compression strengths of reinforced ceramics are often higher than their tensile strengths, presumably because the rigid ceramic matrix affords a much greater level of support to the fibres, allowing their reinforcing potential to be realized, whereas their tensile strengths are often determined by features other than the fibre strength (Harris *et al.*, 1992b). This is shown by the data for the unidirectional SiC-reinforced calcium aluminosilicate composite at the bottom of Table 4.2. In studying compression behaviour, therefore, there is often uncertainty as to whether an apparently low value of the compression strength is a reflection of material behaviour or a consequence of inappropriate testing procedures.

Much effort has been expended on the development of compression test methods to overcome this problem. Devices commonly used include end-constraint techniques to prevent longitudinal splitting and lateral supports to prevent premature buckling, in addition to the more obvious precautionary measures to align the stress axis accurately with the fibres. A recent study by Berg and Adams (1989) suggests that many of the observed discrepancies between published test results for composites occur as a result

Table 4.2 Experimental Tension and Compression Strengths for Various Composite Materials

Material	Lay-up	V_f	Tensile Strength, σ_t GPa	Compression Strength, σ_c GPa	Ratio σ_c/σ_t
GRP	ud	0.6	1.3	1.1	0.85
CFRP	ud	0.6	2.0	1.1	0.55
KFRP	ud	0.6	1.0	0.4	0.40
HTA/913 (CFRP)	$[(\pm45,0_2)_2]_S$	0.65	1.27	0.97	0.77
T800/924 (CFRP)	$[(\pm45,0_2)_2]_S$	0.65	1.42	0.90	0.63
T800/5245 (CFRP)	$[(\pm45,0_2)_2]_S$	0.65	1.67	0.88	0.53
SiC/CAS (CMC)	ud	0.37	334	1360	4.07
SiC/CAS (CMC)	$(0,90)_{3S}$	0.37	210	463	2.20

of test methods. They carried out an evaluation of some of the commonly used anti-buckling test fixtures, and concluded that a method developed at Illinois Institute of Technology (the IITRI fixture), which at the time was an ASTM Standard test method, D 4310, was the most satisfactory method. By such means it is possible to measure compression strengths almost equal to the tensile strength in composites other than those reinforced with Kevlar fibres, as demonstrated by Ewins and Ham (1973) for a series of carbon-fibre composites and, more recently, by Parry and Wronski (1982). Ewins and Ham suggested that the mechanism of failure in CFRP is the same in both compression and tension modes and results from a 45° shear mechanism, but work by Parry and Wronski disagreed with this conclusion on the basis of studies of the effect of superposed hydrostatic pressure on the compression behaviour. It does not follow, however, that for any given composite the 'true' compression strength must be the same as the tensile strength.

Since it is not possible to attach any significance to the notion of 'compression strength' for a 10 μm diameter fibre, when we investigate the effect of reinforcement on compression strength we have to take the tensile strength of the fibre as our point of reference. As a unidirectional composite is loaded in compression, the fibre axis being well-aligned with the stress axis, the elastic models that we have already considered will

Fig. 4.30 Microstructure of a compression failure in a unidirectional glass/polyester composite (the compression axis is vertical).

apply and we would expect the initial stiffness to be the same as the tensile stiffness, as discussed in chapter 3. But the fibres can only be maintained in good alignment by the matrix, which, in resin-based composites usually has low shear stiffness, and when local shear or transverse tensile stresses become greater than the matrix can support, a crack may occur or a shear (kink) band, like that shown in Figure 4.30, may develop and grow as the load increases. Flexible fibres like glass or Kevlar-49 will usually bend or buckle in this kink band, while brittle fibres like carbon will often fracture after the initial microbuckling has occurred. These instabilities then lead to overall failure at stresses below the normal tensile strength of the material. Furthermore, since it is almost impossible to obtain a unidirectional composite with perfectly aligned fibres — the fibres in most practical composites are likely to have small kinks or misalignments arising from manufacture — these defects will act as initiators for the microbuckling mechanisms.

In an early model for compression strength, Rosen (1965) proposed that failure was initiated by matrix instabilities that lead to co-operative, in-plane buckling of the reinforcement. The model was developed for lamellar composites with reinforcing plates, rather than fibres, and, except at low V_f, it predicted failure by a shear mode at a stress:

$$\sigma_{comp} = \frac{G_m}{1 - V_f} \tag{4.30}$$

where G_m is the matrix shear modulus. The deficiency of the model, apart from the fact that it applies to the buckling of plates, is apparent from the fact that $\sigma_{comp} \rightarrow G_m$ as $V_f \rightarrow 0$, and equation 4.30 consistently overestimates the compression strength of real fibre

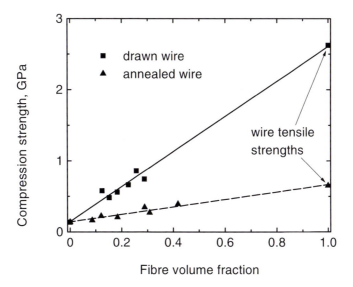

Fig. 4.31 Compression strengths of steel-wire/polyester-resin composites (de Ferran and Harris, 1970).

composites. Lager and June (1969) suggested that out-of plane buckling occurred in fibre composites, and proposed the use of an empirical 'influence factor' to obtain agreement between their results for boron/epoxy composites and Rosen's theory. This was a logical idea, since the stored elastic energy in a fibre which buckles into a helix is bound to be lower than that in one constrained to preserve a planar buckled form. It was later demonstrated by de Ferran and Harris (1970) that helical buckling occurred in fibres extracted from compressed MMCs. They also showed that the compression strengths of steel-wire/resin composites obeyed a mixture rule based on the *tensile* strength of the reinforcing wires, as illustrated in Figure 4.31. These results for two different species of steel wire reinforcements were in good agreement with a model of Hayashi and Koyama (1971) which allowed for the stress-dependence of the shear modulus. The implication was that in this case the fibres were controlling compression failure by virtue of their inherent shear resistance which, in a ductile metal, is the same in compression as in tension.

By contrast, Martinez *et al.* (1981) have shown that in composites reinforced with much finer filaments of carbon, glass and Kevlar-49 the compression strength appears not to be sensitive in any straightforward fashion to either the strength or stiffness of the reinforcing filaments, as can be seen from Figure 4.32. The linear portions of the curves for glass and both varieties of carbon (high-strength and high-modulus) are largely indistinguishable and extrapolate to stress levels at $V_f = 1$ of between 1.3 and 1.4 GPa, reasonable for the tensile failure of glass or HMS carbon, but only half what might have been expected for HTS carbon. The curve for the aramid composites lies well below those of the other composites, and the low strengths are inconsistent with the high tensile strength and stiffness of the Kevlar-49 fibre itself. The curves for all of the composites

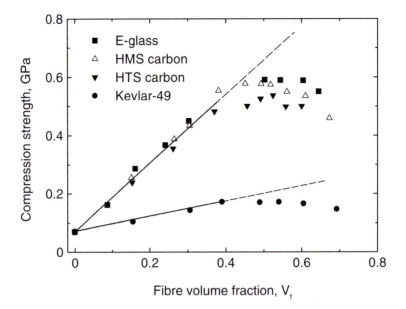

Fig. 4.32 Variation of compression strength with fibre volume fraction in unidirectional composites of polyester resin reinforced with carbon, glass and Kevlar-49 fibres (Martinez *et al.*, 1981).

are linear up to V_f values of only about 0.4, the deviations from linearity and subsequent downward trends implying changes in failure mode.

Early models often assumed that fibre buckling was responsible for the compression yielding of composites and that the Euler buckling theory for a column on an elastic foundation was a suitable basis for calculating the composite compression strength. Since buckling would be controlled by fibre modulus and diameter, the results in Figure 4.32, representing fibres of roughly similar diameter but having a wide range of strengths and moduli, and the earlier results of de Ferran and Harris, representing experiments with wires of different strengths but having identical moduli and diameters, offer no support for these ideas. A corollary of the buckling model was that fibres that were already misaligned or kinked would buckle more readily than misaligned fibres and that the apparently low compression strengths of many composites could be accounted for in this way. A series of experiments was therefore carried out by Martinez *et al.* (1981) with composites in which the fibres were deliberately misaligned with respect to the stress axis by twisting the tows of fibre which were used to reinforce a polyester resin. In this way a known angle of deviation could be introduced. Their results showed that for misalignments up to 10° the compression strengths of brittle-fibre composites actually increased and in some cases did not fall below the aligned-fibre value until nearly twice this degree of misalignment. The increase can be explained in purely geometric terms in the same way as the maximum-stress model of Stowell and Liu (*qv*) explains an initial increase in tensile strength with fibre misorientation.

Although the results described above offer little support for the idea that fibre misalignment controls compressive failure of unidirectional composites, a recent predictive model suggests otherwise. Barbero (1998) has derived an explicit mechanics-based model which is able to predict compression strength with good accuracy on the basis of a non-dimensional function:

$$\chi = \frac{G_{12}}{\Omega \tau_{11}}$$

where G_{12} is the composite shear stiffness and τ_{11} its shear strength, while Ω is the standard deviation (in degrees) of the fibre misalignment. The advantage of this model is that values of all three terms in the definition of χ may be obtained from well-established experimental techniques. Agreement with experimental data was good for a wide range of unidirectional composites, including glass- and carbon-fibre-reinforced polyesters, epoxies and thermoplastics.

The factors other than fibre thickness which affect the compression strength are the matrix strength and rigidity, and the interfacial bond strength. The more rigid the matrix, the greater the compression strength. Thus, ceramic and metal-matrix composites are much stronger in compression than resin-based composites. CMCs, in which the fibres and matrix often have similar levels of stiffness, may have compression strengths that exceed their tensile strengths, as shown in Table 4.2. Most modern reinforcing fibres are only about 10-15 μm in diameter. Boron, on the other hand, although not now much used, is much thicker — about 40 μm — and thicker fibres resist microbuckling much better than thin ones, as predicted by the Euler theory. Boron-fibre-reinforced aluminium is currently the composite material having the highest ratio of compression strength to density. Curtis and Morton (1982) have shown that the compression strengths of multidirectional coupons of an unspecified CFRP material were markedly affected by fibre surface treatment, increasing by some 50% as the level of a commercial treatment was increased from nothing to about 10% of the normal commercial level, and then remaining constant, at which stage the measured strengths were only some 10% less than the tensile strength of the same composite. This discrepancy could be removed, however, by taking into account small degrees of eccentric loading in the test jig.

4.2.1 COMPRESSION STRENGTH AFTER IMPACT

We have seen that tensile loading introduces damage into a composite and that the damage accumulation process is partly responsible for determining the strength of the composite. It is one of the perceived virtues of composite materials that they are tolerant of damage and that tensile damage does not seriously affect tensile load-bearing ability, but a word of caution is necessary here. Composite laminates in many external applications, especially in aircraft, are susceptible to impact damage from low-velocity projectiles travelling normal to the surface. The simplest, and most-often quoted, example is the damage caused by a spanner dropped by a person standing on an aeroplane wing. A tool weighing 1 kg dropped from waist height delivers a blow with about 10 J of energy.

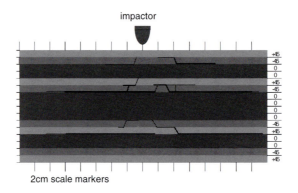

Fig. 4.33 Damage map of the longitudinal section of a $[(\pm45,0_2)_2]_S$ CFRP composite after it had received a low-velocity impact of energy 3J (Beheshty, 1997).

In an impact between a round-ended projectile and a flat plate, tensile stresses may occur in the matrix and across the interlaminar planes of weakness. In a low-velocity impact the rate of build-up of the Hertzian contact stresses is high. Even a toughened epoxy resin may therefore appear to be brittle under such conditions and cracks may occur, as shown in Figure 4.33, as planar cracks within the interlaminar regions and as transverse cracks in an apparently conical distribution spreading outwards through the composite away from the point of contact. The development of this kind of damage has been modelled by Davies *et al.* (1995). If the impacter has sufficient energy, there may be serious surface cracking or spalling damage on the back face of the laminate, even when the damage on the impacted face is scarcely detectable, a condition referred to as BVID — barely-visible impact damage. The problem for users of composites is that although this kind of damage does not markedly affect the tensile strengths of most laminates, it has an adverse effect on the compression strength because the interlaminar cracks are able to extend as delaminations. The CAI, or compression-after-impact, test has become a standard way of assessing the severity of damage induced by low-velocity impact events. An illustration of the effects of such impacts on the tension and compression strengths of a CFRP laminate is shown in Figure 4.34. For this fairly typical composite consisting of a high-strength fibre in a modified epoxy resin, an impact of only 5 J reduces the compression strength to some 32% of its initial value, while reducing the tensile strength by only 5 or 6%. Such a result has a knock-on effect for other properties of the material, such as the flexural response and fatigue behaviour. There are two aspects of the laminate structure that may be considered responsible for this behaviour. First, the fact that the matrix resin is an epoxy — even a so-called 'modified' one — which is naturally rate-sensitive means that the matrix is likely to become brittle at high rates of loading, and if the stresses are concentrated, either by the presence of notches or by the Hertzian contact effect, the rate is increased even further. Second, if the laminate is manufactured from prepreg sheet, the interlaminar planes form natural regions of weakness, as we have seen. Two possible solutions to this problem are a) to use a thermoplastic matrix instead of

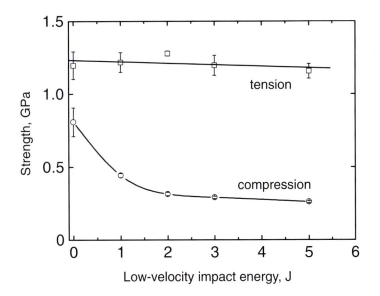

Fig. 4.34 Effect of low-velocity impact damage on the residual tensile and compression strengths of a $[(\pm 45,0_2)_2]_S$ HTA/982 CFRP laminate (Beheshty & Harris, 1997).

a thermoset, and b) to use woven-cloth reinforcement instead of non-woven fibres. An illustration of the power of this argument is given by some results of Ghaseminejad and Parvizi-Majidi (1990) who studied the impact behaviour and damage tolerance of two thermoplastic polyaryl resins, poly(ether ether ketone), known as PEEK, and poly(phenylene sulphide), or PPS, reinforced with woven carbon-fibre cloth. The compression strength after impact of the PEEK-based composite remained at 47% of the value for virgin material even after sustaining an impact of 29 J.

Guild *et al.* (1993) developed a model for the reduction in compression strength of laminates following impact. The virtue of their model is that it is based on the premise that failure in post-impact compression occurs as a consequence of the stress concentration that arises because of the presence of a damaged zone. It does not require the use of fracture-mechanics concepts which, as the authors point out, is ruled out by the observed fact that materials whose fracture behaviour is characterized by different values of the strain-energy release rates G_{Ic} and G_{IIc} often show comparable performances in respect of compression-after-impact. The model predicts, in agreement with experimental results, that the normalized compression-after-impact strength falls linearly with the width of the damage zone, whatever the material.

4.3 SHEAR STRENGTH

Most laminated continuous-fibre composites contain planes of weakness between the laminations and along fibre/matrix interfaces. In shear the composite strength will

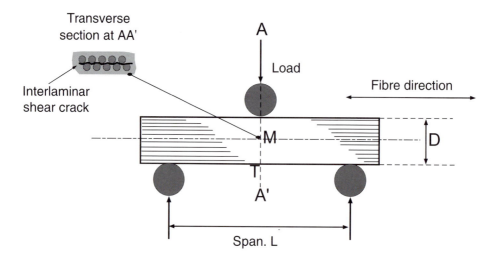

Fig. 4.35 Schematic illustration of interlaminar shear failure in a unidirectional laminate.

be dominated by these weaknesses unless the stress direction intersects the fibres. Shear stresses at the interfaces between the plies can seldom be avoided by lay-up design because of the anisotropies of neighbouring plies, as discussed in chapter 3. These interlaminar shear stresses are usually high at edges, as shown by Pipes and Pagano (1970), and often give rise to delaminations which propagate into the composite from the edges, significantly reducing the laminate tensile strength (Pagano and Pipes, 1971). Delamination is a major cause of failure in laminated composites, and one of the concerns of the designer is to ensure that shear stresses are diffused safely away from stress concentration points.

Interlaminar shear failure is most readily seen in the three-point bending of short beams, a method commonly used to measure what is usually referred to as the interlaminar shear strength, or ILSS, although this is often regarded as unsatisfactory because the state of stress is not pure shear. In the diagram in Figure 4.35, if the level of horizontal shear stress at the mid-plane point M reaches the interlaminar shear strength, τ_{IL}, of the composite before the tensile stress level at T reaches the composite strength, σ_c, then the beam will fail as shown. If the beam is longer than a certain critical length, however, it will fail in a normal bending mode by a tensile failure initiating at the midpoint of the outer face. The variation in failure load as the span-to-depth ratio of the beam increases is shown schematically in Figure 4.36. As long as the failure mode is interlaminar shear, the failure load is independent of length, but when the tensile failure mode predominates, the failure load falls rapidly in inverse ratio to the span of the beam. The critical span-to-depth ratio for the transition can be found by equating the expressions for the tensile and shear stresses in the beam.

The tensile stress at T is 3PL/2BD², while the shear stress at the neutral plane is 3P/4BD (the standard formulae for the bending beam, with B = width, D = thickness, and L = length). Equating these, we have:

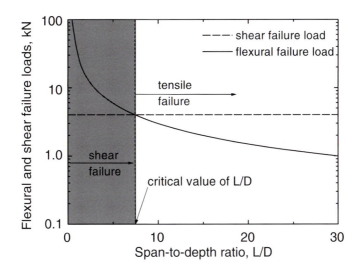

Fig. 4.36 Transition from shear to flexural (tensile) failure of a thin laminate plate in three-point bending as the span-to-depth ratio, L/D, increases. The composite tensile strength for this example is 750 MPa and its interlaminar shear strength is 50 MPa.

$$\frac{\sigma_c}{\tau_{IL}} = \frac{3PL}{2BD^2} \Big/ \frac{3P}{4BD} = \frac{2L}{D} \qquad (4.31)$$

and the beam will therefore fail in interlaminar shear at the neutral plane if $L/D < \sigma_c/2\tau_{IL}$. For a typical unidirectional CFRP laminate with $\sigma_c \approx 1.5$ GPa and $\tau_{IL} \approx 50$ MPa, a shear failure will occur in bending at span-to-depth ratios less than about 15:1, and when the short-beam shear test is used to evaluate the ILSS, it is usually recommended that a span-to-depth ratio of about 5 is used.

The interlaminar shear strength depends in part on the shear strength of the matrix and in part on the shear resistance of the interface. The stronger the interfacial bonding, then, and the greater the amount of interface relative to the amount of matrix, the greater will be the ILSS: the ILSS almost always increases with fibre volume fraction unless the level of fibre/matrix bonding is very low. With modern carbon and glass fibres this rarely occurs, however, because the reinforcing fibres are usually surface treated after manufacture in order to improve the interfacial bond. In carbon-fibre composites it was long ago shown that the ILSS was roughly inversely related to the stiffness of the reinforcing fibres, which reflected the fact that both the fibre mechanical properties and their chemical activity (wettability) are determined by the level of heat-treatment they receive during processing.

Interlaminar weakness is aggravated by the presence of voids or moisture in the resin, as shown in Figure 4.37. In some carbon/epoxy composites, for example, 10vol% of voids could reduce the ILSS by some 25%: moisture absorption will reduce the intrinsic matrix strength and may also cause a deterioration in the interfacial bond strength, both of which will reduce the ILSS. Corten (1968) suggested that the effect of voids on the

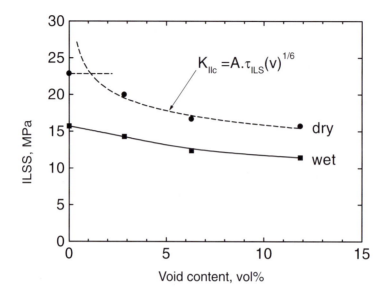

Fig. 4.37 The effect of voids and moisture on the interlaminar shear strength of a unidirectional carbon/epoxy laminate (Beaumont & Harris, 1972). The dashed line represents a suggestion of Corten (1968) that the effect of voids on ILSS may be represented by a constant value of the mode-II critical stress intensity factor and the voids as having the effect of cracks (see text).

shear strength could be treated in terms of a fracture-mechanics approach (see chapter 5 for further information on fracture mechanics). If an increase in void content is notionally equivalent, in terms of stress concentration, to an increase in crack size, and the equivalent crack dimension, a, is roughly proportional to the cube root of the void content, $\sqrt[3]{v}$, then an expression for the mode II critical stress-intensity factor, K_{IIC}, is:

$$K_{IIc} = \tau_{IL}\sqrt{\pi a} = A\tau(v)^{1/6} \qquad (4.32)$$

As the dashed line in Figure 4.37 shows, an expression of this form fits the data very well down to some threshold void content of about 1%. The constant, A, has a value of about 23 when the ILSS is in MPa and v is in %).

Although an appreciation of the reasons for the weakness of composites in shear is of great importance to the designer, it is not necessarily advantageous to use composites with the highest obtainable shear strengths. Cracks running normal to fibres in a tension structure may pass at various times through fibres, resin and interface, but if nothing changes the crack path, a reasonably flat fracture surface may result and the composite will be notch-sensitive, even brittle. If the material has an interface of only moderate strength, however, cracks may deviate through weak regions, such as the interlaminar planes, to become delaminations. As a result, a crack tip is effectively blunted and the material will appear tougher. Materials suppliers are required to maintain a fine balance in controlling the fibre, matrix and interface properties in order to optimise strength (tensile and shear) and toughness. An alternative approach to modifying the interfacial bond strength is to improve the delamination resistance of a laminate by 'stitching' the laminae together with fibres running in the direction perpendicular to the plane of the laminate. This process has been shown (Dransfield *et al.*, 1994) to improve both the ILSS (delamination resistance) and also the toughness of the composite, although Adanur *et al.* (1995) have shown that there is a limiting stitch density beyond which these improvements are reversed.

Another difficulty often faced by designers is that of assessing the reliability of a reported value of the shear strength of a composite, which we have previously termed τ_c. The short-beam shear test described above is very easy to carry out and is very useful as an experimental tool. But although it forms the basis of an international standard test, the complex stress state involved makes it unsatisfactory for obtaining absolute values of the shear strength: τ_{IL} is not necessarily a good measure of τ_c. The ideal method of measuring τ_c is perhaps by the torsion testing of thin-walled, hoop-wound tubes, as described by Puck and Schneider (1969), which gives pure states of shear stress and strain in the tube wall. Alternative solutions have usually been sought, however, because of the time and expense involved in carrying out such tests. One such solution is the measurement of the *intralaminar* shear strength by the tensile testing of unidirectional test coupons at an angle to the fibres. Chamis and Sinclair (1972) showed, from a consideration of the relative stress and strain magnitudes in tensile test samples at various angles, θ, to the fibre direction, that the in-plane shear strain is a maximum when $\theta = 10°$ for typical CFRP laminates. A tensile test on a unidirectional coupon cut at $10°$ to the fibres should therefore provide an accurate measure of the intralaminar shear strength. The tensile stress normal to the fibres is not zero at this angle, however, and locally the composite is effectively in a biaxial stress state. Hart-Smith (1996) argues that a tensile test on a $\pm45°$ laminate provides a more appropriate value of the in-plane shear strength, and there is much current interest in the Iosipescu test (Iosipescu, 1967), a method involving transverse shear loading of deeply notched test beams which, when carried out with adequate controls, generates a state of pure shear and provides good values of shear stiffness, but which still leaves something to be desired when used to obtain strength data (Pierron *et al.*, 1996).

4.3.1 THE SHEAR STRENGTH OF THE INTERFACIAL BOND

At several points in this book we refer to the importance of the fibre/matrix bond in determining important physical properties of composites. The economic and safe exploitation of modern high-performance composites depends directly on our ability to employ the fibres in resin and other matrices in such a way as to permit the optimization of the major mechanical properties of the composite. Our ability to tailor composites to a given specification, and to predict accurately the properties of manufactured materials, calls for an understanding of the nature of the individual components and of the manner in which they interact when working in combination.

Since much of this understanding hinges on characterization of the strength of the interface, it seems appropriate to digress briefly from dealing with macroscopic properties in order to consider some of the methods that are used for determining this all-important characteristic.

One of the difficulties in making valid assessments of the interfacial shear strength, τ_i, is that the simplest experimental methods involve interactions between single fibres and blocks or drops of resin rather than realistic composite structures. Some of the earliest experiments with single fibres embedded in resin were carried out by Broutman and by McGarry (see Broutman, 1967). These were carried out on specially shaped specimens which, when loaded in compression, caused shear or tensile debonding of the fibre from the matrix. These methods have largely been superseded, possibly because of the difficulty of preparing the samples. A second problem is that the analysis of experimental data often involves oversimplified assumptions about the stress distribution along a fibre and about the manner of interfacial failure. Some of these limitations will be apparent in the following discussion of several of the more familiar test methods.

4.3.1.1 Values From Macroscopic Shear-Strength Measurements

It might be anticipated from the schematic illustration of an interlaminar shear failure shown in Figure 4.35 that, since this kind of failure is a mixed fracture passing partly through the matrix and partly through the interface, the ILSS would be a rule-of-mixtures sum of the shear strengths of the matrix and the interface. It will not be a linear function of V_f, however, because the interface failure is associated with a non-planar surface area. As shown by Hancock and Cuthbertson (1970), the ILSS for a unidirectional composite would be given by:

$$\tau_{IL} = a_i\tau_i + (1 - a_i)\tau_m \tag{4.33}$$

where τ_m is the shear strength of the resin and a_i is the fraction of the fracture surface area which consists of fibre/resin interface. Since $a_i \to 1$ in a composite of high V_f, this relationship may provide a useful first estimate of τ_i, but this simplified model is likely to be inadequate because of the shear-stress concentrations that result from the inhomogeneity of the material. The results of Iosipescu or other macroscopic shear tests could, of course, be analysed in a similar manner to that described above.

4.3.1.2 Values From Crack-Spacing Measurements

When dealing with brittle-matrix composites like CMCs, estimates of the value of τ_i may be made from observations of crack spacings in damaged composites. In such a composite the matrix is usually of lower extensibility than the fibres, so that when the composite is loaded in tension, multiple cracking of the matrix occurs until a limiting crack density is reached. The Aveston, Cooper and Kelly theory (Aveston et al., 1971, 1974, 1979), which is discussed in more detail in section 4.6, shows that the limiting crack separation in a brittle-matrix composite is between x' and 2x', where

$$x' = \left(\frac{V_m}{V_f}\right) \cdot \left(\frac{\sigma_{mu}r}{2\tau_i}\right) \tag{4.34}$$

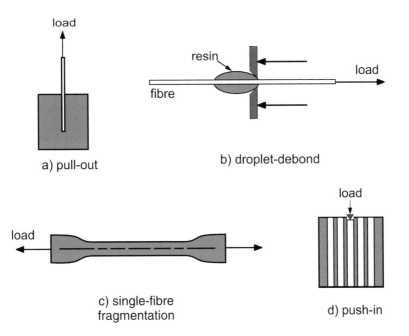

Fig. 4.38 Schematic illustration of four of the most common experimental methods for determining the interfacial shear strength.

σ_{mu} being the failure stress of the unreinforced matrix and r the fibre radius. Aveston *et al.* (1974) show that the mean crack spacing, s = 1.364x'. Thus, for a typical mean crack spacing of the order of 0.2 mm in a SiC/CAS composite of $V_f = 0.4$, (Harris *et al.*, 1992b), x' is about 0.15 mm, and the τ_i must therefore be about 3.5 MPa if σ_{mu} is 100 MPa. This low value is typical of many CMCs.

4.3.1.3 Fibre Pull-Out Tests

There are two main variants of this method, as illustrated in Figure 4.38a and b. In the first, single fibres of different lengths are embedded in resin and loads are applied. If the embedded length, ℓ, is less than $\ell_c/2$, the bond will be broken and the fibre extracted. When $\ell > \ell_c/2$, the fibre will break outside the resin (see problem 14, Chapter 8). When the end of a fibre with $\ell < \ell_c/2$ is loaded by pulling, the load increases to some maximum level at which the induced shear forces acting at the interface are able to initiate a debond and the load drops to some lower level as the debond rapidly propagates, as illustrated in Figure 4.39. The equilibrium state identified by the bottom of the load drop relates to a length of fibre separated from the surrounding resin and an end 'plug' which maintains contact and exerts an opposing frictional drag as the applied force withdraws the fibre from the matrix. The peak load must reflect the strength of the combined chemical and physical bond, while the portion of the curve after the drop relates to the work of fibre pull-out.

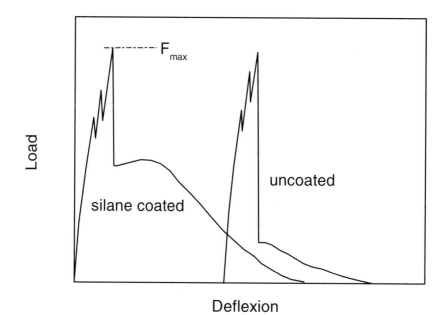

Fig. 4.39 Load/deflexion curves obtained from single-fibre pull-out experiments on glass/polyester composites (Harris *et al.*, 1975). The peak loads are not directly comparable because the pulled lengths of fibre were different.

The interpretational problem associated with this test is that we have no easy means of knowing exactly what shear/debond condition the force F_{max} or the load at the bottom of the load-drop shown in Figure 4.39 relate to. We do not know what the relative lengths of the plug and the 'free' portion of the fibre are, and we do not know how the shear stress varies in the plug region, nor how it changes as the pull-out process progresses. F_{max} is often used to obtain an indirect estimate of the maximum interfacial shear stress required for debonding:

$$\tau_i = F_{max} / \pi d \ell \qquad\qquad (4.35)$$

and the area under the curve after the load drop has been equated with the work of fibre pull-out, which provides an estimate of the interfacial frictional stress. Outwater and Murphy (1969) and Gao *et al.* (1988) adopted fracture-mechanics-based models to describe the debonding process, and while these may well offer more realistic models of the pull-out process, their implementation as tests for determining τ_i depends on knowing the critical strain-energy release rate for debonding.

Pitkethly and Doble (1989) pointed out that if equation 4.35 is applied to a long fibre then the shear strength will be underestimated because a part of the fibre is included over which no stress transfer is taking place. Ideally, the embedded fibre length should be as near to zero as possible, but for more manageable fibre lengths they took the value of the shear stress obtained from equation 4.35 as an average value of the shear debond stress and applied a correction derived from shear-lag theory to obtain a more realistic estimate of the maximum interfacial shear strength. A recent critique of the test and its analysis has been given by DiFrancia *et al.* (1996). They cite values of the debonding shear strength for surface-treated carbon/epoxy combinations of about 70 MPa, compared to about 25 MPa for untreated fibres. Qiu and Schwartz (1993) have tried to overcome the unrealistic condition of pull-out from a plain resin by designing micro-composite samples in which the central fibre is withdrawn from a 'composite' consisting of seven fibres in resin.

4.3.1.4 Droplet-Debond Tests

In the droplet-debond test (Miller *et al.*, 1991) a small droplet of resin is suspended on the fibre and, after curing, the fibre is withdrawn from the droplet (Figure 4.38b). Because the interfacial contact is small, there is a high probability that debonding will occur before fibre rupture or matrix deformation. The interfacial shear strength is again calculated from the maximum debonding force, F_{max}, from equation 4.35. It is usually found that there is a high degree of variability in these tests (cv $\approx$ 20%), often as a consequence of the inherent non-uniformity of fibre surfaces. An accurate assessment of the area of the fibre perimeter within the droplet is required, and this is made either by wettability measurements or fibre diameter measurements. An accurate assessment of the embedded length is also sometimes difficult to obtain if there are marked menisci at the ends of the droplet and these break off at an early stage in the test. Good instrumentation is needed if these dimensional measurements are not to constitute a major source of error.

4.3.1.5 Micro-Indentation (or Push-In) Tests

This test was first proposed by Mandell *et al.* (1980), and described for use with CMCs by Marshall and Evans (1985). Analyses have also been given by Chen and Young (1991) and Ho and Drzal (1996). A cross-section is taken through a real composite and polished by normal metallographic procedures. A very small diamond indenter, such as that in a micro-hardness testing machine, is then used to push one of the fibre ends into the body of the composite (Figure 4.38c). By choosing the load so that the indenter just contacts the matrix, an estimate of τ_i can be obtained from the relationship:

$$\tau_i = F^2 / \left(4\pi^2 u r^3 E_f\right) \tag{4.36}$$

where r is the fibre radius and E_f is the fibre elastic modulus. F is the force applied to the fibre, and this is calculated from the relationship:

$$F = 2a^2 H,$$

H being the fibre hardness which is obtained by indenting fibres under lower loads where no sliding occurs, and a is the length of the diagonal of the indentation impression in the fibre. The term u is the distance by which the fibre has been depressed at maximum load and is calculated from the dimensions of the indentation in the fibre and matrix. Two common difficulties with this test are that it is hard to assess the effect of the sideways force exerted by the indenter normal to the direction of sliding, and it is often difficult to obtain a realistic value of fibre 'hardness' because the structure (and therefore the properties) of common fibres are often very variable through the cross section. More sophisticated purpose-built testing machines provide force/displacement information directly from sensitive transducer read-outs.

4.3.1.6 Single-Fibre Fragmentation Tests

This test has become one of the principal methods for assessing fibre/matrix interfacial characteristics. The method, introduced by Kelly and Tyson (1965), consists in embedding a single fibre in a clear resin sample (Figure 4.38d) which is loaded in tension so as to cause the fibre to break down into successively shorter lengths until all fragments have lengths shorter than or equal to ℓ_c. To ensure that this process can be completed it is usually necessary to ensure that the resin has a high failure strain, thus imposing a second element of artificiality on the test. The fragment lengths are measured with an optical microscope focused on the fibre within the resin (for detailed descriptions, see, for example, Drzal et al., 1982; Netravali et al., 1989; Feillard et al., 1994).

The principle of the test depends on the force-balance relationship between the critical aspect ratio of the fibre, ℓ_c/d, the fibre strength, σ_f, and the interfacial shear strength, τ_i, given by equation 3.30:

$$\frac{\ell_c}{d} = \frac{\sigma_f}{2\tau_i} \qquad\qquad (4.37)$$

for rigid/perfectly-plastic behaviour of the composite. In the use of the fragmentation test it is assumed that the value of critical length obtained from the distribution of broken fibre lengths and the single-filament strengths obtained from single-filament tests may be used in this equation to obtain a value for the interfacial shear strength. The situation is less simple than it appears, however. The distribution of broken fibre lengths is not independent of the fibre strength distribution, and assessment of a true value of ℓ_c is not straightforward.

Assuming a uniform interfacial strength and a fibre strength distribution with little dispersion about the mean value, the shortest broken length that will be found should be equal to $\ell_c/2$, and the longest observed length will be just shorter than ℓ_c, the range of lengths being uniformly distributed between these two extremes, provided the test has been continued to sufficiently large strains. Thus, it is frequently assumed that the average observed broken length will be $3\ell_c/4$, and the value of the critical length derived from such measurements is $4\ell_{mean}/3$. Two typical distributions of aspect ratios, ℓ/d, of carbon-fibre fragments in epoxy resin are shown in Figure 4.40 (Harris et al., 1992a). They fit two-parameter Weibull distributions (see Appendix 2) with wide spreads. The shape factor, m, is about 6, which is similar to values obtained for the fibre strength distributions, as we saw in section 4.1.2. It can be seen that the effect of carrying out a commercial surface oxidation treatment on the fibre is to shift the distribution of fragment lengths to the left, so that the characteristic length, obtained from the scale parameter, b, falls from

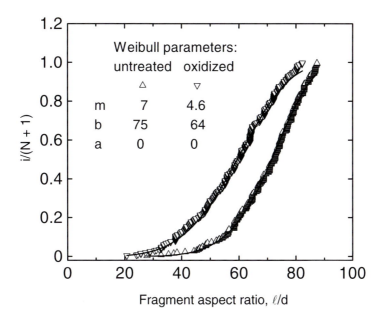

Fig. 4-40 Probability plot of fibre fragment aspect ratios from a single-fibre fragmentation test. Untreated and oxidized ENKA HTA carbon fibres.

about 525 mm to about to 450 mm as a result of this oxidation. Since both the fibre strengths and the fragment lengths are distributed according to two-parameter Weibull models, it is therefore necessary to use a statistical approach in order to determine a cumulative distribution function for the interfacial shear strength, from which an average *effective* interfacial shear strength can be obtained.

In obtaining an estimate of the interfacial shear strength from equation 4.37, the usual assumption is that the value of strength to be used in the equation is that which is characteristic of the fibre at a gauge length equivalent to the critical length, ℓ_c. If we consider the process of fragmentation, either in a single-fibre composite or in a real composite, we see from the weakest-link model that each time a fibre fractures, the remaining pieces of the fibre must become stronger. During the fragmentation process, therefore, there will be a continuing change in the value of fibre strength that is actually affecting the fragmentation process and therefore a change in the balance of forces implied by equation 4.37. The final fracture in a given piece of fibre, will occur when the fibre length, ℓ, is somewhat greater than ℓ_c and somewhat less than $2\ell_c$, *i.e.* $\ell_c < \ell < 2\ell_c$. Thus, the appropriate value of strength to use in equation 4.37 will be that which is characteristic of a length of about $1.5\ell_c$, and this can be obtained from the strength/ length relationship of equation 4.10. Some typical experimental results for values of the interfacial shear strength in carbon/epoxy systems are given in Table 4.3. These are taken from published papers, and full references may be found in Harris *et al.*, 1992a.

Table 4.3 Some Published Values of Carbon/Epoxy Interfacial Shear Strength Values Measured by the Single-Fibre Fragmentation Test

Carbon-Fibre/Resin System	Fibre Surface Treatment	Reported Value of τ_i, MPa
HMU/828	None	6-14
HMS/828	Commercial	20-47
AU/828	None	24
AS/828	Commercial	74
AU4/828	None	29-36
AS4/828	Commercial	47, 48
T300/5208	None	39
T300/5208	Commercial	49

It is unlikely that the rigid/perfectly-plastic assumption is truly valid for resin-based composites. Lacroix *et al.* (1992) have made a detailed critique of the method, arguing that ℓ_c may not be considered to be a material constant, since it depends on the shear-stress-transfer model used. The Kelly and Tyson model(1965) which yields equation 4.37 is valid only for the condition of total debonding and fails to distinguish between the interfacial bond strength and the frictional shear strength. Predicted values of ℓ_c may be lower than those characteristic of complete debonding if the interfacial bond strength is sufficiently larger than the frictional shear stress. Despite the obvious experimental difficulties with this test, the results in Table 4.3 are reasonably self-consistent, and the method can presumably be used for comparative measurements, even though the absolute values of τ_i obtained from it may not be accurate.

4.3.1.7 Direct Measurement by Laser-Raman Spectroscopy

This is essentially a means of measuring strain by monitoring the frequency of atomic vibrations in a crystalline material which can be probed by spectroscopic techniques such as Raman or IR spectroscopy. As atomic bond lengths increase under load, frequency shifts occur which are directly related to the applied load. The use of the laser Raman method, which is widely applicable to materials such as commercial polymers and ceramics, does not require physical contact, unlike other strain-measuring techniques, and it is highly selective, being capable of probing areas as small as 1 or 2 μm. Pioneering work in this field has been carried out by Galiotis and co-workers (1984, 1991).

It is commonly observed that the results obtained from many of the above tests show considerable variability and there are also usually significant differences between

Table 4.4 A Comparison of Results of Interfacial Shear-Strength Measurements Obtained by Different Test Methods for AS4 Carbon Fibres in 828 Epoxy Resin (Herrera-Franco & Drzal, 1991)

Experimental Method	τ_i, MPa
Short-beam shear test (ILSS)	84.0
Iosipescu shear test	95.6
$[\pm 45]_{3s}$ tension test	72.2
Droplet-debond test	50.3
Micro-indentation test	65.8
Single-fibre fragmentation test	68.3

the experimental estimates obtained from the different tests, as shown in Table 4.4 by some results published by Herrera-Franco and Drzal (1991).

4.4 FLEXURAL STRENGTH

As discussed in section 4.3, we might expect that when a composite is loaded in bending it would fail when the stress in the outer fibres on the tensile face reaches the normal composite tensile strength. However, early flexural tests often gave strength values that were considerably lower than the tensile strength. One reason for this was that the loading points in flexural test rigs designed for metals often caused localized damage which initiated premature failures. But even when this defective test procedure was rectified, low-stress failures still occurred. It was then considered that this was because in materials with poor in-plane shear resistance and with compression strengths that were lower than the tensile strengths, it was shear and/or compression damage modes that initiated the premature flexural failures. Controlled fibre surface treatments have now considerably improved both the interlaminar shear and compression responses of many commercial materials, and it is usually possible to measure flexural strengths that are at least equal to the normal tensile strengths.

Direct comparison of tension and flexural strength is somewhat complicated, however, by virtue of the statistical effect of the different stressed volumes in tensile and flexural test pieces. As discussed in Appendix 2 and illustrated by equation 4.10, the effective strength of a sample of small volume is greater than that of a larger sample. Thus, if two identically sized composite test pieces are tested in axial tension and in bending, the effective strength of the latter will be higher, even though the failure mode in each case is simple tension, because the volume of the bend-test sample subjected to the highest stress level is much less than that in a tensile sample. The effect of such geometrical/statistical effects in ceramics has been demonstrated by Davidge (1979). He shows that, for samples of similar volume, the ratio of strengths measured in four-point bending and tension is a function only of the Weibull shape factor, m, thus:

$$\frac{\sigma_{4B}}{\sigma_T} = \left[\frac{4(m+1)^2}{m+2} \right]^{\frac{1}{m}} \tag{4.38}$$

and for a given value of m the ratio of the three-point bend strength to the tensile strength is even greater.

Measurements on a $[(\pm45,0_2)_2]_S$ HTA/913 carbon/epoxy laminate gave values of 1.17±0.05, 0.81±0.02, and 1.00±0.12 GPa, respectively, for the axial tensile and compression strengths and the 4-point flexural strength (Hardiman, 1997). The Weibull shape factor for most good-quality CFRP materials is usually high, between 20 and 30, by comparison with common values for ceramics which are less than 10. A mean value obtained at Bath for five other CFRP laminates of the same lay-up as the above is 22, and the ratio given by equation 4.38 is therefore about 1.23. Thus, whereas the measured tensile and flexural strengths are statistically almost identical, this apparent equivalence is accidental since, on this basis, we would have expected the flexural strength to be 23% higher than the tensile strength, *i.e.* about 1.44 GPa, if the flexural failure is determined by the tensile strength. As it happens, the same factor applied to the experimental compression strength gives a value of 1.0 GPa, which is equal to the measured value of the flexural strength, even though the flexural failure of these samples was not apparently initiated on the compression face. We note, in passing, that for a group of SiC/CAS CMC laminates, the ratios of three-point-bend strength to tensile strengths were between 1.5 and 2.5, consistent with Weibull m values between 5 and 10. Attempts to reconcile the results of measurements of this kind with the simple statistical model are of course based on the assumption that the characteristics of the critical flaws on the surface and in the volume of the material are identical, which may not be true.

4.5 FAILURE CRITERIA FOR COMPLEX STRESSES

For most practical purposes, designers require models of behaviour that can predict failure under realistic combinations of stresses, rather than for the idealised uniaxial stress conditions under which most laboratory tests are carried out. There has been a great deal of research on complex-stress failure criteria (twenty or more models have been proposed, although the differences between many of them are quite small: see the review by Owen and Found, 1972), and designers throughout the world remain in dispute as to the 'best' method. A recent meeting organized jointly by the Institute of Mechanical Engineers and the Science and Engineering Research Council (now the Engineering and Physical Sciences Research Council) in the U.K. (1991) exposed some of the areas of disagreement, and the Elsevier journal *Composites Science and Technology* has published a Special Issue devoted to evaluating a wide range of popular failure criteria against validated experimental results.[*] The main difficulty in accepting one or other of the common methods is that their validity can usually only be tested over limited ranges of combined stress because of the complexity and cost of the test samples and test procedures for such experiments — tubes under combined tension, torsion and internal pressure, for example. Most of the existing failure criteria are in fact restricted to conditions of plane stress (thin plates) and some are only applicable to orthotropic materials.

We have noted in section 4.1.6 the differences between the maximum-stress criterion and the maximum-strain-energy criterion in predictions of the orientation-dependence of the strength of a single lamina. The former takes into account that a unidirectional laminate will fail by different mechanisms at different angles, but is not able to allow for interactions between failure modes or stresses. The disadvantage of the strain-energy approach is that it does not take account of different failure modes. In its formal form it is an interactive model, although the interaction usually has to be allowed for empirically. The virtue of the strain-energy criterion in the eyes of many designers is that it is represented by a single equation instead of three (or five if compression failure is included). One of the earliest applications of the von Mises model was by Norris and McKinnon (1946) for GRP in plane stress (*i.e.* loading by stresses σ_1, σ_2, τ_{12}, with all other stresses = 0) which predicts that failure will occur when:

$$\left[\frac{\sigma_1}{\sigma_c}\right]^2 + \left[\frac{\sigma_2}{\sigma_t}\right]^2 + \left[\frac{\tau_{12}}{\tau_c}\right]^2 = 1 \qquad (4.39)$$

the stresses σ_c, σ_t, and τ_c being the composite failure stresses in tension, compression and shear, as defined in section 4.16. In this form no interaction effects are allowed for. The later model of Azzi and Tsai (1965), a development of which we have already met, took into account interaction effects between the in-plane normal stresses:

$$\left[\frac{\sigma_1}{\sigma_c}\right]^2 + \left[\frac{\sigma_1\sigma_2}{\sigma_c\sigma_t}\right]^2 + \left[\frac{\sigma_2}{\sigma_t}\right]^2 + \left[\frac{\tau_{12}}{\tau_c}\right]^2 = 1 \qquad (4.40)$$

although in many situations the second term is relatively small and is often ignored, when the equation reverts to equation 4.39. Like the maximum-stress theory, these quadratic models are also extendible into the second, third and fourth quadrants of the (σ_1,σ_2) plane-stress framework, provided appropriate values of the compression strengths are used. Predictions of strength for off-axis loading can be obtained by substituting the stress definitions used for the maximum-stress model (equations 4.16-4.18) into equation 4.39, when the form of the maximum-strain-energy criterion given in equation 4.19 is recovered.

Eckold *et al.* (1978) have developed a laminate theory for the prediction of failure envelopes for filament-wound structures under biaxial loads. They predict failure, from a maximum-stress criterion rather than an interaction model, in a ply-by-ply analysis which allows for progressive failure of the laminate, for shear non-linearity, and for different properties in tension and compression. The composite is assumed to be composed of a series of homogeneous orthotropic plies, as is usual in laminate theory calculations. Good agreement between predictions and experimental results was obtained.

---

[*] 'Failure Criteria in Fibre-Reinforced-Polymer Composites,' M. J. Hinton, P. D. Soden and A. S. Kaddour, eds., *Composites Science and Technology*, **58**, July 1998, 999–1254.

Although single-expression models like equation 4.39 are easy to work with mathematically, they are subject to the very serious criticism that they are truly appropriate only for homogeneous materials and do not take account of the different physical failure modes that occur in composites as the relative orientations of the stress and symmetry axes change. Homogenisation methods of this kind result in single elliptical failure envelopes which can apparently be defined from only three or four data points obtained under straightforward experimental conditions. But they are incapable of distinguishing between fibre and matrix failures, or allowing for alternative matrix damage processes, such as cracking and ductile shear failure. Hart-Smith has been particularly vociferous in condemning the quadratic models used by Tsai and his collaborators (and embodied in most of the currently available computer design software). He maintains (1991) that it is scientifically incorrect to use polynomial interaction failure models whenever the mechanism of failure of the laminate changes with stress and maintains that separate failure criteria are needed for the fibres, the matrix and the interface. He has argued (1991, for example) for the adoption of a maximum-shear-stress (Tresca) model because it automatically introduces truncations related to changes in failure mode which the quadratic models do not, and satisfactorily explains in-plane laminate test data in all of the stress quadrants.

4.6 THE STRENGTH OF CERAMIC-MATRIX COMPOSITES

Because of the brittleness of the matrices of CMCs, their response to stress differs considerably from that of reinforced plastics and metals. The matrix cracks at low stresses relative to the final failure stress, gradually transferring all responsibility for load bearing to the reinforcing fibres. Stress/strain curves for some unidirectional and cross-plied $(0,90)_{3s}$ laminates of SiC/CAS are shown in Figure 4.41. The right-hand section of this figure shows the normal axial stress/strain curves, while the left-hand section shows the variation of the *transverse* strains during loading.

The axial stress/strain relationships for the unidirectional and $(0,90)_{3s}$ laminates show similar features, with a predominantly linear initial portion, a marked knee, an almost linear second stage with a much reduced slope, followed by a third non-linear region in which the average slope increases again. The knee indicates the onset of major matrix cracking.

The transverse strains exhibit unusual variations: they begin in the normal manner by showing a reduction in lateral dimension that is proportional to the load — normal Poisson contraction — but the direction of this strain is subsequently reversed and eventually becomes positive. This type of behaviour has been observed for a number of CMCs and although various explanations have been put forward it seems likely that the reversal is due to a gradual release of residual thermal stresses as matrix cracking develops.

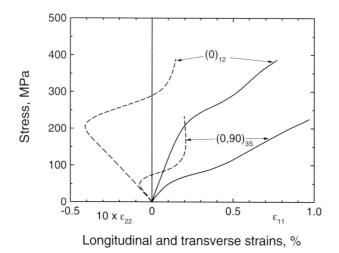

Fig. 4.41 Longitudinal and transverse strains as a function of stress for unidirectional and cross-plied composites of SiC/CAS (Harris *et al.*, 1992b). The transverse strains are multiplied by ten to increase visibility on the graph.

From an energy-based model, Aveston *et al.* (1971) and Budiansky *et al.* (1986) show that the stress at which matrix cracking should occur, σ_{mc}, is given by:

$$\sigma_{mc} = \left[\frac{12\tau_i \gamma_m V_f^2 E_f \left(1 - v_c^2\right)^2}{r V_m E_c E_m^2} \right]^{1/3} \frac{E_c}{\left(1 - v_c^2\right)} \tag{4.41}$$

where τ_i is the interfacial shear strength, r is the fibre radius, and v_c is the composite major Poisson ratio. The two unknowns in this equation are γ_m, the matrix fracture energy, and τ_i, the interfacial shear strength. Estimates of the value of τ_i are often found from observations of crack spacing in damaged composites which, in a brittle/brittle system, is inversely related to the strength of the interfacial bond. When the fibres have a greater failure elongation than the matrix, which is almost always the case in CMCs, multiple cracking of the matrix occurs during loading until a limiting crack density is reached where the crack spacing is between x' and 2x'. The condition for multiple fracture is that:

$$\sigma_{fu} > \sigma_{mu} V_m + \sigma_f' V_f \tag{4.42}$$

where σ_{mu} is the matrix tensile strength and σ_f' is the fibre stress at the matrix failure strain. The crack spacing is obtained from a force balance between tensile force in the fibres and shear force at the interface as:

$$x' = \left(\frac{V_m}{V_f} \right) \cdot \left(\frac{\sigma_{mu} r}{2\tau_i} \right) \tag{4.43}$$

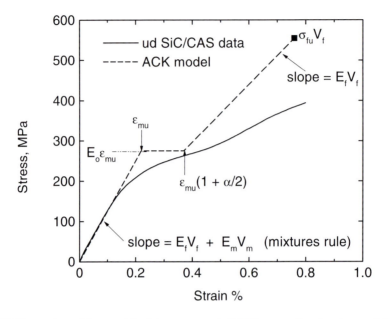

Fig. 4.42 Illustration of the model of Aveston *et al.* (1974) to predict the shape of the stress/strain curve for a ceramic-matrix composite α is the ratio $E_m V_m / E_f V_f$.

σ_{mu} being the failure stress of the unreinforced matrix and r the fibre radius. The cracking stress predicted by equation 4.41 is often very low by comparison with the normal composite tensile strength. For example, Evans *et al.* (1991) quote a range of 140-160 MPa for SiC/CAS composites for which the tensile strength may well be of the order of 1GPa, and acoustic emission studies have shown that substantial sub-critical matrix cracking occurs at stresses well below the level predicted by equation 4.41 (Harris *et al.*, 1992b).

The ACK model was used to predict the stress/strain curves for a number of fibre-reinforced cement-matrix composites with a considerable degree of accuracy (Aveston *et al.*, 1994). The predicted curve follows three distinct, linear stages as shown in Figure 4.42. The first stage has a slope which is given by the usual rule of mixtures, $E_c = E_f V_f + E_m(1 - V_f)$, and rises to a stress level σ_{mc} at the first matrix cracking strain, ε_{mc}. The second stage is horizontal and corresponds to a strain range during which matrix cracking continues at constant stress until a saturation level is reached at a strain $\varepsilon_{mc}(1 + 0.66\alpha)$, α being the weighted modulus ratio, $E_m V_m / E_f V_f$. The plateau joins the third section which is of slope $E_f V_f$, representing a fibre-only contribution, and the failure point is reached at a stress level $\sigma_{fu} V_f$.

Not all CMC stress/strain curves are well predicted by the ACK model, however. Figure 4.42 also includes the stress/strain curve for the unidirectional SiC/CAS laminate referred to in Figure 4.41. It can be seen that matrix cracking, signalled by the deviation from linearity of stage 1 of the curve, starts well below the predicted level and the stress continues to rise as the cracking proceeds, rather than remaining constant. The slope of the third stage of the curve falls well below the predicted fibre-only contribution, and final failure occurs below the predicted failure stress. Possible explanations of this are that the fibres have been damaged by the manufacturing process, or sustain damage as matrix cracking proceeds, perhaps as a result of dynamic crack-growth effects not envisaged in the ACK model.

4.7 REFERENCES

S. Adanur, Y.P. Tsao and C.W. Tam: *Composites Engineering*, **5**, 1995, 1149-1158.

K.E. Andrews: Unpublished Results, *Department of Materials Science and Engineering,* University of Bath, 1998.

J. Aveston, G. A. Cooper and A. Kelly: *Proceedings of NPL Conference on Properties of Fibre Composites,* NPL Teddington, IPC Science and Technology Press, Guildford, U.K., 1971, 15-26.

J. Aveston, R.A. Mercer and J.M Sillwood: *Proceedings of NPL Conference on Composites Standards, Testing and Design*, IPC Science and Technology Press, Guildford, U.K., 1974, 93-102.

J. Aveston and A. Kelly: *Proceedings of Royal Society, London,* **A366**, 1979, 599-623.

J. Aveston, A. Kelly and J.M. Sillwood: Proceedings of ICCM3, Paris, *Advances in Composite Materials,* Pergamon, Oxford, **1**, 1980, 556-568.

V.D. Azzi and S.W. Tsai: *Experimental Mechanics,* **5**, 1965, 283-288.

M.G. Bader: *Science and Engineering of Composites Materials,* **1**, 1988, 1-11.

J.E. Bailey, P.T. Curtis and A. Parvizi: *Proceedings of Royal Society*, **A366**, 1979, 599-623.

E.J. Barbero: *Journal of Composites Material*, **32,** 1998, 483-502.

J.S. Berg and D.F. Adams: *Journal of Composites Technology and Research*, **11**, 1989, 41-46.

W.H. Bowyer and M.G. Bader: *Journal of Materials Science*, **7**, 1972, 1315-1321.

L.J. Broutman: *Modern Composite Materials,* L.J. Broutman and R.H. Krock, eds., Addison-Wesley, Reading, Mass, 1967, 391 *et seq.*

B. Budiansky , J.W. Hutchinson and A.G. Evans: *Journal of Mechanics and Physics of Solids,* **34**, 1986, 167-189.

A.R. Bunsell and B. Harris: *Proceedings of ICCM1: First International Conference on Composite Materials,* E. Scala, E Anderson, I. Toth and B.R. Noton, eds., AIME New York, 1976, 174-190.

C.C. Chamis and J.H. Sinclair: *Experimental Mechanics*, **17**, 1977, 339-346.

E.J.H. Chen and J.C. Young: *Composites Science and Technology*, **42,** 1991, 189-206.

G.A. Cooper and A. Kelly: *Interfaces in Composites* STP 452, American Society for Testing and Materials, Philadelphia, U.S.A., 1969, 90-106.

H.T. Corten: *Fundamental Aspects of Fibre Reinforced Plastics,* R.T. Schwartz and H.S. Schwartz, eds., Wiley/Interscience, New York, 1968, 89-128.

A.H. Cottrell: *The Mechanical Properties of Matter*, John Wiley and Son, NY., 1964.

P.T. Curtis and J. Morton: *The Effect of Fibre Surface Treatment on the Compressive Strength of CFRP Laminates,* RAE Technical Report 82047, Ministry of Defence Procurement Executive, U.K., 1982.

P.T. Curtis: *Composites Science and Technology*, **27**, 1986, 63-86.

R.W. Davidge: *Mechanical Behaviour of Ceramics*, Cambridge University Press, Cambridge, 1979.

G.A.O. Davies and X. Zhang: *International Journal of Impact Engineering,* **16**, 1995, 149-170.

E.M. de Ferran and B. Harris: *Journal of Composites Material* **4**, 1970, 62-72.

R.F. Dickson: *The Strength of Fibre Bundles*, Unpublished Results, School of Materials Science, University of Bath, 1980.

C. DiFrancia, T.C. Ward and R.O. Claus: *Composites,* **A27,** 1996, 597-612.

K. Dransfield C. Baillie and Y.W. Mai: *Composites Science and Technology*, **50**, 1994, 305-317.

L.T. Drzal, M.J. Rich and P. Lloyd: *Journal of Adhesion*, **16,** 1982, 1-30.

G.C. Eckold, D. Leadbetter, P.D. Soden and P.R. Griggs: *Composites*, **9**, 1978, 243-246.

P.D. Ewins and A.H. Ham: *The Nature of Compressive Failure in Unidirectional CFRP,* Technical Report 73057, Ministry of Defence Procurement Executive, U.K., 1973.

P. Feillard, G. Désarmot and J.P. Favre: *Composites Science and Technology,* **50,** 1994, 265-279.

H. Fukuda and K. Kawata: *Fibre Science and Technology,* **10**, 1977, 53-63.

C. Galiotis, R.J. Young, P. Yeung and D.N. Batchelder: *Journal of Material Science,* **19**, 1984, 3640-3684,

C. Galiotis: *Composites Science and Technology,* **42,** 1991, 125-150.

Y.C. Gao, Y.W. Mai and B. Cotterell: *Zeit für Angewandte Mathematics and Physics,* **39,** 1988, 550-572.

M.N. Ghaseminejhad and A. Parvizi-Majidi: *Composites*, **21**, 1990, 155-168.

F.J. Guild, B. Harris and A.J. Wills: *Acoustic Emission,* **1**, 1982, 244-250.

F.J. Guild, P.J. Hogg and J.C. Prichard: *Composites,* **24**, 1993, 333-339.

P. Hancock and R.C. Cuthbertson: *Journal of Materials Science,* **5,** 1970, 762-768.

S.J. Hardiman: Unpublished results, School of Materials Science, University of Bath, 1997.

B. Harris and D. Cawthorne: *Plastics and Polymers*, **42**, 1974, 209-216.

B. Harris and S.V. Ramani: *Journal of Materials Science*, **10**, 1975, 83-93.

B. Harris, F.J. Guild and C.R. Brown: *Journal of Physics D, Applied Physics*, **12**, 1979, 1385-1407.

B. Harris, O.G. Braddell, C. Lefebvre and J. Verbist: *Plastic Rubber and Composites: Processing and Application,* **18**, 1992a, 221-240.

B. Harris, F.A. Habib and R.G. Cooke: *Proceedings of Royal Society*, London, **A437,** 1992b, 109-131.

L.J. Hart-Smith: Proceedings of a Meeting on *Failure of Polymeric Composite Structures: Mechanisms and Criteria for the Prediction of Performance,* St. Albans, U.K., SERC/IMechE, London, 1991, 19-31.

L.J. Hart-Smith: *Aerospace Materials,* **5** (part 2), 1993, 10-16.

T. Hayashi and K. Koyama: *Proceedings of International Conference on Mechanical Behaviour of Materials,* Kyoto, Japan, *Japanese Society for Materials Science,* **5,** 1971, 104-112.

P.J. Herrera-Franco and L.T. Drzal: *Composites,* **23,** 1992, 2-27.

R. Hill: *Proceedings of Royal Society,* London, **A193,** 1948, 281-297.

H. Ho and L.T. Drzal: *Composites:* **A27,** 1996, 961-971.

J.D. Hughes, H. Morley and E.E. Jackson: *Journal of Physics D*: *Applied Physics,* **13,** 1965, 921-936.

N. Iosipescu: *Journal of Materials,* **2,** 1967, 537-566.

R.D. Jamison: *Composites Science and Technology,* **24,** 1985, 83-99.

A. Kelly and W.R. Tyson: *Journal of Mechanics and Physics of Solids,* **13,** 1965, 329-350.

J.A. Kies: *Maximum Strains in the Resin of Fibre-Glass Composites*: Report 5752, U.S. Naval Research Laboratory.

T. Lacroix, B. Tilmans, R. Keunings, M. Desaeger and I. Verpoest: *Composites Science and Technology,* **43,** 1992, 379-387.

J.R. Lager and R.R. June: *Journal of Composites Materials,* **3,** 1969, 28-56.

J.F. Mandell, J.H. Chen and F.J. McGarry: *Adhesion and Adhesives,* **1,** 1980, 40-44.

P.W. Manders and M.G. Bader: *Journal of Material Science,* **16,** 1981, 2246-2256.

D.B. Marshall and A.G. Evans: *Journal of American Ceramic Society,* **68,** 225-231.

G.M. Martinez, M.R. Piggott, D.M.R. Bainbridge and B. Harris: *Journal of Materials Science,* **16,** 1981, 2831-2836.

B. Miller, U. Gaur and D.E. Hirst: *Composites Science and Technology,* **42,** 1991, 207-219.

A.N. Netravali, L.T.T. Topoleski, W.H. Sachse and S.L. Phoenix: *Composites Science and Technology,* **35,** 1989, 13-29.

C.B. Norris and P.F. McKinnon: *US Forest Products Laboratory Report no. 1328,* U.S. Forest Products Laboratory, 1946.

J.O. Outwater and M.C. Murphy: *Proceedings 24th Annual Technical Conference of Reinforced Plastics/Composites Institute of SPI,* 1969, 11C.

M.J. Owen and M.S. Found: *Solid/Solid Interfaces: Faraday Special Discussions,* Chemical Society, London, **2,** 1972, 77-89.

J. Pabiot: *Composite Materials*, Conference Proceedings Number 63, AGARD/NATO Neuilly Paris, 1971, Paper 6.

N.J. Pagano and R.B. Pipes: *Journal of Composites Material,* **5,** 1971, 50-57.

N.J. Parratt, *et al.*: *New Technology (May issue),* Department of Trade and Industry, London, 1971.

N.J. Parratt and K.D. Potter: *Advances in Composite Materials,* Proceedings of ICCM3, A.R. Bunsell, C. Bathias, A. Martrenchar, D. Menkes and G. Verchery, eds., Pergamon,

Oxford, **1**, 1980, 313-326.

T.V. Parry and A.S. Wronski: *Journal of Materials Science,* **17**, 1982, 893-900.

F. Pierron and A. Vautrin: *Realising their Commercial Potential:* Proceedings of 7th European Conference on Composite Materials, ECCM7, Woodhead Publishing, Abington, U.K. and EACM, Bordeaux, **2**, 1996, 119-124.

R.B. Pipes and N.J. Pagano: *Journal of Composites Material,* **4**, 1970, 538-548.

M.J. Pitkethly and J.B. Doble: Proceedings of an International Conference on *Interfacial Phenomena in Composite Materials* '89, F.R. Jones and Butterworths, eds., London, 1989, 35-43.

A. Puck and W. Schneider: *Plastics and Polymers,* **37**, 1969, 33-34.

Y.P. Qiu and P. Schwartz: *Composites Science and Technology,* **48,** 1993, 5-10.

B.W. Rosen: *Fibre Composite Materials: Proceedings ASM Conference,* (American Society for Metals, Ohio), 1965, 37-87.

E.Z. Stowell and T.S. Liu: *Journal of Mechanics and Physics of Solids,* **9**, 1961, 242.

C.T. Sun and J. Luo: *Composites Science and Technology,* **22**, 1985, 121-134.

W. Weibull: *Journal of Applied Mechanics,* (Trans. ASME), **73**, 1951, 293-297.

C. Zweben: *Journal of Materials Science,* **12**, 1977, 1325-1337.

5. Fracture and Toughness of Composites

5.1 STRUCTURAL ASPECTS OF COMPOSITES FAILURE

Many reinforced plastics consist of brittle fibres, such as glass or carbon, in a weak, brittle polymer matrix, such as epoxy or polyester resin. An important characteristic of these composites, however, is that they are surprisingly tough, largely as a result of their heterogeneous nature and construction. During deformation, microstructural damage is widespread throughout the composite, as we saw in chapter 4, but much damage can be sustained before the load-bearing ability is impaired. Beyond some critical level of damage, failure may occur by the propagation of a crack which usually has a much more complex nature than cracks in homogeneous materials. Crack growth is inhibited by the presence of interfaces, both at the microstructural level between fibres and matrix, and at the macroscopic level as planes of weakness between separate laminae in a multiply laminate. The fracturing of a composite therefore involves not only the breaking of the load-bearing fibres and the weak matrix, but a complex combination of crack deviations along these weak interfaces. For similar reasons, even all-brittle ceramic/ceramic composites can also exhibit toughnesses which are orders of magnitude higher than those of unreinforced ceramics.

The microstructural inhomogeneity and anisotropy of fibre composites are together responsible for the fact that the fracture of these materials is rarely a simple process. Although the complex combination of micro-failure events that leads to deterioration of load-bearing ability, or to destruction, can often give rise to surprisingly high levels of toughness, the same complexity makes it difficult or impossible to use procedures based on fracture mechanics for design purposes. There have been many theoretical and experimental studies of cracking in composites and of the mechanisms by which toughening is achieved, but there is still a large measure of disagreement about the contributions to the overall toughness of the various processes by which cracks are stopped or hindered. This is probably because of the great diversity of composite types that have been studied. The toughness of a composite is derived from many sources, and the relative magnitudes of the separate contributions depend not only on the characteristics of the components, but also on the manner in which they interact. There is thus no simple recipe for predicting the toughness of all composites. The lay-up geometry of a composite strongly affects crack propagation, with the result that some laminates appear to be highly notch-sensitive whereas others are totally insensitive to the presence of stress concentrators. The selection of resins and fibres, the manner in which they are combined in the lay-up, and the quality of the manufactured composite must all be carefully controlled if optimum toughness is to be achieved. Furthermore, materials requirements for highest tensile and shear strengths of laminates are sometimes incompatible with requirements for highest toughness. Final selection of a composite for a given application is often therefore a matter of compromise.

In this chapter we examine the cracking of a composite at various structural levels. We begin by considering how the behaviour of a crack in the matrix phase is modified by the presence of reinforcing filaments and how the mechanical characteristics of the fibres themselves affect cracking of the composite. These effects are distinct from those related to true composite action in which it is the interfacial discontinuity between fibres and matrix that controls crack propagation. At the macroscopic level there are other discontinuities, the interfaces between laminae for example, or the resin-rich zones around the boundaries of fibre tows, and these discontinuities also affect crack growth.

5.1.1 FRACTURE FORMALISMS

The Griffith model of fracture of cracked brittle solids (1920) has had far reaching consequences for designers, specifically in having laid the foundations of modern fracture mechanics as couched in terms of the familiar stress-based stress-intensity factor, K, and the associated energy-based parameter, G, the strain-energy release rate. With appropriate modifications to the basic concepts and definitions of Griffith (see, for example, Atkins and Mai, 1985; Broek, 1986), we arrive at design methods that can reasonably account for modest levels of non-linear, inelastic behaviour in a variety of homogeneous brittle and semi-brittle materials — ceramics, metals, and even plastics. It was natural, therefore, that many attempts should have been made to apply the same concepts to fibre composites in order to obtain design data that could be directly compared with those already in use for conventional engineering materials. And, despite the limitations imposed by the microstructural complexity of composites, by the scale of those structural features relative to the crack geometry, and by the distributed patterns of damage, there has been a measure of success. However, it is now apparent that the stress-intensity factor is not an appropriate design parameter for use with fibre composites, except for situations where a delamination crack spreads between the plies of a conventional laminate (see Hashemi $et\ al.$, 1990, for example).

The stress approach is dubious because it is unable to account for mechanistic input in the way that an energy-based approach can, and perhaps the most successful early studies of composite toughness have been those based on the Orowan fracture-energy development of the Griffith model (see McGarry and Mandell, 1972, and Phillips, 1972, for example). Measurements of the fracture energy, or work of fracture, γ_F, provided information which related in a fairly obvious way to the observed fracture processes, permitted comparisons of the performance of different materials, and encouraged deductions to be made as to the likely effects of materials/processing parameters on the behaviour of a given composite system. Unfortunately, although some early attempts to correlate γ_F with the conventional fracture toughness parameters were moderately successful, γ_F cannot be thought of as a design parameter although it fulfils an important rôle in furthering our understanding of composites fracture processes. The fracture energy idea has also led to the development of the concept of crack-resistance curves (R curves) which have particular relevance to materials like composites where sub-critical cracking and progressive failure are common features (Gurney and Hunt, 1967; Mai *et al.*, 1976). In this book, in order to preserve the analogy with the Griffith surface energy, γ_S, we define γ_F as the total energy absorbed during the complete failure process divided by twice the sample cross-sectional area: it is thus half the value of R or G, as usually defined.

5.2 FRACTURE PROCESSES IN COMPOSITES

As a starting point, we take the simple, Griffith planar-crack model and the knowledge that during crack extension an irreversible process takes place in which the net change in the stored elastic energy of the whole system (testing machine plus sample) caused by an increment of crack growth serves to provide the energy, γ_S, needed to produce new surfaces plus the energy, γ_F, required to activate any other deformation or damage mechanism that must operate in order for the crack to grow. The second of these contributions is minute in bulk glass where crack-tip plastic deformation is very limited (though still an order of magnitude greater than γ_S,) but is much greater in metallic or polymeric solids where extensive plastic or viscoelastic deformation may lead to the development of a sizeable plastic zone ahead of the crack.

This concept of a 'process' zone near the crack tip in which we encounter a variety of deformation or failure events all drawing on the energy of the system is familiar in ceramics where the events in question may include a distribution of subcritical micro-cracks, crack deviation, and crack blunting. Ceramic composites are designed to add other, perhaps more effective, damage mechanisms. An example of a particulate ceramic composite is the family of partially stabilized zirconias, or PSZ materials which, on account of their improved intrinsic toughness, may provide a better starting point as a matrix for subsequent fibre reinforcement in CMCs (Evans *et al.*, 1981; Garvie, 1985).

5.2.1 MATRIX EFFECTS

In practical fibre composites the proportion of fibres present will vary between about 10% (in DMCs, for example) and 70% (in high-performance CFRP). Crack-resistance effects directly attributable to the matrix material may not therefore be significant except perhaps in composites reinforced with relatively low volume fractions of short fibres. Three specific matrix effects might be listed, however:

i.   The effective toughness of a non-brittle matrix, such as a metal or thermoplastic, will be *reduced* by the presence of a high V_f of rigid, brittle fibres or particles as a result of plastic constraint which leads to triaxial tensile stress components in the matrix. A cermet, or metal-bonded ceramic, is a good example of this.

ii. The effective toughness of a soft, flexible matrix may be *increased* by the presence of a low V_f of rigid particles or fibres because a high overall stress level on the composite may then be required to generate a critical degree of crack opening in the stiffened matrix (Harris, 1980).

iii. The effective toughness of a low-toughness matrix may be *increased* by the presence of fibres or particles as a result of the slowing up of cracks in the neighbourhood of the filler. The toughness of certain brittle plastics for example, shows a strong dependence on the crack speed, often linked to changes in the crack face roughness — the slower the crack, the rougher the crack face and the higher the associated work of fracture (Harris, 1980).

5.2.2 FIBRE EFFECTS

Certain types of fibres such as the drawn steel wires used for reinforcing concrete, the polyester textile fibres used to reinforce rubber tyres, or the aromatic polyamide fibres used in high-performance laminates, may be classified as 'tough' because they are capable of extensive non-elastic deformation after yielding. Familiar mechanisms of deformation in such fibres are plastic necking in metal wires and drawing or interfibrillar splitting in polymer fibres. Such fibres often show high tolerance of defects and surface damage, by contrast with brittle fibres like glass or carbon. A bundle of such fibres possesses a large fracture energy, and a substantial portion of this fracture energy can be transferred into a composite containing the bundle.

Fibres such as glass, carbon and boron, which have high breaking strengths and failure strains (when undamaged) have very low intrinsic fracture energies, of the order of only 10-100 Jm^{-2}, and their failure is governed by the statistical nature of flaw distributions. These low fracture energies make no direct useful contribution to the composite work of fracture, even though each fibre may break many times (multiple cracking). But as a result of this multiple cracking, damage is widespread, crack paths in composites become complex, and it is this that results in very high values of the work of fracture.

5.2.3 SIMPLE FIBRE/MATRIX ADDITION EFFECTS

When fibres are incorporated into a matrix of any kind the separate phases may be able to contribute their individual levels of toughness to the composite in an additive fashion, or they may not, depending on the kind of interaction that occurs and on the level of constraint that is set up as a result of the differences in their properties. Early studies (Cooper and Kelly, 1969; McGuire and Harris, 1974) of metal-matrix composites in which the reinforcements were also metallic, showed that the specific fracture energy

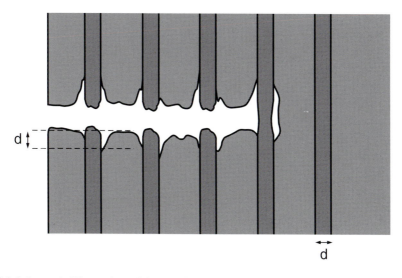

Fig. 5.1 Schematic illustration of the way in which fibres in a plastically deforming metal-matrix composite constrain the plastic deformation of the matrix.

of the composite, $(\gamma_F)_c$, could be obtained as a mixture-rule sum of the fracture energies of the individual components:

$$(\gamma_F)_c = (\gamma_F)_m (1 - V_f) + (\gamma_F)_f V_f \qquad (5.1)$$

where the subscripts c, m and f refer, as usual, to the composite, matrix and fibres, provided account was taken of the fact that the presence of the stiffer fibres caused the extent of matrix plastic deformation to be localized in the neighbourhood of the crack plane. Typically, a metal filament like tungsten or cold-drawn steel undergoes plastic deformation which is concentrated within a volume of the order of one fibre diameter, d, in height on either side of the final break, as illustrated schematically in Figure 5.1. If the fibre failure stress and strain are σ_{fu} and ε_{fu}, respectively, the fibre fracture energy is therefore roughly equal to $2d.\sigma_{fu}\varepsilon_{fu}.\pi d^2/4$ per fibre, which amounts to $2dV_f\sigma_{fu}\varepsilon_{fu}$ for the total fibre contribution. The matrix contribution is likewise obtained from the stress/strain product for the matrix, $\sigma_{mu}\varepsilon_{mu}$, and the deforming volume which is determined by the geometry of fibre/matrix distribution and the size of the deformation zone in the fibre. An estimate of this component (Cooper and Kelly, *op cit*) gives $d(1 - V_f)^2 V_f^{-1}\sigma_{mu}\varepsilon_{mu}$, and the total work of fracture for the composite is then:

$$(\gamma_F)_c = 2dV_f\sigma_{fu}\varepsilon_{fu} + d(1-V_f)^2 V_f^{-1}\sigma_{mu}\varepsilon_{mu}, \qquad (5.2)$$

The terms in d and V_f modify the plastic work terms (stress x strain) to take into account the localization of the plastic deformation. Both matrix and fibres are subject to large local deformation, and debonding may also occur. This decohesion relaxes some

Fig. 5.2 Tensile failure surface of a tungsten-wire/aluminium composite, showing plastic deformation of both fibres and matrix (McGuire, 1972).

of the triaxiality that would otherwise prevail at the interface. Failure to debond would cause conditions approaching plane strain throughout the composite and failure surfaces would have the appearance of a flat brittle-looking failure rather than the tougher, pseudo-plane-stress tensile failure of the locally separated components shown in Figure 5.2. There is also evidence that the strain to failure of a metal wire embedded with good wetting in a ductile metal matrix is rather higher than that of an identical isolated wire, perhaps because a strong fibre/matrix bond delays the onset of localized necking in the fibre. Estimates given by equation 5.2 are in reasonable agreement with measured values in composites where the interfacial bond is not very strong so that decohesion is able to occur and the two components can deform in an unconstrained fashion.

Modern composites rarely contain metallic reinforcements, however, and the more usual situation is that the stiffness ratio, E_f/E_m, is very high. This reduces the deforming volume of even a highly deformable matrix and limits the extent to which it can contribute its plastic work to the toughness of the composite. When we consider that a composite such as a GRP, in which both fibres and matrix are brittle (with fracture energies of the order of 0.5 and 100 Jm^{-2}, respectively), may exhibit a work of fracture of the order of 10^5 Jm^{-2}, it is clear that estimates based on the intrinsic toughnesses of the components have no relevance to real behaviour. It is in the special nature of composite action that the behaviour of individual components is dramatically modified; new damage mechanisms are introduced which drain energy from the system and in doing so contribute to the composite toughness, or crack-growth resistance. This is one reason for the success of GRP and other high-performance reinforced plastics, and it is the driving force for the intensive search for toughness in ceramic-matrix composites.

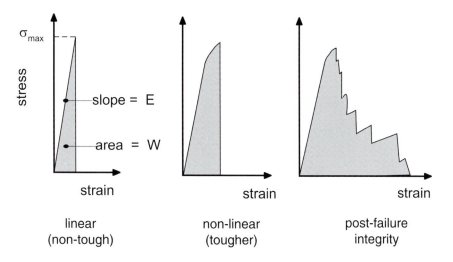

Fig. 5.3 Schematic stress/strain curves illustrating different classes of brittle/tough behaviour.

5.3 TOUGHNESS DERIVING FROM COMPOSITE ACTION

5.3.1 Toughening Mechanisms in Fibre-Reinforced Plastics

We have seen, for a simple case, how estimates of the energy required for deformation mechanisms associated with the matrix and fibres provide reasonable models of macroscopic toughness. In real fibre composites, however, the microstructural inhomogeneity and anisotropy cause the failure process to be very complex, and the combination of microfailure events — sometimes interacting, sometimes not — that leads ultimately to 'destruction' can often give rise to high levels of fracture energy. We should be clear that by using the γ_F definition of toughness we are begging a number of questions. The engineer used to dealing with K or G will usually be interested only in events up to maximum load. Unlike metals, most high-performance composites do not exhibit work-hardening phenomena and the load-bearing ability of a component may fall quite rapidly after peak load, even though the deterioration process may be gradual, and much energy may be required to complete the destruction process, as illustrated schematically in Figure 5.3. Such a composite would have a high work of fracture, and be valuable for many applications where energy absorption rather than load-bearing ability (or strength) is called for. There are also composites in which, as in plastically deforming metals, a process zone develops and increases in size as a single crack grows and the 'toughness', now defined as a crack-growth resistance, R, increases.

Low-energy failures of the kind implied by a rule-of-mixtures equation (5.1) for a brittle/brittle composite rarely occur in modern engineering laminates except, for example,

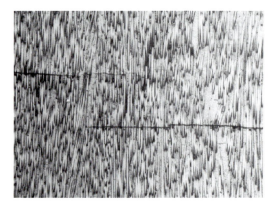

Fig. 5.4 Brittle fibre-to-fibre failure in a high-performance GRP pultruded rod exposed to stress-corrosion conditions in dilute hydrochloric acid solution.

in certain types of GRP under the combined action of stress and an aggressive environment (stress corrosion, as in Figure 5.4, for example), although it was a common feature of early CFRP in which the commercial fibre surfaces were often over-treated. Such failures are rare, partly because the interface will usually modify the mode of crack propagation and partly because of the statistical variation of fibre strengths. At any fibre break the load is shed back, *via* the matrix, to the neighbouring fibres, so that the stresses in these adjacent fibres will be concentrated somewhat above the average fibre stress level, as illustrated by Figure 4.8 in chapter 4. It is statistically unlikely, however, that there will be weak spots at the same points in these neighbouring unbroken fibres, and in the early stages of deformation fibre failures are widely distributed throughout the stressed volume. If the fibre/matrix bond is not too strong some relaxation of the stresses around the broken fibre ends can occur and this, together with local creep relaxation in the matrix, will help to reduce the level of stress concentration in neighbouring fibres. Thus, in many modern composites a great deal of fibre damage may occur before the number of fibre breaks in any cross-section reduces the load-bearing ability of that cross-section below the current level of applied load. At this point we have, effectively, a crack nucleus of critical size which could propagate rapidly to cause failure of the composite, and the simple cross-section argument should be modified to include the effect of the crack-tip stress concentration. If the fibre/matrix bond is too strong, however, this prevents relaxation, so that failure of whole bundles of fibres can occur, the crack spreading from one fibre to the next *via* the matrix with little or no deviation at the interface, in the brittle mode described earlier. A detailed study of tensile failure due to inter-fibre stress transfer within the axial plies of notched laminates has been made by Potter (1978).

Practical fibre-reinforced plastics consist of relatively brittle resins (γ_F values of a few hundred to a few thousand Jm^{-2}) reinforced with strong, brittle fibres which may be very stiff (*e.g.* carbon) or less stiff (*e.g.* glass or aramid polymers). The failure strains of these fibres are generally less than that of the matrix. The fibres may be continuous and straight, continuous and woven, or chopped and randomly dispersed. When a continuously reinforced composite laminate is loaded, the weaker fibres in the distribution begin to break at relatively low loads, and this damage is distributed throughout the material, as illustrated in chapter 4. A brittle fibre fracture is a process which requires an amount of energy $\frac{1}{2}\sigma_{fu}\varepsilon_{fu}$ per unit volume. If we allow all of the fibres in a 0.7 V_f carbon-fibre composite to break once, with a deforming volume of the order of d^3, their net contribution to the composite fracture energy can only be of the order of 1 Jm^{-2}. Because of the statistical distribution of fibre strengths, each fibre may break many times before the composite fails, but even if, in a 10 cm long sample, say, each fibre could be completely broken down to pieces of near critical length (about 0.5 mm in a CFRP), the total available fracture work from this source still amounts to only a few hundred Jm^{-2}.

In order to distinguish the separate micro-mechanisms of toughening it is convenient to consider a rather artificial model in which a crack travelling in the matrix approaches an isolated fibre (Figure 5.5). The crack is effectively halted by the fibre (picture b), first, because the higher stiffness of the fibre inhibits further opening of the matrix crack, and second, because the strength of the fibre is too high to permit of its being broken by the level of stress currently concentrated at the tip of the matrix crack. Further crack opening may occur if the local shearing force acting at the fibre/matrix interface is sufficiently high to allow the fibre to become locally debonded. This debonding is also encouraged by the fact that there may be a significant level of transverse tensile stress concentration ahead of the crack tip, parallel with the crack plane, as shown by the work of Cook and Gordon (1964). The overall consequence is that debonding may occur, the debonded fibre will extend elastically, and subsequent further crack opening can take place during which the matrix slides relative to the fibre (stage c): all of these processes require energy. The crack may progress past a number of fibres in this way, leaving them unbroken in its wake and for a given fibre/matrix/interface system there may be some equilibrium state where a stable number of these bridging fibres continue to support part of the load, modifying the thermodynamic driving force for crack growth. This fibre bridging is thus a further toughening mechanism. Its effect is illustrated in Figure 5.6. These are results of a simple double-cantilever beam (DCB) experiment on a polyester resin sample containing a single array of glass fibres spaced as indicated in the figure (Harris and Ankara, 1978). The crack propagates through the array, and the unbroken fibres behind the crack tip provide a closure force which opposes the applied stress, so reducing the driving force for crack growth, an idea proposed very early on by Romualdi and Batson (1963) for the toughening effect in fibre-reinforced cement. The apparent toughness is defined in the graph in terms of a notional 'stress-intensity factor', $K = \sigma\sqrt{a}$, 'a' being the length of the crack, and σ, the applied stress.

The extent of this increase in crack resistance as the crack extends, which is commonly termed 'R-curve behaviour', may be limited or extensive, depending on the material. At

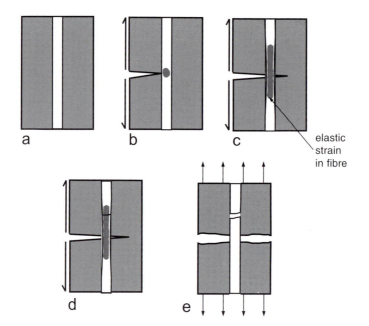

Fig. 5.5 Schematic illustration of the stages in crack growth in a fibre composite (see text for explanation).

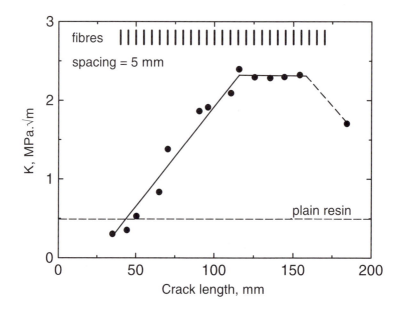

Fig. 5.6 Plot of K against crack length, during a test on a double-cantilever beam sample of polyester resin containing a linear array of glass fibres 5 mm apart (Harris and Ankara, 1978).

Fig. 5.7 Photographs of fibre pull-out in a carbon-fibre-reinforced epoxy composite (bottom) and a SiC-fibre-reinforced glass-ceramic. The magnifications may be judged from the fibre diameters, carbon 7 μm and SiC 15 μm.

some point, however, the rising load on a bridging fibre must reach a level at which the fibre will break (stage d in Figure 5.5). During loading, the fibre will have been released by Poisson contraction from contact with the matrix over a debonded length that will depend on the interfacial bond strength, and a weak spot may be encountered anywhere within this length (or just outside it, where contact and frictional grip is being re-established). When the fibre failure occurs away from the crack plane, the stored elastic energy is released and the fibre expands to regain contact with the matrix. Many of the broken fibres will therefore remain bridging the crack, and the embedded ends must be pulled out of the matrix holes in order for further crack opening to occur (stage e). This fibre 'pull-out' makes further demands on the energy of the system, and leads to the characteristic brushlike failures of many types of composite, as shown in Figure 5.7.

Since the earliest years of research on fracture-energy mechanisms in composites, there have been strong disagreements about the major contributing factors. McGarry and Mandell (1972), for example, were categoric that the primary source of toughening in glass-fibre/resin composites was the work necessary to break the fibres in a damage zone at the crack tip, and that contributions from the matrix toughness, fibre debonding or splitting, and fibre pull-out were negligible. Outwater and Murphy (1969) were equally convinced that it was the debonding process that was the major contributor, while other researchers believe that fibre pull-out is dominant (Harris *et al.*, 1975). The truth is not easy to unravel, especially when several mechanisms, which may be interdependent, may be making similar contributions, and it is in any event clear that different situations occur in different kinds of composites. We now examine the major mechanisms in a little more detail in order to see how simple energy considerations may offer guidance in designing composite materials.

5.3.1.1 Fibre/Resin Debonding

Glass fibres are coated prior to incorporation in the resin matrix with a complex mixture that includes a coupling agent, usually an organo-silane, the purpose of which is to encourage wetting and the establishment of a direct chemical bond between the glass surface and the resin molecular network. No such direct bond can easily be produced between a carbon fibre and the resin, so that the coating which is put on carbon fibres is merely a size which assists handling and encourages wetting-out during composite manufacture. However, carbon-fibre surfaces are also usually treated, by chemical, electrochemical, or plasma methods, to modify their surface chemistry and so further enhance wetting. Depending on the type of fibre, then, and the type and level of coating or surface treatment, the degree of actual chemical bonding between fibre and resin and the efficacy of any purely physical bond may vary immensely from system to system. There is uncertainty about the exact nature of the interface in fibre/resin composites (Drzal *et al.*, 1982, 1983; Kardos, 1985) and in this discussion of debonding we shall therefore adopt a very simple approach.

As discussed in section 4.3.1, when the end of a fibre, of length less than half the critical transfer length, ℓ_c, which is embedded in a block of cured resin, is loaded by pulling (Figure 5.8a), the load increases to some maximum level at which the induced shear forces acting at the interface are able to initiate a debond and the load drops to some lower level as the debond rapidly propagates. The peak load must reflect the strength of the combined chemical and physical bond, and the shaded triangular area on the diagram in Figure 5.8b represents the elastic energy that was needed to cause the debond, *i.e.* the debond energy. The fibre is not free, however, because the cured resin has shrunk onto the fibre during the curing process and during cooling from the cure temperature. As a consequence, further work is needed to withdraw the fibre completely from the resin block, and the full loading curve is as shown in Figure 5.8b.

The magnitude of the debonding energy can be roughly assessed (Harris *et al.*, 1975) by assuming that it is equivalent to the energy released when the fibre finally breaks. If a fibre debonds to a total distance $\frac{1}{2}\,\ell_{db}$ on either side of the crack face, the stored elastic energy released at fibre fracture is likely to be somewhat greater than $\frac{1}{2}(\sigma_{fu}^2/E_f).(\pi d^2/4).\ell_{db}$. If N fibres break in the cross section of the composite, the debonding contribution to the composite fracture work, w_{db}, normalized with respect to the sample cross-sectional area, 2A, is:

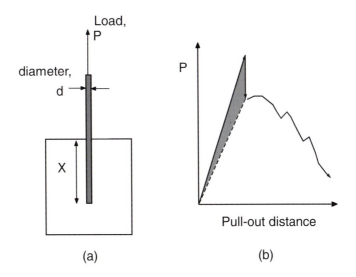

Fig. 5.8 A simple pull-out experiment. As the embedded fibre is loaded (a), the resulting load/deflexion curve frequently resembles that shown in (b).

$$\frac{W_{db}}{2A} = \frac{N\pi d^2 \sigma_{fu}^2 \ell_{db}}{16AE_f} \tag{5.3}$$

and since the fibre volume fraction $V_f = N\pi d^2/4A$, we obtain the debond component of the work of fracture:

$$(\gamma_f)_{debond} = \frac{V_f \sigma_{fu}^2 \ell_{db}}{4E_f} \tag{5.4}$$

This is an underestimate, since it assumes that the fibre/resin interfacial shear strength falls to zero after debonding and that the fibre is not strained outside the debonded region, neither of which is likely to be valid.

In their early attempt to assess the work of debonding, Outwater and Murphy (1969) treated the problem as a fracture-mechanics case of the propagation of a mode-II shear crack along a cylindrical interface. For a long fibre of diameter d embedded in a block of matrix, as in Figure 5.8a, and debonded over a distance x from the surface, the stress needed to continue the debonding process is given by:

$$\sigma_{fu} = \frac{4\tau x}{d} + \sqrt{\frac{8E_f G_{IIi}}{d}} \tag{5.5}$$

where τ is the sliding friction stress operating between fibre and matrix in the region already debonded, and the second term is the energy required to break the bond, G_{IIi}

Fig. 5.9 Flat surface topography of a typical brittle composite fracture. The material is a unidirectional carbon/epoxy composite containing plasma-treated fibres.

being the fracture surface energy of the interface in shear. The condition for the start of debonding is given, when $\tau = 0$, as:

$$\sigma_{fu} = \sqrt{\frac{8E_f G_{IIi}}{d}} \qquad\qquad (5.6)$$

It is evident that there is a limiting fibre diameter, $8E_f G_{IIi}/\sigma^2_{fu}$, equivalent to equation 5.3 modified so as to relate to the debonded surface area of the fibre, above which the fibres will break before they can debond. In such circumstances a crack would propagate cleanly through fibres and matrix, with no deviation, resulting in a flat fracture face, such as that shown in Figure 5.9, and a very low fracture energy. Such a situation could occur if the fibre/matrix bond were excessively strong, and fibre surface treatments therefore need to be carefully controlled so that they are neither too weak (low G → low composite toughness) nor too strong (no debonding → low toughness). The interface would be likely to have a fracture energy intermediate between that for a resin fracture and that for the failure of a glass/resin adhesive bond, say a few hundred Jm^{-2}. But Outwater and Murphy measured values of about 4 kJm^{-2} for G_{IIi}, for glass/polyester models, which suggests that simple mode-II cracking is not occurring during the debonding process. Kelly (1970) considered the debonding energy to be small, although more recent work by Wells and Beaumont (1985) suggests that it could be a major contributor to toughening.

The effect of debonding is to permit a large increase in the volume of fibre that is under stress and, therefore, the amount of stored elastic energy that is subsequently released when the fibre fails. Harris (1980) suggested that it may well be this consequence of debonding rather than the debonding itself that adds substantially to the composite fracture energy. It has also been suggested that in glass/polyester model composites the overall fracture energy is relatively little affected by the ease or difficulty of debonding (Harris and Ankara, 1978), presumably because the area under the tail of the load/deflexion curve in Figure 5.8b is usually substantially greater than that in the debonding triangle indicated.

5.3.1.2 Frictional Work Following Debonding

As implied by equation 5.5, after debonding has taken place the fibres and matrix move relative to each other as crack opening continues, and work must be done against frictional resistance during the process. The magnitude of this work is hard to evaluate because we do not know with any certainty what level of friction force is operating. The work expended will result from the action of the interfacial frictional force over a distance equal to the differential displacement of fibre and resin which Kelly (1970) suggested was roughly the product of the debonded length and the differential failure strain, $\ell_{db}\Delta\varepsilon$. In early composites, the matrix failure strains were usually much less than those of the fibres, and $\Delta\varepsilon$ could be taken as being approximately equal to ε_{fu}. If the initial friction force, $\tau_i \pi d.\frac{1}{2}\ell_{db}$, acts in each direction from the crack face over a distance $\frac{1}{2}\varepsilon_{fu}\ell_{db}$, the post-debond friction work is (Harris *et al.*, 1975):

$$W_{pdb} = \left(\frac{2\tau_i \pi d\ell_{db}}{2}\right)\left(\frac{\varepsilon_{fu}\ell_{db}}{2}\right) \tag{5.7}$$

per fibre. Normalizing again with respect to the sample cross-sectional area, the contribution to the work of fracture is therefore:

$$(\gamma_F)_{post-debond} = \frac{N\tau_i \pi d\ell_{db}^2\varepsilon_{fu}}{4A} \tag{5.8}$$

for N fibres in the cross-section, and again converting to V_f, we have:

$$(\gamma_F)_{post-debond} = \frac{V_f \tau_i \ell_{db}^2\varepsilon_{fu}}{d} \tag{5.9}$$

Substituting reasonable values for the parameters ($V_f = 0.6$, $\tau_i = 40$ MPa, $\ell_{db} \approx 10d$ to $50d$, $d = 7$ μm, and $\varepsilon_{fu} \approx 1.5\%$), the contribution varies between 250 Jm^{-2} to 6.3 kJm^{-2}. For modern fibre/resin combinations with toughened resins, the resin failure strain may well be much closer to that of the fibre, and the net contribution to toughening from this source would then become almost insignificant. It is interesting to note that Marston *et al.* (1974) had some success in attempting to account for the fracture energies of various types of composite by associating the surface energy of the debonded interface with the actual debonded surface area. They generalized the Outwater debonding model and showed how interfacial shear strength might be tailored to obtain reasonable levels of toughness while maintaining rule-of-mixtures strengths in boron/epoxy and carbon/epoxy composites.

5.3.1.3 Fibre Pull-Out

The pull-out work has long been considered to be of great significance in providing toughness in composites, and the earliest solutions were given by Cottrell (1964) and Kelly (1973). When a fibre end of length x is pulled from the matrix, the effective friction force, $\tau_i \pi d(x/2)$, acts over the distance x to do work $\frac{1}{2}\tau_i\pi dx^2$. Since x is normally considered to vary between 0 and $\frac{1}{2}\ell_c$ (where ℓ_c is the critical fibre length), the mean work done in pulling N fibres out to a maximum distance $\frac{1}{2}\ell_c$ over the composite cross section is:

$$W_{pull-out} = \frac{1}{2}N\pi d\tau_i \frac{\int_0^{\ell_c/2} x^2 dx}{\int_0^{\ell_c/2} dx} \qquad (5.10)$$

The pull-out contribution to the work of fracture is then:

$$(\gamma_F)_{pull-out} = \frac{N\pi d\tau_i}{48A} = \frac{V_f\tau_i\ell_c^2}{12d} \qquad (5.11)$$

The relationship given in section 4.1.9 of chapter 4:

$$\frac{\ell_c}{d} = \frac{\sigma_{fu}}{2\tau_i}$$

between fibre strength, σ_{fu}, effective interfacial bond strength, τ_i, fibre critical length, ℓ_c, and fibre diameter, d, which was derived by Kelly and Tyson (1965) for metallic composites without interfacial chemical bonding, may be used to write equation 5.11 in slightly different form:

$$(\gamma_F)_{pull-out} = \frac{V_f\sigma_{fu}\ell_c}{24} \qquad (5.12)$$

For typical values ($V_f = 0.6$, $\sigma_{fu} = 3$ GPa, $\ell_c \approx 10d$ to $50d$), this amounts to between 5 and 25 kJm^{-2}, clearly a potentially powerful source of toughening. We see from equations 5.11 and 5.12 that control of the interface characteristics in a composite of given fibre/matrix composition is vitally important in designing toughness into a system. An illustration is shown in the results of some crack growth experiments of the kind referred to earlier (cf. Figure 5.6) in which the surfaces of glass fibres were treated in various ways prior to embedding in polyester DCB samples. The curves in Figure 5.10 again show the increase in apparent toughness as a crack passes through an array of fibres (R curves), and it can be seen the results are not entirely in accord with conventional ideas. It is true that all of the fibre-containing samples develop crack resistances greater than that of the plain resin. But the toughening effect is considerably greater when the fibre surfaces are pristine than when they are coated with a silane coupling agent, applied

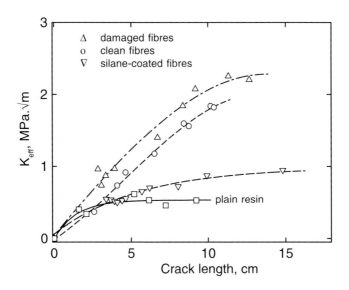

Fig. 5.10 Notional stress-intensity factor, K_{eff}, as a function of crack length for plain polyester resin and three composite samples (DCBs) containing glass fibres with different surface treatments (Harris *et al.*, 1975).

in practice to *improve* the fibre/matrix bond, and the highest toughness is obtained with a fibres that have been damaged by environmental contamination prior to embedment in the resin.

The parameters σ_{fu}, τ_i and ℓ_c are not independent, and there are limits to the extent to which the interfacial bond strength and the pull-out length (which is related to ℓ_c) can be varied. It has been shown, for example, that as the fibre/resin bond strength is gradually increased by fibre surface treatments, the work of fracture first increases and then decreases as the initial improvement with increasing τ_i is replaced by a deterioration as ℓ_c becomes very small. This is illustrated in Figure 5.11 by some early results for carbon/ polyester composites in which the fibres have been exposed to a range of chemical treatments which have modified the strength of the interfacial bond. The effects of those treatments are also reflected in changes in the interlaminar shear strength of the composite (*qv*, chapter 4), and this property is used as the abscissa in the graph. Early commercial fibre surface treatments were sometimes carried to excess in the drive to raise the composite shear resistance, with the consequence that many brittle structural failures were encountered (the datum point at the extreme right of the graph was for a commercial treatment). Modern treatments are designed to optimise strength and toughness. Beaumont and Phillips (1972) showed very early on that in carbon-fibre-reinforced plastics, untreated fibres produced composites that were notch-insensitive, whereas treated fibres, especially the somewhat over-treated fibres of that time, led to notch sensitivity.

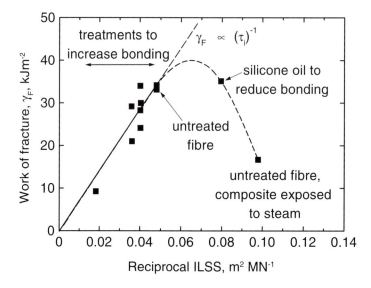

Fig. 5.11 Effect of fibre surface treatments in modifying the work of fracture and the interlaminar shear strength of carbon-fibre/polyester composites (Harris *et al.*, 1971).

There have been many studies of the effects of interface control in modifying the toughness of composites, and it has often been observed that attempts to improve the work of fracture (as opposed to the notional fracture-mechanics K_C value) often resulted in a reduction in strength. Atkins (1974) showed, however, that if reinforcing filaments were made to have random alternating sections of high and low interfacial bond strength (giving locally low and high interfacial toughness) a desirable combination of strength and toughness could still be achieved. The strongly bonded regions ensured that the fibre strength could be utilized and the weakly bonded regions, randomly arranged in the paths of running cracks, provided the source of debonding and pull-out energy. In fact, it was possible to achieve fracture energies of the order of 300 kJm^{-2} in boron/ epoxy composites containing 'barber's pole' release-agent coatings without loss of strength relative to composites with uncoated fibres.

The simple models for debonding and pull-out described above both make the incorrect assumption that the fibre strength has a unique value, σ_{fu}, having been originally derived, like many of the early theoretical studies of composites behaviour, for metal-fibre/metal-matrix systems. Wells and Beaumont (1985) and Thouless and Evans (1988) have recognized that the statistical distribution of fibre strengths must be taken into account in attempting to determine the location of a fibre failure (and hence the pull-out length) and its relationship with the fibre, matrix and interface properties.

In a composite containing fibres at angles to a propagating crack, extra work may be needed to withdraw the fibres (Harris and Ankara, 1978; Harris 1980; Helfet and Harris,

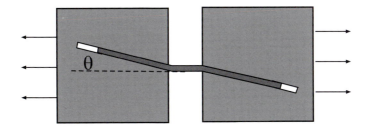

Fig. 5.12 Model for the plastic shearing during pull-out of a fibre lying at an angle to the crack plane.

1972; Piggott, 1970 and 1974). The sources of this extra work will depend on the constituents of the composite. Ductile or tough fibres, like metal wires or polymer filaments, will be capable of deforming inelastically in order to follow the more complex pull-out path without breaking. A metal filament that is capable of work-hardening, like a steel wire in concrete, may contribute a considerable level of plastic deformation work to the overall composite toughness by being forced to deform first one way and then another, as illustrated in Figure 5.12. Results of Morton and Groves (1974) support those of Helfet and Harris in demonstrating that a higher angle of misorientation leads to a higher pull-out stress. They developed plastic-hinge models to determine extra plastic work and showed that the pull-out force was a maximum at an angle, θ, of 45°.

Piggott (1970, 1974) suggested that the apparent strength of *brittle* fibres would be reduced by the bending process, although the extent of the strength loss was smaller the softer (or more ductile) the matrix resin since the bending fibres can cause the matrix to yield at the corners formed by the intersections of the fibre holes and crack faces. However, in recent work, Khatibzadeh and Piggott (1997) have carried out more detailed studies of the effect of loads applied at an angle, and have found that the apparent strength of brittle fibres is actually initially increased as the angle between fibres and crack plane increases. At larger angles, though, the combination of geometric effects and mechanical damage to the fibres will offset the gains described above and the overall work of pull-out will be reduced.

For a brittle-fibre composite, Piggott (1974) proposed that the work of fracture of a composite containing fibres at an angle θ to the crack plane was given by:

$$(\gamma_F)_\theta = (\gamma_F)_0 (1 - B \tan \theta) \tag{5.13}$$

where $(\gamma_F)_0$ is the work of fracture of a composite with the fibres normal to the crack plane and the parameter B, which is determined by the force exerted by the fibres on the matrix, is related to the ratio of the matrix yield strength to the fibre tensile strength, τ_{my}/σ_{fu}. For ductile fibres, the fracture work increases with increasing angle provided the breaking stress of the fibre is not exceeded. Some examples of the effect

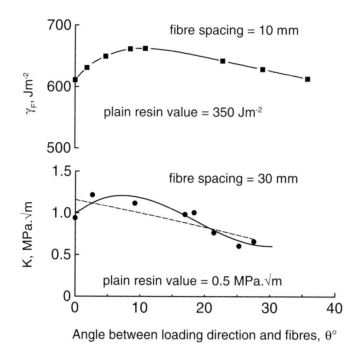

Fig. 5.13 Effect of the angle, θ, between the fibre direction and the crack plane on the resistance to crack propagation for (a) tungsten wires in PMMA, and (b) glass fibres in polyester resin. The test samples were double cantilever beams with a single row of spaced fibres (Harris, 1980). The dashed curve in the lower plot represents the Piggott model of equation 5.13, with $(\gamma_F)_0 = 1.16$ and B = 0.78.

of fibre orientation for metallic and glass fibres in model composites are shown in Figure 5.13. The results for the steel-wire/PMMA composites are presented as work-of-fracture versus angle, while those for the glass/polyester, following the style of Figure 5.6, are given as K ($= \sigma \sqrt{c}$). It can be seen that for both cases the composite toughness first increases with θ and then falls as the angle increases. Although the Piggott model is for work of fracture rather than K, the data agree reasonably well with equation 5.13, as shown, with values of $(\gamma_F)_0$ of 1.16 and B = 0.78. The disagreement at small angles clearly relates to the issue of whether or not the apparent strength of the glass fibres increases for small deviations of alignment.

    If the fibres are discontinuous, as in a DMC, for example, the situation is changed since the length of the fibre and its orientation may affect the pull-out behaviour. In a composite containing aligned, short fibres with lengths shorter than ℓ_c, the toughness increases with increasing length since the fibres cannot be broken by the external forces and increasing their length increases the extent to which they can be pulled from the matrix. From equation 5.11, it can be seen that the fracture work will be proportional to

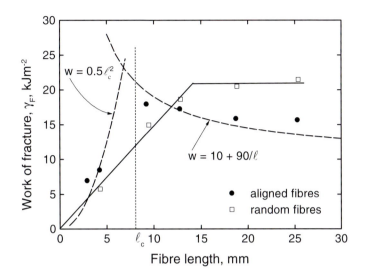

Fig. 5.14 Variation of work of fracture with fibre length in random and aligned composites of chopped steel wires in polyester resin. The fibre volume fraction is 0.1. (Helfet and Harris, 1972).

the square of the fibre length for $\ell < \ell_c$. Beyond the critical length, however, a proportion of those bridging a matrix crack will have an end within ½ ℓ_c of the plane of fracture and these will pull out: the rest can fracture and, as a result, the amount of pull-out work required is reduced (Cooper and Kelly, 1969; Harris and Ramani, 1975), the fracture work being inversely proportional to the fibre length. Thus, for a given system, the toughest composites will be those with fibres with lengths equal to ℓ_c, but, in order to achieve this benefit, strength and stiffness would have to be sacrificed. If the fibres are randomly oriented instead of being aligned, the effects described above are less clear cut. The combined effects of fibre length and orientation on the work of fracture of some model steel-wire/polyester composites are shown in Figure 5.14. The simple model for the effect of fibre length is not entirely satisfactory, since the proportionality constant for the experimental ℓ^{-1} relationship shown in this figure is lower than that calculated by Helfet and Harris (*op cit*), and the two curves do not intersect at precisely the experimental value of the fibre critical length, but the argument is at least mechanistically reasonable.

We would normally expect composite toughness to be raised by increasing the fibre volume fraction, by using fibres of larger diameter, or by using stronger fibres. Contrary to expectation however, improving the fibre/matrix bond will usually reduce the toughness because it inhibits debonding and therefore reduces pull-out (Beaumont and Harris, 1972).

Detailed analysis of the individual energy-absorbing mechanisms that are believed to contribute to the overall toughness of reinforced plastics is useful, as we have said, for

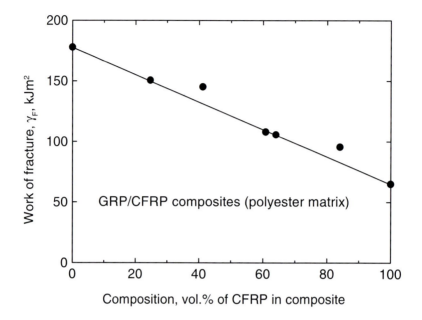

Fig. 5.15 Work of fracture, measured by notched Charpy impact tests, for hybrid CFRP/ GRP composites (Harris and Bunsell, 1975).

guiding us in the design of composites that may be expected to show improved mechanical properties, particularly in respect of combined strength and toughness. But it is easy to develop an oversimplified view of composite behaviour, especially when the analysis is carried out on model systems which are deliberately designed to prevent interaction of microfailure events.

5.3.2 TOUGHNESS OF HYBRID COMPOSITES

When more than one species of fibre is embedded in a single resin, the behaviour of the resulting hybrid composite may not be easy to predict from simple models. Early experiments with glass/carbon/resin composites led Harris and Bunsell (1975) to conclude that the work of fracture measured by notched impact followed a simple mixture-rule variation with composition (proportions of CFRP and GRP), as shown in Figure 5.15, regardless of the distribution of the two constituent fibres. But more detailed work on carbon/glass and carbon/Kevlar hybrids suggested that this may not always be the case. Anstice and Beaumont (1983), for example, showed that the presence of the carbon fibres, as an 'environment', as it were, for the glass filaments, effectively modified the debond and pull-out lengths of the glass so as to reduce the overall hybrid fracture energy below that predicted by the mixture rule. Stefanidis *et al.* (1985) reported similar results from impact-energy values for carbon/Kevlar hybrids. These are examples of a negative 'hybrid effect'.

An interesting effect that is often overlooked is that since cracks propagating in laminated composites may be forced to travel simultaneously through two materials — *e.g.* two plies of different orientation, or two similarly oriented plies of different composites (in a hybrid) — which have different intrinsic fracture toughnesses, this will constrain the crack and modify the material's apparent resistance to crack growth. When a crack is forced to move with a common front through a sandwich consisting of a well-bonded high-toughness and low-toughness pair of layers, since the composite fails at a common displacement, the failure strain for the tougher ply will be reduced and that of the less-tough ply will be increased. Guild *et al.* (1978) showed by a simple mechanics analysis that the fracture energy value for the composite, $(\gamma_F)_c$, would be the mixture-rule average, *i.e.*:

$$(\gamma_F)_c = \frac{1}{2}[(\gamma_F)_L + (\gamma_F)_H] \qquad (5.14)$$

of the toughness of the low- and high-toughness components, $(\gamma_F)_L$ and $(\gamma_F)_H$. Rewriting this in terms of the strain-energy release rate, G $(= 2\gamma_F)$, given by the compliance relationship:

$$G = \frac{P^2}{2}\left(\frac{dC}{dA}\right)_A \qquad (5.15)$$

where $(dC/dA)_A$ is the rate of change of compliance with crack area, A, it is then easy to show that the cracking load for the composite beam, P_C, is given by:

$$P_c^2 = \frac{1}{2}\left(P_L^2 + P_H^2\right) \qquad (5.16)$$

from which we conclude that the composite cracking *load*, and therefore the related critical stress-intensity factor, K_C, must always be greater than a simple mixture-rule sum such as that given for $(\gamma_F)_c$. Atkins and Mai (1976) showed that the cracking load is given by the volume-fraction sum only when there is no bond between the components. This is an example of an apparent synergistic effect, as discussed above and in chapter 4, where the composite appears to perform better than would be expected from simple mixture-rule considerations.

5.3.3 TOUGHNESS OF LAMINATED STRUCTURES

5.3.3.1 Delamination

In discussing these microstructural failure events we have taken a simplified view of failure, but frequently there are further complications which may swamp the effects just described. The debonding process is effectively a localized crack-blunting and crack-deviation mechanism, and the growing crack described above may in fact change planes rather than always continuing on the same plane. Multiple cracks may also be distributed throughout the process zone. But in most practical composites, there is the added complexity of laminate structure. A laminate is constructed from groups of individual unidirectional plies which are laid at various angles, depending upon design requirements, or from layers of woven cloth laid at various angles to the main stress axes. The tension/ shear coupling effects discussed in chapter 3 cause shear stresses to be developed in the plane of the laminate, especially near free edges, when the material is stressed. And since interlaminar planes in non-woven composites are always planes of weakness, the interlaminar shear stresses may easily become large enough to disrupt or delaminate a composite well before its overall tensile strength is reached. A crack travelling through a given ply may therefore find it energetically favourable to deviate along an interlaminar plane: under such circumstances, considerations of fibre debonding and pull-out are of little significance.

Although the specific fracture energy for splitting is low (of the same order as that of the resin, *i.e.* a few hundred Jm^{-2}), the total area of delaminated surfaces created may be very large, so that the net contribution of a delamination to the overall composite γ_F may be substantial. Aveston (1971) has analysed the work of fracture that may be available when the destruction of a bending beam occurs by a process of repeated fracture arrest and delamination. From strain-energy considerations, he has shown that the total work for a single delamination may be three times the normal stored elastic energy to peak load, even in all-brittle composite systems. Most prepreg laminated composites exhibit multiple delamination, with consequent further increases in fracture work. But delaminations of this kind are not always able to travel unhindered through fibre-free planes of weakness in modern composites because the manufacturing process has usually forced the original prepreg layers into reasonably good contact, and it is often found that, after some degree of splitting, fibres lying partially across the splitting plane exert a bridging effect in the form of a tied zone. Crack propagation then becomes increasingly difficult as the delamination extends. An indication of the relative degrees of resistance to impact damage can be seen in Figure 5.16. Test coupons of a quasi-isotropic carbon-fibre laminate and a woven-roving glass/epoxy composite were subjected to repeated impacts by a round-nosed tup, the samples being supported by a solid steel backing during the impacts, and the residual levels of toughness were determined as a function of cumulative incident energy. It can be seen that the prepregged CFRP material begins to lose toughness following the initial impact, whereas the woven GRP composite retains a very high level of toughness after many impacts. This is of course one of the reasons why crash helmets made of this kind of GRP composite perform so much better than cheaper, unreinforced polycarbonate helmets.

Since delamination appears to be advantageous in improving the overall toughness of composites, it is natural that much attention should have been paid to methods of increasing the work of delamination, as characterised by both the mode-I (tensile) strain-energy release rate, G_{IC}, and the mode-II (shear) characteristic, G_{IIC} (O'Brien, 1982; Hashemi *et al.*, 1990). Results obtained from these two types of measurement sometimes appear to give contradictory indications. Various methods have been tried to increase the work of delamination, including:

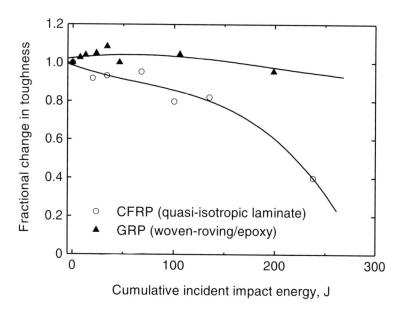

Fig. 5.16 Effect of repeated impact on the notched tensile strength of a quasi-isotropic carbon/epoxy laminate and a woven-roving glass/epoxy laminate (Harris *et al.*, 1991).

1. Modification of the matrix resin chemistry to increase the basic toughness of the resin. Alternatively, the use of intrinsically tough thermoplastic matrix polymers like PEEK and other polyaryls is advantageous, but calls for somewhat more complex manufacturing processes.
2. Modification of the matrix by the incorporation of 'flexibilisers' — a process akin to that used in the production of high-impact polystyrene or ABS. Liquid CTBN rubber added to an epoxy resin results in a tougher, phase-separated structure containing rubber particles, but the hot-wet properties of a material are often reduced by 'flexibilisation'.
3. Modification of the resin by the incorporation of particles, whiskers or chopped fibres into the resin, usually in the interlaminar regions of a laminate
4. Interleaving with inert, usually soft, polymer films.
5. Through-thickness stitching of the laminate.

The first three of these bring about some improvement, but the degree of improvement is rarely sufficiently great to make much difference to the overall composite toughness. Srivastava and Harris (1994), for example, showed that by the incorporation of a number of types of particulate material into the interlaminar planes of a cross-plied CFRP laminate both the mode-I and mode-II delamination toughnesses could be increased. G_{IC} was approximately doubled, from about 0.6 kJm^{-2} to about 1.3 kJm^{-2}, and G_{IIC} was increased from about 120 Jm^{-2} to about 200 Jm^{-2}. These improvements, which were a maximum at particle addition levels of about 3%, falling back thereafter towards the original levels, were greater the higher the rigidity of the particulate additive. The resultant improvements in laminate toughness were much smaller, however, amounting to only about a 25% increase in the un-notched impact strength.

If the composite laminae are deliberately separated by interleaving with inert films which can act as delamination initiators, the tied zone caused by fibre bridging can no longer form, but the extra energy of delamination contributes to toughening instead. Favre (1977) showed that the fracture energy of CFRP laminates may be increased in proportion to the number of deliberately introduced delamination initiators such as films of aluminium foil or polyimide resin bonded between the carbon/epoxy laminae. Altus and Ishai (1990) also demonstrated the value of this process of 'interleaving' by observing the effects of soft interleaved layers on delamination. They considered that delamination was triggered by transverse ply cracking, and that, up to a certain level of thickness, the presence of soft layers reduces stress associated with transverse cracks and hence delays the start of delamination. Sela and Ishai (1989), in a useful review of interlaminar toughening methods, showed that interleaving markedly improved the compression-after-impact strength (CAI) of laminates, but has the disadvantage that it brings a weight penalty because of the concomitant reduced stiffness. The net V_f is lower than in an equivalent composite with no interleaving, so that additional plies have to be used to maintain design requirements. They also recommend selective interleaving as a useful compromise. Mai (1993) shows that the use of plastic interleaving in which the film is perforated so as to allow intermittent separation of the composite laminae leads to significant increases in toughness without the concomitant loss in tensile strength which accompanies the incorporation of non-perforated interleaving films. This effect is seemingly analogous to that of intermittent fibre coating described in section 5.3.1.3.

We saw in section 4.3 of chapter 4 that third-direction stitching was a useful means of improving the interlaminar shear strengths of laminates by its effect in improving the delamination resistance. Dransfield et al. (1994) showed that stitching was able to improve some mechanical properties such as the delamination resistance and the compression-after-impact strength, but that it may cause degradation of other properties as a result of the creation of pockets of resin or the breakage of load-bearing fibres. Adanur et al. (1995) had recognised the problem of the fibre damage caused by the stitching process, and they describe an improved stitching method to reduce it. Materials produced by the improved method then exhibited reduced delamination and increased shear strength and impact energy up to a critical level of stitch density. It is interesting to note that Mouritz et al. (1997) found that stitching increased the mode-I delamination resistance but not the mode-II resistance. They also noted that the fibre damage caused by stitching led to reductions in the flexural properties, including the ILSS, especially under impact conditions.

The interlaminar fracture toughness of composites often shows R-curve behaviour — an increase in crack resistance as an interlaminar crack grows — as a result of the effect of fibres bridging the growing crack behind the moving crack tip. Truss et al. (1997) demonstrated that the misalignment of short fibres in a discontinuously reinforced

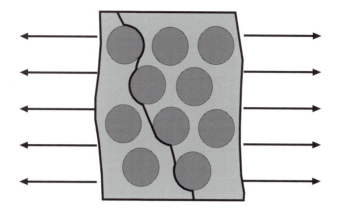

Fig. 5.17 Schematic illustration of the crack path during transverse cracking through the matrix and interface.

laminate added to this fibre-bridging effect, increasing the fracture toughnesses for both initiation and growth of interlaminar and intralaminar cracks.

5.3.3.2 Transverse-Ply Cracking

In the case of an orthogonally laminated composite (*e.g.* of $[(0,90)_n]_S$ lay-up, or similar), loading along one of the principal axes will cause cracking of the transverse plies at relatively low strains. In a typical GRP laminate, for example, this cracking, which is easily detected by techniques such as acoustic emission monitoring, may begin at strains as low as 0.2%, depending on the thicknesses and distribution of the plies and the resin toughness (Garrett and Bailey, 1977; Harris *et al.*, 1979) and continue to strains of about 1% — perhaps half the overall composite failure strain — leaving the transverse plies completely cracked into blocks of lengths of the order of the ply thickness, as shown in Figure 4.15 of chapter 4. This cracking occurs partly through the weak fibre/matrix interface and partly through the matrix resin, as shown schematically in Figure 5.17 and it reduces the composite stiffness. If a 10 cm x 2 cm x 3 mm strip of a six-ply $(0,90)_{3S}$ composite were loaded until each of the 90° plies was fully cracked, with a crack spacing of about 0.5 mm, and if the γ_F value for transverse-ply cracking were of the order of 200 Jm^{-2}, there would be about 0.012 m^2 of crack surface, and this would require 2.4 J of energy. Normalized with respect to twice the sample cross-sectional area, this provides about 20 kJm^{-2}. This is no doubt a vast overestimate, but it certainly suggests that transverse-ply cracking may not be a negligible contribution to the composite work of fracture.

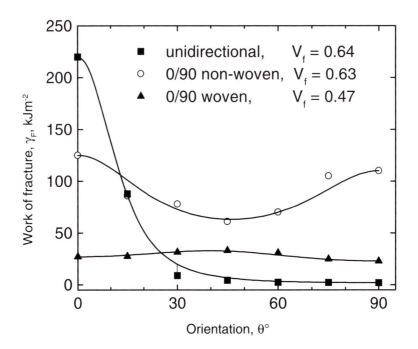

Fig. 5.18. Orientation dependence of the work of fracture of some GRP laminates. The data were obtained from notched Charpy impact tests (Harris, 1980).

5.4 TOUGHNESS OF PRACTICAL REINFORCED PLASTIC LAMINATES

The fracture energy of laminated composites depends strongly on both the ply stacking sequence and on the arrangement of the reinforcing fibres in the individual plies. The degree of anisotropy of the impact fracture energy of some typical high-quality E-glass/epoxy laminates is shown in Figure 5.18 (Harris, 1980). The high degree of anisotropy of the unidirectional composite is indicated by the ratio of about 100:1 between the extreme values for cracking normal to and parallel with the fibres. The fracture energy for unconstrained cracking in the transverse direction, approximately 2 kJm^{-2}, is some four or five times higher than typical values for cracking of unreinforced resin, an increase partly due to the greater complexity of the crack path and partly to the higher fracture surface energy of the fibre/resin interface. For the 0/90 non-woven laminate, which has about the same V_f as the unidirectional composite and contains approximately equal proportions of fibres in the two directions, the fracture energy in the two orthogonal directions is roughly half that of the unidirectional composite. This implies that it is the crack/fibre interaction in the longitudinal plies that determines the overall toughness of

Fig. 5.19 Damage accompanying fracture of the notched Charpy samples of the GRP laminates for which the fracture energies are shown in Figure 5.18. In sequence, left to right, unidirectional, 0/90 non-woven, and 0/90 woven.

the 0/90 laminate, although this may be an over-simplification since we observe relatively little reduction in γ_F at angles other than 0° and 90° where the fracture within the plies must be more complex. The woven-cloth laminate, which has a somewhat lower V_f has a work of fracture only one quarter that of the non-woven 0/90 laminate and it is more nearly isotropic, showing a slight *increase* in γ_F at 45° to the fibre directions. It is interesting to note that the orientation-dependence of the fracture energy, like those of the elastic modulus and tensile strength, can be represented by an expression of the form:

$$\frac{1}{(\gamma_F)_\theta} = \frac{a\cos^4\theta}{(\gamma_F)_0} + \frac{b\sin^4\theta}{(\gamma_F)_{90}} + c\sin^2\theta\cos^2\theta \qquad (5.17)$$

and the fitted curves in Figure 5.18 have this form, although we make no attempt to interpret the values of the constants a, b and c in terms of physical mechanisms.

The differences in γ_F levels between these three materials and the degree of anisotropy exhibited by each of them cannot be explained on the basis of either fibre V_f or elastic modulus variations. Perhaps the most significant difference between the macroscopic fracture patterns of these three composites relates to the volume of material comprising the fracture zone (Figure 5.19). In the unidirectional and non-woven 0/90 composites the continuity of fibres results in an extensive damage zone around the notch tip, whereas in a woven-cloth laminate the damage zone is much more restricted. Fracture of a 0/90 non-woven laminate at other than a right angle to the fibre direction occurs by transverse fracture of the two sets of plies, interlaminar shearing failure between the plies, and final separation by a type of pulling-out of the interlocked plies. The more localized failure of a woven-cloth laminate involves fracture and pulling-out of stubby bundles of fibres, mostly less than 10 fibre diameters in length, the shortness of these pull-outs presumably being due to the high local curvature of the filaments in the woven cloth. These effects of style of reinforcement on fracture energy are ascribed to the fact that woven reinforcement constrains the extension of the damage zone and serves to transfer the stress in a manner analogous to that which occurs with thin interspersed plies in a non-woven laminate. The tighter the weave, the lower the toughness.

McGarry *et al.* (1976) have discussed in detail the factors affecting the toughness of practical laminates. They describe the development of a damage zone ahead of a crack which like the plastically-yielded zone in a metal, helps to relax the high stresses concentrated at the crack tip. For practical reinforced plastics the damage zone is a region containing intra-ply cracks propagating parallel to the fibres and inter-ply delaminations, and its size and shape exert an important influence on the resistance of the laminate to crack propagation. Sharp cracks in a ply are effectively blunted by the intra-ply cracks so that failure requires a higher applied stress, but final failure is still determined by the stress distributions outside the damage region. Growth of a crack through the thickness of a laminate is inhibited by the development of the delaminated region between cracked and uncracked plies. As a consequence, certain types of laminate construction may be highly notch-sensitive whereas others may be completely insensitive to the presence of notches. Thus, notched CFRP laminates with ± 45 lay-up fail at net section stresses higher than their un-notched failure stress, σ_c, whereas 0/90 laminates fail at net section stresses much lower than σ_c. The ± 45 laminates are, of course, much weaker than the 0/90 laminates despite their lack of notch sensitivity. Grouping of the plies so as to reduce the effective number of interfaces between differently-oriented plies reduces the efficiency of stress transfer from ply to ply and dramatically increases the fracture toughness of the laminate. An illustration of the differing degrees of notch-sensitivities of some practical CFRP laminates is shown in Figure 5.20 (Lee and Phillips, 1981). In this diagram, the straight line represents the behaviour of a material that is completely insensitive to the presence of a notch, in this case a composite of lay-up [(± 45)$_2$]$_S$. The extent to which the experimental measurements of notched strength (shown here normalized with respect to the un-notched strength) fall below the straight line is then an indication of notch sensitivity. A more systematic analysis of the effect of lay-up in one specific material, a T800/924 carbon/epoxy system (Soutis, 1994), shows quite clearly how increasing the percentage of 0°plies increases the notch sensitivity (Figure 5.21). The test pieces used by Soutis contained circular holes rather than notches, but the effect is similar to that shown by Lee and Phillips. In Figure 5.21 the straight line again represents notch-insensitive behaviour, and the bottom curve, which is effectively a lower bound to the experimental curves, identifies the fully-notch-sensitive definition of the stress concentration for a circular hole, K_t (equal to 3 for a circular hole in an infinite plate).

5.4.1 SHORT-FIBRE COMPOSITES

Thermoset and thermoplastic moulding compounds containing randomly arranged chopped fibres, usually glass, are cheap materials which are finding wide-spread use in mainly non-load-bearing applications. Although the stiffnesses of these materials are

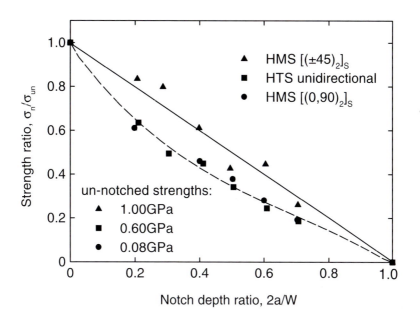

Fig. 5.20 Notch sensitivity of some CFRP laminates (Lee and Phillips, 1981). The stresses are the nominal failure stresses, σ_n (peak load divided by sample cross-section area) normalized with respect to the un-notched strength, σ_{un}. They are plotted as a function of ratio of the notch length, 2a, to the specimen width, W. Two types of carbon fibre, high-strength (HT) and high-modulus (HM), both surface treated, are represented.

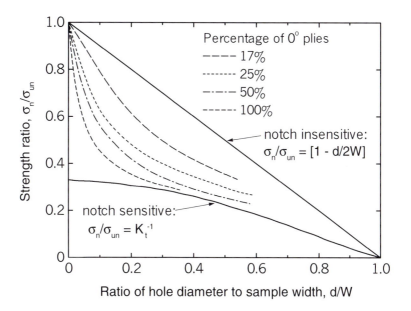

Fig. 5.21 Results of Soutis (1994) for the compression strengths of T800/924 CFRP laminates of various lay-ups containing holes. $K_t = 3$ for a circular hole in an infinitely large plate.

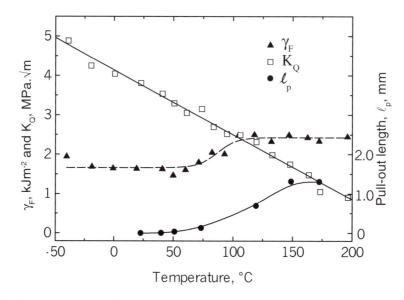

Fig. 5.22 Temperature-dependence of the fracture energy, critical stress intensity, and fibre pull-out length for DMC (Harris and Cawthorne, 1974).

usefully greater than those of the base resins from which they are compounded, their strengths are rarely much higher than those of the plain (filled) matrices. An important virtue of such materials, however, apart from their cheapness and mouldability, is that they are substantially tougher than unreinforced polymeric materials on account of the toughening mechanisms described above. They are roughly isotropic, and the lack of obvious planes of weakness such as occur in prepreg laminates means that the cracking process resembles that in conventional homogeneous materials more than that in anisotropic laminates. Because of their structure and composition their behaviour is more matrix- and interface-dominated than that of higher-performance composites. Polyester dough-moulding compounds provide a useful example of the behaviour of this kind of composite. Some measurements of the work of fracture, γ_F, and the candidate critical stress-intensity factor, K_Q, for a typical DMC are shown in Figure 5.22 (Harris and Cawthorne, 1974). The work of fracture of the compound is of the order of 2 kJm^{-2}, much greater than that of the polyester/chalk-filler material which constitutes the matrix. It is interesting to note that the two fracture parameters behave quite differently as functions of temperature, and are clearly not measuring the same property even though we use both to refer to toughness. The K_Q values fall linearly with increasing temperature, presumably because K_Q is influenced by the material strength which falls markedly over the temperature range shown. On the other hand, γ_F shows classical tough-brittle transformation behaviour, the transition occurring roughly in the region of the matrix

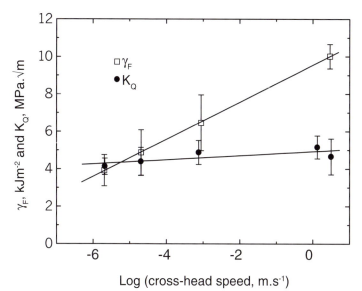

Fig. 5.23 Variation of work of fracture and critical stress intensity with testing speed (three-point bend tests) for a polyester dough-moulding compound (DMC) at room temperature (Harris and Cawthorne, 1974).

glass-transition temperature. Thus, the fact that the matrix is becoming less brittle is dominating the variation of this property. The figure also shows the observed fibre pull-out lengths, and it can be seen that these vary with temperature in roughly the same way as the γ_F values. It should be recalled, with reference to Figure 5.3, that K_Q relates only to behaviour as far as the peak load, whereas γ_F includes the work required for the whole failure process. In matrix-dominated materials the measured values of toughness parameters are also more rate-dependent than those in fibre-dominated composites. Figure 5.23 shows, for example, how the two fracture parameters vary with testing speed in notched three-point bend tests. In this case, γ_F is much more sensitive than K_Q, but both increase with testing speed.

5.4.2 DESIGN OF DAMAGE-TOLERANT COMPOSITES

Lee and Phillips (1981) have given some rules-of-thumb for the design of damage-tolerant composites. The requirements are:

i.   The fibre/matrix bond strength is most important and must be optimized: too strong a bond results in brittle composites.

ii. The composite needs to store large amounts of strain energy, $\frac{1}{2}\sigma_f\varepsilon_f$, so high-strength, intermediate-modulus, or high-failure-strain fibres are required.

iii. Thick plies delaminate more easily and reduce interactions between adjacent cracked and uncracked plies, thereby increasing toughness. But there is a need to compromise because thick plies increase the susceptibility to transverse cracking, and hence reduce fatigue and environmental resistance.

iv. The stacking sequence needs to be arranged so as to encourage delaminations between shear cracks in 45° plies adjacent to load-bearing 0° plies (*e.g.* as in $(\pm 45,0_2)_s$ structures) in order to increase notched strength by protecting the 0° plies. By contrast, if delamination is suppressed (as in [0, ± 45, 0]$_s$ structures) then the shear cracks which form in 45° plies at low stresses have a detrimental effect on the 0° fibres, leading to brittle behaviour.

v. The style of reinforcement is important: cloth limits the size of the damage zone, rendering the composite more brittle.

vi. 45° plies are weak but notch insensitive. They can therefore be used as crack arrestment strips in a laminate normally reinforced with 0° load-bearing plies.

Mai (1993) has also given a valuable summary of the idea of developing high fracture toughness in composites by engineering the interfaces in the composite at several levels which encapsulates some of the main ideas presented in the foregoing sections.

5.5 TOUGHNESS OF CERAMIC-MATRIX COMPOSITES

The current high level of interest in ceramic-matrix composites has largely been generated by users of materials in high-temperature applications for which existing metallic alloys are increasingly becoming border-line candidates. In most of the CMCs that have been studied hitherto, both fibre and matrix are refractory in nature and therefore brittle. Practical applications consequently depend on the ability of these brittle/brittle combinations to exhibit energy-absorbing mechanisms which are more efficient than those associated with simple failure of the matrix and the fibres.

In typical CMCs, by contrast with reinforced plastics, the stiffnesses of the reinforcing and matrix phases are often quite similar, although the strength and failure strain of the fibre will almost always be higher than those of the matrix. Many of the mechanisms of energy absorption that have already been described for polymer composites will also occur in CMCs, but the consequence of the extreme brittleness of the matrix — *i.e.* multiple matrix cracking — is to modify in some respects the extent to which these other toughening mechanisms can operate. Hillig (1987), following Cottrell (1964), has summarized the situation for CMCs with reference to the expression for the fracture energy (in terms of the critical strain-energy release rate, G):

$$G_c = 2 \int_0^{U_{max}} \sigma dU \qquad (5.17)$$

where U is the non-elastic displacement relative to the bulk material at the crack plane under critical conditions for crack propagation, and σ is the corresponding stress level in that plane. A large value of G_c results from a combination of a high crack-tip stress and a broad process zone, which sets the guiding principles for the development of successful CMCs. The extent of the process zone can be modified, as with reinforced plastics, by the provision of a group of energy-absorbing mechanisms which, in ceramic materials, will include incipient microcracking, crack deflexion and branching, fibre/matrix debonding, and fibre pull-out.

5.5.1 MATRIX CRACKING

One of the seminal treatments of the failure of CMCs was that of Aveston and co-workers (1971, 1973, 1974) which led to what is now almost universally referred to as the Aveston-Cooper-Kelly, or ACK, model. It is an energy-balance model set up to explain the multiple fracture of CMCs and is frequently used to predict the occurrence of the first crack as a design parameter. In composites of this kind first cracking may be more important than the notional strength. The ACK model starts from the assumption that a crack cannot form in the matrix unless the work done by the applied stress is greater than the increase in elastic strain energy of the composite plus the fracture surface work of the matrix per unit area of cross section. The model takes into account the work done by the applied stress on the composite body, the work expended in debonding the fibres from the matrix, the matrix fracture energy (usually negligible), the friction work expended as the matrix and fibres slide past each other, the decrease in the elastic strain energy of the matrix, and the increase in the strain energy of the fibres. The friction stress controlling the interfacial sliding is assumed to be constant and characteristic of the system. After cracking of the matrix, the bridging fibres are then assumed to carry the entire load formerly shared with the matrix. A brief introduction to the derivation of the ACK model is given in section 4.6 of chapter 4. We refer again, briefly, to matrix cracking in section 5.5.3.

5.5.2 FIBRE BRIDGING

In order for crack bridging to occur and contribute to composite toughening, it is essential that the fibres debond at the interface in preference to fracturing at the crack front. When this condition is satisfied, the rate of load transfer from fibres to matrix is determined by the frictional resistance to sliding at the interface. If τ_i is high, load transfer is efficient, and the load in the fibre decays rapidly with distance from the crack face. The statistical nature of the strengths of brittle fibres then results in the occurrence of fibre failures mainly in locations close to the crack face, thereby reducing the potential pull-out contribution to toughening. A low level of frictional resistance thus promotes toughness and, as in polymer-matrix composites, control of the interface is therefore of paramount importance in determining the composite fracture resistance. In a review of the mechanical behaviour of CMCs, Evans and Marshall (1989) point out that in order for debonding to occur instead of fibre fracture, the ratio of the interface fracture energy,

G_{IC}, to that of the fibre, G_{fc}, should be less than about 0.25, although, as they pointed out, there was at the time no experimental validation of this requirement. The effect of bridging fibres on the composite toughness can be modelled (Romualdi and Batson, 1963; Marshall *et al.*, 1985) by considering the stresses in the fibres as crack-surface closure tractions which reduce the stresses at the crack tip.

5.5.3 FIBRE PULL-OUT

The fabrication of CMCs almost always involves the use of high temperatures and/ or pressures in order to achieve good levels of density and uniform microstructures, and under conventional processing conditions this may lead to interfacial reactions which can result in reductions in the strength of the reinforcing filaments, increases in the interfacial bond strength, and the build-up of differential thermal stresses. These effects, as we have already seen, should result in marked reductions in the level of toughness that can be contributed to the composite through the pull-out and other mechanisms. Further exposure to elevated temperatures during service will also lead to deterioration. Results that have been reported range from composites where the bond is so strong that little or no pull-out can occur to systems where the bonding is so weak that the matrix can completely disintegrate, leaving dusty bundles of bare fibres. Neither of these extremes offers a useful engineering composite, and it is clear that control of the interface is therefore just as important an issue in CMCs as in reinforced plastics and metal-matrix composites.

A good deal of early research on CMCs related to composites with glass or glass-ceramic matrices reinforced with carbon or SiC fibres which were manufactured by slurry impregnation and hot-pressing. In much of this work, the composites produced exhibited works of fracture of the order of 3-10 kJm^{-2}. Examples were the work of Phillips (1972, 1974a) on borosilicate glass and a low-expansion lithium alumino-silicate (LAS) glass-ceramic reinforced with high- and low-modulus carbon fibres. The fracture energies of the composites were related to differences in the fibre/matrix bond strength and the simple pull-out model described earlier appeared to account for the greater part of the measured toughness values. Subsequent CMC development, particularly in the U.S.A., concentrated on glass and glass-ceramic systems reinforced with Nicalon SiC fibres which have been reported on extensively by Prewo and his co-workers (1980, 1982a, 1982b) who gave fracture energy values of about 30 kJm^{-2}. Values of γ_F for SiC/Pyrex composites were also published by Ford *et al.* (1987) who showed that, depending on heat-treatment conditions, the same composite could have a fracture toughness as low as 2 kJm^{-2}, with little pull-out, or as high as 25 kJm^{-2} when the interfacial bond strength was optimised. When the same material was heat treated so as to cause extensive delamination, γ_f values over 40 kJm^{-2} were obtained.

One of the difficulties in using a simple model such as that of equation 5.12 for CMCs is that it is often impossible to know exactly what is the strength of the fibre after it has undergone the rigours of the manufacturing process. In resin-based or glass-based composites it is possible to dissolve away the matrix and carry out strength measurements

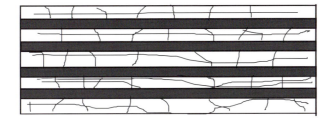

Fig. 5.24 Schematic illustration of a unidirectional CMC composite after loading so that the matrix is cracked into small blocks surrounded by an array of longitudinal and transverse cracks. See text for explanation.

on the filaments so as to obtain their true strength *in situ*, but in composites with more resistant matrices it is impossible to do this without the fibre sustaining further damage. To remove glass or a silica-based ceramic from a Nicalon-fibre composite, for example, it is necessary to use hydrofluoric acid or similar aggressive chemicals, and since Nicalon itself contains some SiO_2 it is likely that some attack of the fibres will also occur. For a SiC/CAS composite of V_f about 0.4 manufactured by Corning, experimental measurements gave values of γ_F of about 50 kJm^{-2} (Harris *et al.*, 1992). Scanning electron microscope pictures showed pull-out lengths of about 2-3 fibre diameters, so the contribution of fibre pull-out to the toughness, from equation 5.12, is only about 5 kJm^{-2}, if the fibre strength is about 1.5 GPa (a value assumed from the measured strengths of extracted fibres). Harris *et al.* showed that a substantial part of the remaining fracture energy of the SiC/CAS composite could have been contributed by the multiple transverse and longitudinal matrix cracking which occurs during deformation, reducing the matrix to a series of roughly equi-axed blocks as shown schematically in Figure 5.24. From an estimation of the total crack surface area in a typical cracked sample, it was deduced that the contribution of matrix cracking to the composite work of fracture must be about 250 γ_m, where γ_m, the specific work of fracture of the matrix is about 50 Jm^{-2}. This works out at about 30 kJm^{-2} which, together with the pull-out work and other small contributions, approximates reasonably well to the measured value.

5.6 APPLICATION OF FRACTURE MECHANICS TO COMPOSITES

5.6.1 STRENGTH OF NOTCHED COMPOSITES AND NOTCH SENSITIVITY

Before discussing the possibility of applying a fracture-mechanics formalism to composites behaviour it is instructive to consider a number of attempts that have been made to model the strength of composites containing holes or notches. In the first place,

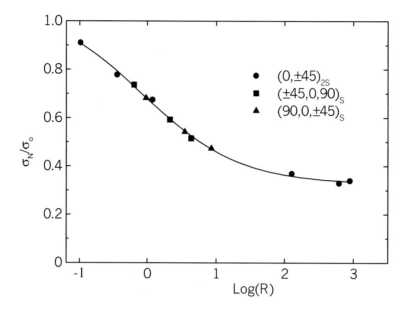

Fig. 5.25 Predicted values of the ratio of notched to un-notched strengths of three types of laminate as a function of hole radius (Pipes *et al.*, 1979). The fitted curve is a sigmoidal (Boltzmann) curve.

experimental work, like that of Waddoups *et al.* (1971) on $(0, \pm45)_{2S}$ CFRP laminates, demonstrated that the notched composite strength is often strongly dependent on the size of the hole. In order to explain observed notch strengths, it is necessary to know the stress distribution in the vicinity of the notch and to assume that failure occurs when the stress at some distance from the notch (known as the *point-stress criterion*) or the mean stress in a given region in the neighbourhood of the notch (the *average-stress criterion*) reach the un-notched tensile strength of the laminate. Whitney and Nuismer (1974) used a two-parameter model, arguing that while the stress-concentration factor is the same for holes of different sizes in an infinite plate, the stress *gradient* is different for each. Thus large stresses are localized more closely to the edge of a small hole than a large hole, and a critical defect is more likely to occur in a region of high stress for a large hole. The 2-parameter point-stress and average-stress models are both capable of predicting observed notch size without recourse to fracture-mechanics methods. A 3-parameter 'classical elasticity' model involving a transposition procedure proposed by Pipes *et al.* (1979) was also successful in being able to predict notched strength versus notch radius relationships for quasi-isotropic laminates of any stacking sequence, given the relationship for a single sequence, as illustrated in Figure 5.25. Potter (1978) considers that microcracks that form in the damage zone near a notch as a result of over-stressing do not inevitably lead to failure, and may even act as stress-relievers. Then, although the composite may

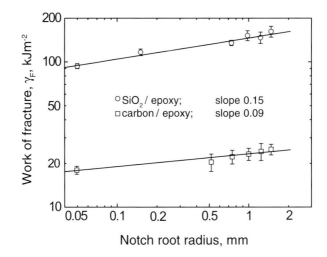

Fig. 5.26 The effect of notch acuity on measured works of fracture for unidirectional carbon/epoxy and SiO$_2$/epoxy composites (Ellis and Harris, 1973).

exhibit linear-elastic behaviour, the notched laminate may fail at a stress greater than that indicated by the linear elastic stress concentration — *i.e.* it will show a notch-size effect. Potter showed that tensile fracture is governed by inter-fibre stress transfer within the axial plies, and that limitations on the extent of this stress transfer imposed by matrix strength and fibre/matrix bond strength result in 'large' and 'small' notch effects. 'Large' notches initiate a sequential fibre failure process in the axial plies, and their effect is predictable by orthotropic laminate analysis, assuming inhomogeneity. In the neighbourhood of 'small' notches on the other hand, fibre failures in the 0° plies do not initiate this sequential fibre failure process, but form stable damage zones such as we have already discussed, which consist of matrix and fibre/matrix interface failures propagating parallel to the fibres and inhibiting inter-fibre stress transfer. 'Small' notches become 'large' when the notch geometry is modified in an appropriate way by the developing damage zone. Early work on unidirectional epoxy-based composites reinforced with high-modulus carbon and with low-modulus SiO$_2$ suggested that there is a logarithmic relationship between the fracture energy and the radius of the notch tip in Charpy impact tests, as shown in Figure 5.26.

5.6.2 THE APPLICATION OF FRACTURE MECHANICS

Fracture mechanics is universally accepted as a useful discipline for characterising the toughness of materials that are homogeneous and isotropic on the macroscopic scale — materials like metals and alloys. In the field of composites, however, there is serious doubt about whether the additional complexities of heterogeneity and anisotropy do not preclude the practical application of fracture mechanics. Early questions that were asked in this context (Sih, 1979) are:

- Is it feasible to use the same test procedures (such as critical stress-intensity and compliance measurements) to characterize composite toughness?
- Can the basic equations of fracture mechanics be modified to include inhomogeneity and anisotropy?
- To what extent is crack propagation (when it occurs in an appropriate manner) controlled by initial crack geometry, loading, and material orientation?
- What kinds of theoretical and experimental investigations are needed to develop predictive techniques to forecast composite behaviour?

Much experimental and theoretical work has indeed been published about a wide variety of composite types. But it is still far from clear whether fracture mechanics is or is not a useful investigative tool and predictive model (see, for example, Sendeckyj, 1975, and Sih and Tamuzs, 1979).

It is well known that certain composite or laminate types give results which can apparently be treated by linear-elastic fracture mechanics, or extensions of it, whereas others cannot. Several distinct types of composite stress/strain behaviour may be observed, for example, as shown in Figure 5.3. The linear, or non-tough, kind of behaviour, undesirable in engineering structures, is most likely to allow the application of fracture mechanics whereas the toughest kind of behaviour, in which the material retains a substantial degree of 'post-failure' integrity involving highly complex deformation and cracking mechanisms, seems least likely to obey fracture-mechanics principles. And in the failure of such a desirable engineering material, we have to ask whether the more important parameter is that representing the initial crack-growth resistance — G or K — or the total fracture work, γ_F. The spreading of a delamination crack is perhaps the simplest mode of cracking in a laminate, and is satisfactorily modelled, both for simple crack growth and for fatigue, by linear-elastic fracture mechanics, as shown by Vancon *et al.* (1984). On the other hand, the rupturing of a notched unidirectional GRP rod, which may involve the total disintegration of the resin matrix over a large part of the sample volume, will be too complex to treat by the usual models. Furthermore, as we have seen, for any given laminate it is sometimes possible to change from one extreme type of fracture behaviour to another simply by arranging the plies in different ways.

5.6.3 APPLICATION TO SPECIFIC COMPOSITES

In seeking to apply fracture mechanics to composite materials we are, in essence, attempting to define a fracture criterion that concerns only the macroscopic behaviour of the material in the presence of notches, flaws or other design features that are large by comparison with the fibre diameter. At first sight this does not appear to be feasible because the fracture behaviour of composites can be considerably modified by the presence of weak interfaces, and the mode of crack propagation does not seem to satisfy the conditions laid down for the application of fracture mechanics. Experimental results seem to suggest however that under some circumstances fracture-mechanics ideas can be used. Wu (1968) showed that a fracture-mechanics approach to the propagation of cracks in homogeneous, isotropic plates could be applied in a limited way to unidirectionally-reinforced plastics. He determined the critical stress for the propagation of cracks of various lengths parallel with the fibres under tension and shear loading conditions, and showed that for each case a relationship of the form $k = \sigma_\infty \sqrt{(\pi a)}$ accurately fitted the results, k being a constant, a the crack length, and σ_∞ the fracture stress. He also showed that the interaction of stress-intensity factors for combined conditions (tensile and shear mode) was well described by a relationship of the form

$$\frac{K_I}{K_{IC}} + \left(\frac{K_{II}}{K_{IIC}}\right)^2 = 1 \tag{5.18}$$

This empirical relationship was not in agreement, however, with a general theoretical one given by Paris and Sih (1965) for anisotropic materials:

$$\left(\frac{K_I}{K_{IC}}\right)^2 + \left(\frac{K_{II}}{K_{IIC}}\right)^2 = 1 \tag{5.19}$$

although both could be considered to be specific cases of a more general power-law model. Sanford and Stonesifer (1971) measured the fracture toughness and strain-energy release rate, G, of several types of glass-fibre/epoxy composites and showed that G is a bulk property of the material, independent of the mode of loading. For combined tension and shear they give

$$G = G_I + G_{II} \tag{5.20}$$

so that fracture toughness properties for the in-plane forward-shear mode, which are experimentally much more difficult to measure than those in tension, could be deduced in a straightforward manner from opening-mode results. Since G and K are not independent properties of the material, and can be written

$$G_c = CK_c^2 \tag{5.21}$$

C being an elastic compliance property. Equation 5.21 can then be given in the alternative form

$$C_1 K_1^2 = C_2 K_2^2 = \text{constant} \tag{5.22}$$

which is more in keeping with the expression of Paris and Sih. C_1 and C_2 are functions of the elastic compliances for the material, *viz.*

$$\left(\frac{C_1}{C_2}\right)^{\frac{1}{2}} = \left(\frac{E_1}{E_2}\right)^{\frac{1}{4}} \tag{5.23}$$

where E_1 and E_2 are the elastic moduli in the longitudinal and transverse directions, respectively. The full form of equation 5.21 for opening-mode cracks in a rectilinearly anisotropic body (as opposed to an inhomogeneous composite) has been given by Sih, Paris and Irwin (1965):

$$G_{IC} = K_{IC}^2 \left(\frac{S_{11}S_{22}}{2} \right)^{\frac{1}{2}} \left[\left(\frac{S_{22}}{S_{11}} \right)^{\frac{1}{2}} + \left(\frac{2S_{12}+S_{66}}{2S_{11}} \right) \right]^{\frac{1}{2}} \qquad (5.24)$$

On making the appropriate substitutions for the S_{ij} in terms of the engineering elastic constants (see chapter 3), and ignoring a small term involving Poisson's ratio, we obtain:

$$G_{IC} = K_{IC}^2 \left(\frac{1}{2E_1E_2} \right)^{\frac{1}{2}} \left[\left(\frac{E_1}{E_2} \right)^{\frac{1}{2}} + \frac{E_1}{4G_{12}} \right]^{\frac{1}{2}} \qquad (5.25)$$

This equation can be presented in a form analogous to the conventional fracture-mechanics relationship for isotropic materials, *viz.*:

$$G_c = \frac{K_{IC}^2}{E} \qquad \text{(isotropic solids)} \qquad (5.26)$$

$$G_c = \frac{K_{IC}^2}{E_{effective}} \qquad \text{(orthotropic materials)} \qquad (5.27)$$

the effective modulus thus being the reciprocal of the bracketed terms in equation 5.25. It was suggested in the First Edition of this book (page 75) that for orthotropic composites with a strong anisotropic axis, such as CFRP, the effective modulus could be approximated by $2\sqrt{(E_1 G_{12})}$. Closer inspection of the function defined in equation 5.25, however, suggests that this is invalid and that for most practical purposes the value of the transverse modulus, E_2, is a better approximation.

From their work on stress concentrations in flawed $(0, \pm 45)_{2S}$ carbon/epoxy laminates referred to in the previous section, Waddoups *et al.* (1971) concluded that the experimental behaviour of their laminates, when tested without deliberately introduced notches, was consistent with predictions of a fracture-mechanics analysis for a crack dimension of approximately 1 mm, even though no macroscopically detectable crack was present. This and other treatments, including that of Whitney and Nuismer (*op cit*), therefore gives rise to the concept of an energy-intense region at each crack tip which leads to an empirical extension of isotropic linear-elastic fracture mechanics in which the fracture toughness, K_c, for an infinite body under tensile load containing an internal crack of length 2a is given by

$$K_c = \sigma \sqrt{[\pi(a+r_\rho)]} \qquad (5.28)$$

where r_ρ is the dimension of the energy-intense region at the crack tip, referred to in other formalisations as the *process zone*: r_ρ is treated as a disposable parameter and is evaluated from experimental data. Similar approaches have been followed by Owen and Rose (1973) for glass/polyester composites, and by Beaumont and Phillips (1972) who suggested that the damage zone consists largely of material in which the fibres were debonded from the resin. Owen and Bishop (1972) have found that as far as the fracture-mechanics approach is concerned this zone can be treated in the same way as the plastic zone in metallic materials, and they used measured values of the zone size to correct K_c results obtained from a variety of GRP laminates. By determining the fracture toughness of plates up to 105 mm wide in this way they were able to show that K_c could be considered to be a genuine material parameter for design purposes. The work of Owen and his co-workers has largely been carried out by testing wide, notched plates in tension. One of the difficulties in comparing results obtained by different investigators, however, is that the various test methods are sometimes found to give quite different data, and it is not clear what type of test can be said to give absolute values.

Phillips (1974b) measured G and K values for 0/90 carbon/epoxy and carbon/glass composites and concluded that K_C was independent of crack size and could therefore be used to predict composite strength. A fracture-mechanics analysis gave fracture energies that agreed quite well with those obtained by other methods, indicating the validity of the Griffith-Irwin criterion for crack-propagation normal to fibres. He obtained K_C values of 6-25 MPa√m for various CFRP materials, G_C being between 1 and 20 kJm^{-2}. For the carbon-fibre/glass material, he obtained a K_C value of about 10 MPa√m and a G_C of about 2 kJm^{-2}.

In attempting to use the concept of K_C for a unidirectional GRP, Barnby and Spencer (1976) showed that the compliance calibration is significantly different from that for a continuous elastic medium, and is non-conservative (i.e. the error goes the wrong way). With an experimental calibration involving a fourth-order polynomial they obtained K_C values of about 15 MPa√m, similar to those of Owen and Rose (1973), but observed that the failure mode was not transverse to the fibres.

An example of what can be achieved by a fracture-mechanics type of analysis applied to a series of carbon-fibre laminates is illustrated by some work reported by Dorey (1977). He shows that when an attempt is made to compare a range of laminate lay-ups having different notch sensitivities, measurements of the apparent fracture toughness, $K_C = \sigma_c \sqrt{(\pi a)}$, give results that were strongly dependent on the test-panel width and the notch length and acuity, and which do not therefore permit of easy evaluation of the relative toughness of different laminates. By contrast, when the results are adjusted by the application of a sample finite-width correction factor, α, and the damage-zone size correction factor, r_ρ, of equation 5.28:

$$K_c = \alpha\sigma\sqrt{[\pi(a + r_\rho)]} \tag{5.29}$$

the resulting values become independent of notch and sample geometry and can therefore be used to make comparisons of materials characteristics such as resin type, ply distribution and layer thickness. His results illustrate the wide variations in fracture toughness that may be observed in any given class of composites as a result of the use of different resins and alternative arrangements of the plies in (0, ±45) laminates.

Bishop and McLaughlin (1979), working with laminates similar to those of Dorey, summarized the materials characteristics that affect the toughness of this type of composite. They showed that the notch-sensitivity of a $(0, \pm 45)$ CFRP laminate in tension can be reduced by avoiding too high a bond strength, since in general all lay-ups tend to become brittle if the bond strength is too high. At modest levels of interfacial bonding, increases in toughness can be obtained by selecting ply angles and stacking sequences, and by increasing the thickness of the plies. Thicker plies tend to delaminate more readily than thin plies. Thus, increasing the laminate thickness by repeating the basic stacking sequence has little effect on failure characteristics, but increasing the thickness of the layers leads to more delamination and shear cracking, so that toughness is increased. Laminates with four-ply layers, for example, had notched tensile strengths some 50% greater than similar laminates with single-ply layers. The load-carrying ability of 0° plies is increased by shear splitting (in the 0° plies), but is reduced by cracks in adjacent 45° plies: 0/45 ply interactions are of course reduced when delamination occurs. Thus, even if fibres, resins, and bond strengths are pre-selected on the basis of some optimisation procedure for strength and stiffness, further optimisation for improvements in toughness can still be done in terms of fibre angles, stacking sequences and ply thickness.

Some conclusions from work by Wang *et al.* (1975) on through-thickness cracks in cross-plied laminates may serve as a summary of these ideas. It is clear that the stress field round a crack is severely modified by the introduction of damage consisting of sub-cracks parallel to the fibres in each ply. A core zone can be defined around the main crack tip such that outside this zone the stress field is undisturbed while inside the zone the stress field is relaxed by high levels of inter-laminar stress that are present. Mixed-mode crack propagation is therefore automatically implied. These sub-cracks serve a function similar to the plastic zone in metals, and crack resistance results from reduction of the in-plane stress field at the main crack tip. The relaxed core zone is embedded in a stress field at the main crack tip which is unaltered by the sub-cracks, and the in-plane stresses follow a classical $r^{-1/2}$ stress dependence in the near-crack-tip region.

5.7 CONCLUSION

Under some circumstances, linear-elastic fracture mechanics can be used directly. Typically this applies to brittle composites with high fibre/resin bond strengths and brittle matrices. In such materials crack paths may be planar, or nearly so, and damage zones small (*e.g.* ≤ 1 mm), and the effects of composite heterogeneity are not apparent. Similarly, it is likely that in random composites like chopped-strand-mat laminates or in short-fibre moulding compounds like DMC and SMC, anisotropy will not be a problem and fracture mechanics may be used. It has also been observed that for specific cases of cracks running parallel with the fibres (in a unidirectional laminate) or delaminating cracks in a complex laminate, crack growth appears to obey fracture-mechanics requirements. In such cases as these the modified fracture mechanics can often be used for cracking under both monotonic and cyclic loading conditions.

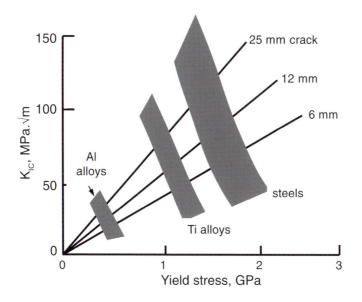

Fig. 5.27 Schematic illustration of the general relationship between fracture toughness and strength for families of metallic materials.

By contrast, in tough composites damage is complex and single-crack failure may not occur. In cases where a single crack does propagate, a plastic-zone or process-zone correction factor may work, but very large samples are often needed to obtain sensible results. Low V_f composites are also likely to cause difficulties because the assumption of structural uniformity is no longer valid.

An important generalisation is that generalisations should not be made! There is always likely to be uncertainty over the question of validity of LEFM for any given case.

Designers dealing with metallic materials recognise that, in general, high toughness and high strength are mutually exclusive characteristics of materials, and that there is roughly an inverse relationship between toughness and strength. This is illustrated schematically in Figure 5.27. Markedly contrasting behaviour in composites was shown by Harris *et al.* (1988) who examined many published and unpublished results, obtained over a 15-year period, for a wide variety of composite types. The range of materials included moulding compounds, cross-plied and quasi-isotropic laminates, unidirectional composites, glass, carbon, Kevlar-49 and hybrid laminates, and glass- and ceramic-matrix composites. They showed that it was possible to plot measured values of K (the candidate fracture toughness, K_Q), measured in tension, against the tensile fracture stress and show, as seen in Figure 5.28, that there is a reasonably good linear relationship between the two. Clearly a substantial amount of scatter is to be expected, given the different experimental and analytical methods used by the large number of authors whose work

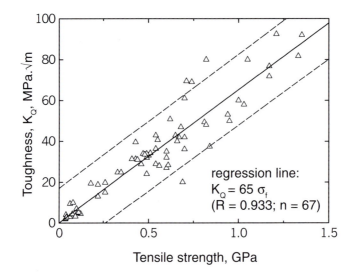

Fig. 5.28 Relationship between apparent toughnesses, K_Q, of a wide variety of composite materials and their tensile strengths. The dashed lines are prediction bands at the 90% level (Harris *et al.*, 1988).

appears on this diagram, but the correlation is statistically significant. A crude analysis on the basis of the familiar approximate relationship:

$$K_Q \approx 1.2.\sigma_c \sqrt{[\pi a]} \tag{5.30}$$

indicates that the intrinsic crack dimension for all of this diverse range of materials is about 1 mm, which is identical with the result of Waddoups *et al.* (1971) for one specific laminate. A similar result was also shown by Wetherhold *et al.* (1986). We may suppose then that the conclusion of Waddoups *et al.* is generally valid and a crack-tip process zone of the order of 1 mm across forms near cracks in most composites prior to failure. This is of the order of 100 fibre diameters for composites like GRP and CFRP, but it is clear from the results of Jamison referred to in Chapter 4 that crack nuclei of this size are never found in practice. It appears, nonetheless, that regardless of the apparent notch-sensitivity or otherwise of fibre composites, the micro-structural features which determine their resistance to crack propagation also control their fracture behaviour in the absence of deliberately introduced notches.

5.8 REFERENCES

S. Adanur, Y.P. Tsao and C.W. Tam: *Composites Engineering*, **5**, 1995, 1149-1158.

E. Altus and O. Ishai: *Composites Science and Technology*, **39**, 1990, 13-27.

P.D. Anstice and P.W.R. Beaumont: *Journal of Materials Science Letters*, **2**, 1983, 617-622.

A.G. Atkins and Y.W. Mai: *International Journal of Fracture*, **12**, 1976, 923-924.

A.G. Atkins: *Nature*, **252**, 1974, 116-118.

A.G. Atkins and Y.W. Mai: *Elastic and Plastic Fracture*, Ellis Horwood, Chichester, U.K., 1985.

J. Aveston: *Proceedings of NPL Symposium on The Properties of Fibre Composites*, IPC Science and Technology Press, Guildford, U.K., 1971, 63-73.

J. Aveston, G.A. Cooper and A. Kelly: *Proceedings of NPL Conference on The Properties of Fibre Composite*, IPC Science and Technology Press, Guildford, U.K., 1971, 15-26.

J. Aveston and A. Kelly: *Journal of Materials Science*, **8**, 352-362.

J. Aveston, R.A. Mercer and J.M. Sillwood: *Proceedings of NPL Conference on Composites - Standards, Testing and Design*, IPC Science and Technology Press, Guildford, U.K., 1974, 93-102.

J.E. Bailey, P.T. Curtis and A. Parvizi: *Proceedings of Royal Society, London*, **A366**, 1979, 599-623.

J.T. Barnby and B. Spencer: *Journal of Materials Science*, **11**, 1976, 78-82.

P.W.R. Beaumont and B. Harris: *Journal of Materials Science*, **7**, 1972, 1265-1279.

P.W.R. Beaumont and D.C. Phillips: *Journal of Composite Materials*, **6**, 1972, 32-46.

S.M. Bishop and K.S. McLaughlin: *RAE (Farnborough) Technical Report TR*, 1979, 79051.

D. Broek: *Elementary Engineering Fracture Mechanics*, Martinus Nijhoff, Dordrecht, Netherlands, 1986.

J. Cook and J.E. Gordon: *Proceedings of Royal Society, London*, **A282**, 1964, 508-520.

G.A. Cooper and A. Kelly: *Journal of Mechanics and Physics of Solids*, **15**, 1967, 279-298.

G.A. Cooper and A. Kelly: *Interfaces in Composites STP 452*, American Society for Testing and Materials, Philadelphia, U.S.A., 1969, 90-106.

G.A. Cooper: *Journal of Materials Science*, **5**, 1970, 645-654.

A.H. Cottrell: *Proceedings of Royal Society, London*, **A282**, 1964, 2-9.

G. Dorey: *Damage Tolerance In Advanced Composites Materials, RAE (Farnborough) Technical Report TR*, 1977, 77172.

K. Dransfield, C. Baillie and Y.W. Mai: *Composites Science and Technology*, **50**, 1994, 305-317.

L.T. Drzal, M.J. Rich and P. Lloyd: *Journal of Adhesion*, **16**, 1982, 1-30.

L.T. Drzal, M.J. Rich, M.F. Koenig and P. Lloyd: *Journal of Adhesion*, **16**, 1983, 133-152.

C.D. Ellis and B. Harris: *Journal of Composite Materials*, **7**, 1973, 76-87.

A.G. Evans, D.B. Marshall and N.H. Burlingame: *Advances in Ceramics*, **3**, 1981, 202-216.

A.G. Evans, F.W. Zok and J. Davis: *Composite Science and Technology*, **42**, 1991, 3-25.

A.G. Evans and D.B. Marshall: *Acta Metallurgica*, **37**, 1989, 2567-2583.

J.P. Favre: *Journal of Materials Science*, **12**, 1977, 43-50.

B.A. Ford, R.G. Cooke and S. Newsam: *British Ceramic Proceedings*, **39**, 1987, 229-234.

K.W. Garrett and J.E. Bailey: *Journal of Materials Science*, **12**, 1977, 157-168 and 2189-2193.

R.C. Garvie: *Journal of Materials Science*, **20**, 1985, 3479-3486.

W.W. Gerberich: *Journal of Mechanics and Physics of Solids*, **19**, 1971, 71-88.

A.A. Griffith: *Proceedings of Royal Society, London*, **A221**, 1920, 163-198.

F.J. Guild, A.G. Atkins and B. Harris: *Journal of Materials Science*, **13**, 1978, 2295-2299.

C. Gurney and J. Hunt: *Proceedings of Royal Society, London*, **A299**, 1967, 508-524.

B. Harris, P.W.R. Beaumont and E.M. de Ferran: *Journal of Materials Science*, **6**, 1971, 238-251.

B. Harris and D. Cawthorne: *Plastics and Polymers*, **42**, 1974, 209-216.

B. Harris and A.R. Bunsell: *Composite*, **6**, 1975, 197-199.

B. Harris and S.V. Ramani: *Journal of Materials Science*, **10**, 1975, 83-93.

B. Harris, J. Morley and D.C. Phillips: *Journal of Materials Science*, **10**, 1975, 2050-2061.

B. Harris and A.O. Ankara: *Proceedings of Royal Society, London*, **A359**, 1978, 229-250.

B. Harris, F.J. Guild and C.R. Brown: *Journal of Physics D: Applied Physics*, **12**, 1979, 1385-1407.

B. Harris: *Metal Science*, **14**, 1980, 351-362.

B. Harris, S.E. Dorey and R.G. Cooke: *Composites Science and Technology*, **31**, 1988, 121-141.

B. Harris, A.S. Chen, S.L. Coleman and R.J. Moore: *Journal of Materials Science*, **26**, 1991, 307-320.

B. Harris, F.A. Habib and R.G. Cooke: *Proceedings of Royal Society, London*, **A437**, 1992, 109-131.

S. Hashemi, A.J. Kinloch and J.G. Williams: *Proceedings of Royal Society, London*, **A427**, 1990, 173-199.

J.L. Helfet and B. Harris: *Journal of Materials Science*, **7**, 1972, 494-498.

W.B. Hillig: *Annual Review of Materials Science*, **17**, 1987, 341-383.

J.L. Kardos: *Molecular Characterisation of Interfaces Polymer Science and Technology*, H. Ishida and G. Kumar, eds., Plenum Press, NY., **27**, 1985, 1-11.

A. Kelly and W.R. Tyson: *Journal of Mechanics and Physics of Solids*, **13**, 1965, 329-350.

A. Kelly: *Proceedings of Royal Society, London*, **A319**, 1970, 95-116.
A. Kelly: *Strong Solids,* Second Edition, University Press Cambridge, 1973.
M. Khatibzadeh and M.R. Piggott: *Composites Science and Technology*, **58**, 1997, 497-504.
R.J. Lee and D.C. Phillips: *Proceedings of First International Conference on Composite Structures*, Paisley, Scotland, I.H. Marshall, ed., Applied Science Publishers, London, 1981, 536-554.
Y.W. Mai, A.G. Atkins and R.M. Caddell: *International Journal of Fracture*, **12**, 1976, 391-407.
Y.W. Mai: *Proceedings of 3rd Japan International SAMPE Symposium, Advanced Materials - New Processes and Reliability*, T. Kishi, N. Takeda and Y. Kagawa, eds., Chiba, Japan, Japan Chapter of SAMPE, Tokyo, **2**, 1993, 2099-2107.
D.B. Marshall, B.N. Cox and A.G. Evans: *Acta Metallurgica*, **33**, 1985, 2013-2021.
T.U. Marston, A.G. Atkins and D.K. Felbeck: *Journal of Materials Science*, **9**, 1974, 447-455.
F.J. McGarry and F.J. Mandell: *Proceedings of 27th Annual Technical Conference of Reinforced Plastics/Composite Institute of SPI*, Society for Plastics Industry, New York, 1972, Paper 9A.
F.J. McGarry, F.J. Mandell and S.S. Wang: *Polymer Engineering Science*, **16**, 1976, 609-614.
M.A. McGuire and B. Harris: *Journal of Physics D: Applied Physics*, **7**, 1974, 1788-1802.
J. Morton and G.W. Groves: *Journal of Materials Science*, **9**, 1974, 1436-1445.
J. Morton and G.W. Groves: *Journal of Materials Science*, **11**, 1976, 617-622.
A.P. Mouritz, J. Gallagher and A.A. Goodwin: *Composite Science and Technology*, **57**, 1997, 509-522.
T.K. O'Brien: *Damage in Composite Materials*, K.L. Reifsnider, ed., ASTM STP 775, American Society for Testing and Materials, Philadelphia, U.S.A., 1982, 140-167.
J.O. Outwater and M.C. Murphy: *Proceedings of 24th Annual Technical Conference*, Reinforced Plastics/Composite Institute of SPI, 1969, Paper11C.
J.O. Outwater and M.C. Murphy: *Proceedings of 28th Annual Technical Conference*, Reinforced Plastics/Composites Institute of SPI, 1973, Paper 17A.
M.J. Owen and P.T. Bishop: *Journal of Physics D: Applied Physics*, **5**, 1972, 1621-1636.
M.J. Owen and R.G. Rose: *Journal of Physics D: Applied Physics*, **6**, 1973, 42-53.
P.C. Paris and G.C. Sih: *Fracture Toughness Testing and its Applications*, ASTM STP 381, 1965, 30-81.
D.C. Phillips: *Journal of Materials Science*, **7**, 1972, 1175-1191.
D.C. Phillips: *Journal of Materials Science*, **9**, 1974(a), 1847-1854.
D.C. Phillips: *Journal of Composite Materials*, **8**, 1974(b), 130-141.
M.G. Phillips: *Composites*, **12**, 1981, 113-116.
M.R. Piggott: *Journal of Materials Science*, **5**, 1970, 669-675.

M.R. Piggott: *Journal of Mechanics and Physics of Solids*, **22**, 1974, 457-468.

R.B. Pipes, R.C. Weatherhold and J.W. Gillespie: *Journal of Composite Materials*, **13**, 1979, 148-160.

R.T. Potter: *Proceedings of Royal Society, London*, **A361**, 1978, 325-341.

K.M. Prewo: *Journal of Materials Science*, **17**, 1982(b), 2371-2383.

K.M. Prewo and J.J. Brennan: *Journal of Materials Science*, **15**, 1980, 463-468.

K.M. Prewo and J.J. Brennan: *Journal of Materials Science*, **17**, 1982(a), 1201-1206.

J.P. Romualdi and G.B. Batson: *Journal of Engineering Mechanics Division, Proceedings of American Society Civil Engineering*, **89**, 1963, 147-168.

R.J. Sanford and F.R. Stonesifer: *Journal of Composite Materials*, **5**, 1971, 241-245.

N. Sela and O. Ishai: *Composites*, **20**, 1989, 423-435.

G.P. Sendeckyi ed.: *Fracture Mechanics of Composites*, ASTM STP 593, American Society for Testing and Materials, Philadelphia, U.S.A., 1975.

G.C. Sih, P.C. Paris and G.R. Irwin: *International Journal of Fracture Mechanics*, **1**, 1965, 189-203.

G.C. Sih: *Proceedings of First U.S.A./USSR Symposium on Fracture of Composite Materials*, G.C. Sih and V.P. Tamuzs, ed., Riga 1978, Sijtoff and Noordhoff, Netherlands, 1979.

C. Soutis: *Composite*s Engineering, **4**, 1994, 317-327.

V.K. Srivastava and B. Harris: *Journal of Materials Science*, **29**, 1994, 548-553.

S. Stefanidis, Y.W. Mai and B. Cottrell: *Journal of Materials Science Letters*, **4**, 1985, 1033-1035.

M.D. Thouless and A.G. Evans: *Acta Metallurgica*, **36**, 1988, 517-522.

R.W. Truss, P.J. Hine and R.A. Duckett: *Composites*, **28A**, 1997, 627-636.

M. Vancon, J. Odorico and C. Bathias: *Comptes Rendus des Quatrièmes Journées Nationales sur les Composites*, JNC4, Pluralis, Paris, 1984, 93-120.

M.E. Waddoups, J.R. Eisenmann and B.E. Kaminski: *Journal of Composite Materials*, **5**, 1971, 446-454.

S.S. Wang, J.F. Mandell and F.J. McGarry: *Fracture Mechanics of Composites*, G.P. Sendeckyj, ed., ASTM STP 593, 1975, 36-85.

J. Wells and P.W.R. Beaumont: *Journal of Materials Science*, **20**, 1985, 1275-1279.

R.C. Wetherhold and M.A. Mahmoud: *Materials Science and Engineering*, **79**, 1986, 55-65.

J.M. Whitney and R.J. Nuismer: *Journal of Composite Materials*, **8**, 1974, 253-265.

E. Wu: *Composites Materials Workshop*, S.W. Tsai *et al.*, Technomic Press, Stamford, Conn., U.S.A.

6. Fatigue Behaviour of Fibre Composites

6.1 INTRODUCTION

Our familiarity with fatigue failure is so closely linked with the behaviour of isotropic, homogeneous, metallic materials that we have tended to treat modern fibre composites as though they were metals. The test methods used to study fatigue in metals have also been applied to composites, and the interpretation of the results of such tests has often been clouded by our perception of what constitutes metallic fatigue failure. It is natural that a designer wanting to substitute a composite for a metal component should want to test the new material by applying a cyclic loading régime of the same kind as the component would be required to sustain in service in order to prove that the composite will perform as well as the metal. But it must not be assumed, *a priori*, that there is some universal mechanism by which fluctuating loads will inevitably result in failure at stresses below the normal monotonic failure stress of the material.

Fatigue of metals accounts for a high proportion of engineering failures and has been intensively studied for more than a century. Design data have been accumulated for every conceivable engineering metal and alloy, and the engineer has access to a comprehensive set of rules, some empirical and some based on scientific understanding, with which to cope with any given design requirement, although some designers often choose to ignore these rules. Fatigue in metals progresses by the initiation of a single crack and its intermittent propagation until catastrophic failure. This occurs with little warning and little sign of gross distortion, even in highly ductile metals, except at the final tensile region of fracture. In ordinary high-cycle (low-stress) fatigue, where stress levels away from the crack tip are low, the properties of the metal remote from the crack are only slightly changed during fatigue. It is not a general feature of fatigue in metals and plastics that the strength of the material is reduced by cyclic loading, although work-hardening or work-softening may occur in metals undergoing low-cycle (high-stress) fatigue. The usual effect of fatigue at low stresses is simply to harden the metal slightly. Generally speaking, a stronger material will have a higher fatigue resistance, the fatigue ratio (fatigue limit divided by tensile strength) being roughly constant.

It is not uncommon for users of composite materials, even in the aerospace industry, to express the belief that composite materials — specifically, carbon-fibre-reinforced plastics — do not suffer from fatigue. This is an astonishing assertion, given that from the earliest days of the development of composites, their fatigue behaviour has been a subject of serious study, and what is usually implied is that, because most CFRP are extremely stiff in the fibre direction, the working strains in practical components at conventional design stress levels are usually far too low to initiate any of the local damage mechanisms that might otherwise have led to deterioration under cyclic loads.

The idea of using composites, especially CFRP, only at very low working strains raises two important issues. The first is the obvious one that by using expensive, high-performance materials at small fractions of their real strength, we are over-designing or, in more cost-conscious terms, we are using them uneconomically. The second is that since anisotropy is a characteristic that we accept and even design for in composites, a stress system that develops only a small working strain in the main fibre direction may easily cause strains normal to the fibres or at the fibre/resin interface that may be high enough to cause the kind of deterioration that we call fatigue damage. In designing with composites, therefore, we cannot ignore fatigue. And it follows directly that, in addition to needing to understand the mechanisms by which fatigue damage occurs in composites, we need access to procedures by which the development and accumulation of this damage, and therefore the likely life of the material (or component) in question, can be reliably predicted.

In this chapter, we shall concentrate mainly on the fatigue behaviour of polymer-matrix composites. Since the majority of accessible research studies have concerned these materials, it is inevitable that even a general picture of composites fatigue will be substantially coloured by our knowledge of fibre-reinforced plastics (FRPs). Specific reference to metal- and ceramic-matrix composites will be made after a discussion of FRPs.

6.2 DAMAGE IN COMPOSITES

Unlike metals, composite materials are inhomogeneous (on a gross scale) and anisotropic. They accumulate damage in a general rather than a localized fashion, and failure does not always occur by the propagation of a single macroscopic crack. The microstructural mechanisms of damage accumulation, including fibre breakage and matrix cracking, debonding, transverse-ply cracking, and delamination, occur sometimes independently and sometimes interactively, and the predominance of one or other may be strongly affected by both materials variables and testing conditions.

At low monotonic stress levels, or early in the life of a composite subjected to cyclic loading, most types of composite sustain damage, and this damage is distributed throughout the stressed region, as described in chapter 4, section 4.1.3. The damage does not always immediately reduce the strength of the composite although it often reduces the stiffness. Such strength reductions as might occur (in a process described as *wear-out*) are sometimes off-set in the early stages of life by slight increases in strength, known as *wear-in*. These increases may be caused by such mechanisms as improved fibre alignment which results from small, stress-induced, viscoelastic or creep deformations in the matrix. At a later stage in the life the amount of damage accumulated in some region of the composite may be so great that the residual load-bearing capacity of the composite in that region falls to the level of the maximum stress in the fatigue cycle and failure ensues, as shown schematically in Figure 6.1. This process may occur gradually,

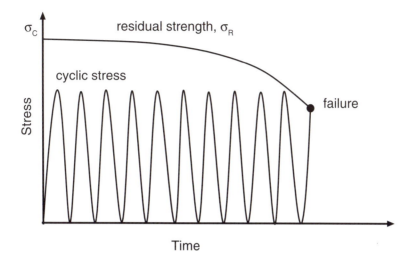

Fig. 6.1 Degradation of composite strength by wear-out until the residual strength, σ_R, falls from the normal composite strength, σ_c, to the level of the fatigue stress, at which point failure occurs.

when it is simply referred to as degradation, or catastrophically, when it is termed *sudden-death*. Changes of this kind do not necessarily relate to the propagation of a single crack, and this must be recognized when attempting to interpret composites fatigue data obtained by methods developed for metallic materials.

When a pre-existing crack is present in a composite it may or may not propagate under the action of a cyclic load, depending upon the nature of the composite. In a metal-fibre/metal-matrix composite the fibre itself will be subject to conventional fatigue processes. Planar (or near-planar) crack propagation can occur with both fibres and matrix exhibiting normal signs of fatigue, as shown in Figure 6.2, and the material behaves simply like a homogeneous metal of higher strength than the unreinforced matrix. In highly anisotropic composites of high V_f the crack will often refuse to propagate normal to the fibres (mode 1) but will be diverted into a splitting mode. In composites like unidirectional CFRP this may result in a brittle, end-to-end splitting failure which simply eliminates the crack (Figure 6.3) while in a metal-matrix composite the splitting will be preceded by a localized but elongated region of intense plastic deformation (McGuire and Harris, 1974). By contrast, in GRP laminates containing woven-roving or chopped-strand mat reinforcement crack tip damage may remain localized by the complex geometry of the fibre array and the crack may proceed through this damaged zone in a fashion analogous to the propagation of a crack in a plastically-deformable metal (Owen and Bishop, 1974).

Fig. 6.2 Fatigue fracture surface of an MMC consisting of tungsten filaments in an Al/4%Cu matrix showing brittle fracture of both fibres and matrix (McGuire and Harris, 1974). Compare with Figure 5.2 which shows a tensile fracture surface of the same composite.

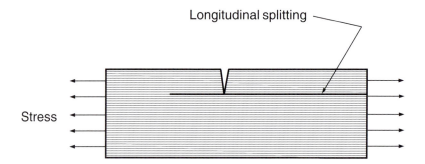

Fig. 6.3 End-to-end splitting parallel to fibres in a unidirectional composite, resulting in effective removal of the stress concentrator.

6.2.1 DEFINITION OF FAILURE AND EXPERIMENTAL SCATTER

Smith and Owen (1968) have demonstrated the extent to which variability affects the results of fatigue tests and have emphasised the value of replicate testing. This variability stems not only from the statistical nature of the progressive damage which leads to fracture of a composite, but more particularly from the variable quality of

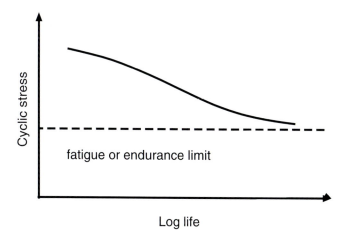

Fig. 6.4 Schematic illustration of the features of an S/N curve (sometimes referred to as a Wöhler diagram).

commercial composite materials. They have also pointed out that in many kinds of composite the number of cycles to complete separation of the broken halves of a sample is a definition of failure that becomes quite meaningless if the sample has lost its integrity and its ability to sustain an applied stress as a result of extensive resin cracking, arguing the need for careful (and relevant) definition of the failure criterion. They have nevertheless shown that designs based on the use of debonding or the onset of resin cracking as failure criteria would drastically impair the economic use of GRP.

6.3 REVIEW OF SOME CLASSICAL FATIGUE PHENOMENA

6.3.1 STRESS/LIFE CURVES

A good deal of the work that has been done on the fatigue of fibre composites reflects the much more extensive body of knowledge relating to the fatigue of metals, and this is not altogether unreasonable since the established methods of accumulating and analysing metallic fatigue data provided a reliable means of describing fatigue phenomena and designing against fatigue. The formidable treatise of Weibull (1959), for example, is a valuable source book for modern workers on composites fatigue. The danger was, and is, in making the assumption that the underlying mechanisms of material behaviour that give rise to the stress/life (S/N or $\sigma/\log N_f$) curve are the same in metals and composites (see Ashby and Jones, 1980, for a description of the various kinds of metallic fatigue).

Before the development of fracture mechanics and its use in treating metallic fatigue as a crack-growth problem, the only available design information on fatigue behaviour was the $\sigma/\log N_f$ curve (Figure 6.4). It represented directly the perceived nature of fatigue in terms of experimental results, but gave no indication of the mechanisms of fatigue damage, of the presence or behaviour of cracks, or of changes in the characteristics of the material as a consequence of the fatigue process. The curve represents the stress to cause failure in a given number of cycles, usually either the mean or median life of a series of replicate tests at the same stress. Despite the anomaly (from a mathematical viewpoint) of plotting the dependent variable on the abscissa rather than *vice versa*, the $\sigma/\log N_f$ curve is nevertheless a useful starting point for the designer, provided due attention has been paid to the statistical aspects of data generation, so that the apparently simple failure envelope which it defines is associated with failure probabilities rather than with some simplistic fail/no-fail criterion. It can then be used, as many designers prefer, without any consideration of the underlying fatigue damage mechanisms, despite the availability of rather better fracture-mechanics methods.

There is logic in using the semi-logarithmic form of the $\sigma/\log N_f$ plot but there is no reason for supposing *a priori* that the failure envelope would be linear on a log-log plot (Basquin's Law for metals). Some stress/life plots are linear on both kinds of plot, some on neither. The knowledge that for bcc metals and alloys such as steels, the $\sigma/\log N_f$ curve flattened out at long lives ($>10^7$ cycles, say) and gave rise to a *fatigue limit*, or *endurance limit*, σ_e, which could be explained in terms of dislocation/solute-atom models, and that the *fatigue ratio*, σ_t/σ_e, (σ_t is the monotonic tensile strength) was roughly constant for a given class of alloys, gave confidence to users of this kind of design data. Finally, there was no general idea that the metallic $\sigma/\log N_f$ curve should necessarily extrapolate back to a stress level which was related to any monotonic strength property of the material (at $N_f = 1$ cycle, for example).

6.3.2 CONSTANT-LIFE DIAGRAMS

In order to use stress/life information for design purposes, a common procedure was to cross-plot the data to show the expected life (or expected stress for some particular probability of failure) for a given combination of the alternating component of stress, σ_{alt}, defined as half the stress range, $\frac{1}{2}(\sigma_{max} - \sigma_{min})$, and the mean stress, σ_m, which is $\frac{1}{2}(\sigma_{max} + \sigma_{min})$. The *stress ratio*, R, is $\sigma_{min}/\sigma_{max}$. In metallic fatigue it was frequently assumed that compression stresses were of no significance because they acted only to close fatigue cracks, unlike tensile forces. Master diagrams of this kind are presented in a variety of forms, all more or less equivalent, but the most familiar is that which is usually referred as the Goodman diagram (Figure 6.5). For design purposes, it was useful to have an equation to represent the fail/safe boundary in this diagram, and it is the linear relationship of Figure 6.5 that is associated with the name of Goodman (1899) although others have been proposed, including the earlier parabolic relationship of Gerber (1874). The linear and parabolic 'laws' have been modified to include safety factors on one or

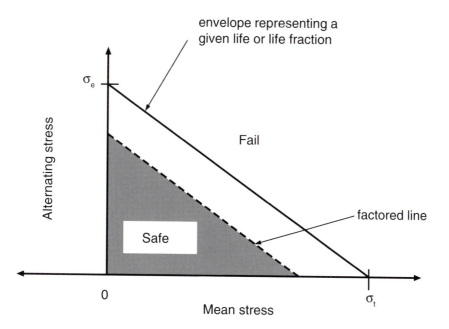

Fig. 6.5 Schematic illustration of a constant-life or Goodman diagram. The stress σ_e is the endurance or fatigue limit, and σ_t is the material tensile strength.

both of the stress components. An important question, for metals as well as for composites, relates to the minimum amount of test data that is needed to define the failure envelope with a level of reliability sufficient for engineering design of critical components.

6.3.3 STATISTICAL ASPECTS OF THE ANALYSIS OF FATIGUE DATA

The mechanical properties of practical composites are almost always variable as a consequence of the variability of the material quality. It is not surprising, therefore, that the variability of the fatigue response of a composite is even greater than that associated with metallic materials. Stress/life data may be obtained by testing single samples at many different stress levels, or by carrying out replicate tests at rather fewer stresses: the latter is usually considered to be the more satisfactory method because it provides statistical information at each stress, and provides probability/stress/life curves in addition to median-life or mean-life curves.

One of the problems is to know how many replicate tests should be done at each stress level since, given the cost of fatigue-testing programmes, the smaller the number of tests that can be used to establish a 'safe' $\sigma/\log N_f$ curve, the better. From a statistical point of view it is usually accepted that at least 20 replicate tests may be necessary before the user can have any confidence in a statistical analysis of results (see, for example,

Lee *et al.*, 1997). But when stress/life curves are required at, say, five different R ratios, even five tests at each stress level may be all that can be provided in a reasonable amount of time, especially at long lives.

A variety of distributions have been used to characterize fatigue lives, but the three-parameter Weibull function, described in Appendix 2, is usually considered to be the most appropriate model for this purpose. The form of the cumulative distribution function used for fatigue is:

$$P(N_f; a, b, m) = 1 - \exp\left[-\left(\frac{N_f - a}{b}\right)^m\right] \tag{6.1}$$

where P is the probability of a life N_f, and a, the location parameter, defines a number of cycles for which there is zero probability of failure. For metallic materials, the value of the shape parameter, m, is often in the range $2 < m < 6$ (Freudenthal and Gumbel, 1953), whereas several recent studies of CFRP fatigue suggest that for these materials $1 < m < 2$ (Whitney, 1981; Harris *et al.*, 1997). It is interesting to note that for an m value of unity the Weibull model reduces to a simple exponential distribution. The exponential distribution, which is sometimes used to model distributions of failure times for the reliability of a product, governs systems where age has no influence on the probability of failure. If a component survives for a length of service t, the probability of failure within a subsequent time interval from t to $(t + \Delta t)$ is $\lambda\Delta t$, where λ is a constant for all t. Thus, failure is a random event, and the system does not deteriorate as a result of service. We would not normally consider this as being applicable to the fatigue failure of reinforced plastics for which it is known that residual performance is certainly reduced as a result of the accumulation of damage. One reason why the data may appear to fit an exponential distribution is that the range of lives at a given stress level is often very wide, spreading sometimes over two decades or more. Whereas when viewed by eye on a logarithmic life axis this may not appear too serious, when considered in the Weibull analysis as a linear function, N_f, rather than as a logarithmic function, $\log N_f$, the range could thus easily be judged to be bounded at the lower extreme at $N_f = 0$, and hence the distribution may appear exponential. In reality, the shape of the Weibull distribution changes from being of classical exponential form to being peaked over a very small range of values of m between 1 and 2.

Although we often work in terms of median or mean lives when discussing fatigue data, it is important to recognize that to the designer of critical structures, such as aircraft components, the requirements are much more restrictive, and it is likely that the desired failure probability for such a structure would be at a very low level, say 5%, instead of the 50% level implied by the median life. An even more appropriate 'life' parameter may be the minimum extreme value obtained from an application of the theory of extreme values (Castillo, 1988). Extreme-value models are appropriate models for describing many engineering phenomena and systems where the relevant parameters are the characteristic largest and characteristic smallest values of a distribution. For example, in the case of the life of a gas-turbine disc fitted with a large number of blades, the turbine life is determined by the life of the shortest-lived blade — a 'weakest-link' model. Such a model, referred to as a Type III asymptote model, is also applicable to the fatigue behaviour of engineering materials.

One of the characteristics of the Weibull model is what is known as its reproductive property (Bury, 1975). A consequence of this property is that for a population of results that is well modelled by the Weibull distribution certain other features of the population, such as the minimum extreme values, will also be described by a Weibull distribution. Put more formally, the exact distribution of the smallest observations in sets of data that are described by a Weibull model also fits a Weibull model. Thus, if we have replicate data sets of fatigue lives at a given stress and R ratio which are described by the two-parameter version of equation 6.1 (see Appendix 2), with shape parameter m and scale parameter b (a being equal to 0), the smallest observations of these data sets also exhibit a Weibull distribution similar to that of the parent distribution and with the same value of the shape factor, m, but rescaled by $n^{1/m}$. The characteristic minimum value for a test sample of n tests will therefore be ($b/n^{1/m}$) and the modal value of the distribution (*i.e.* the most probable value) will be given by:

$$\frac{b}{n^{\frac{1}{m}}}\left(1-\frac{1}{m}\right)^{\frac{1}{m}}$$

If, on the other hand, the desired level of failure probability happens to be the 5% level ($q = 0.05$), then the appropriate value of 'life' is given by:

$$b\left[ln\left(\frac{1}{1-q}\right)\right]^{\frac{1}{m}}$$

Thus, from a series of replicate data sets of fatigue lives at various cyclic stress levels, a stress/life/probability (or S/N/P) diagram resembling that shown in Figure 6.6 may be drawn, provided the distribution parameters, b and m, can be obtained for each data set. From a design point of view, the lower curve in this diagram, representing some given failure probability, is clearly of more interest to the engineer than any other part of the data, but it can only be obtained by statistical means which requires reliable values of the two distribution parameters, b and m. This, in turn, calls for replicate data sets of at least 20 test results for each stress level and R ratio, a requirement which carries with it a serious time/cost disincentive.

However, Whitney (1981) has suggested that where only small numbers of life values are available at a number of different stress levels, the data may be pooled to give an overall value of the Weibull shape parameter, m, this value then being used to obtain working stress/loglife curves for any given failure probability. This is done by normalising each test-stress data set with respect to either the characteristic life, b (the scale factor of the Weibull distribution for the data set) or the median life for the data set, pooling all data sets for *all* stress levels and *all* R ratios, and then re-ranking them in order to allot a new failure probability function to each point (see Appendix 2 for a discussion of ranking

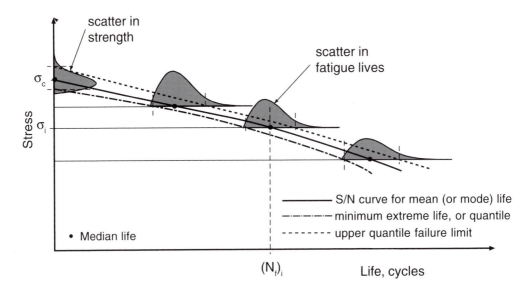

Fig. 6.6 Effects of variability in strength and fatigue life on the definition of the $\sigma/\log N_f$ curve. The scatterband of lives is jointly defined by the upper and lower probability limits of the fatigue data sets and the monotonic strength distribution.

methods). The virtue of this procedure is that a much larger population is being used to derive the value of the Weibull shape parameter and calculations of an expected life based on that m value will be much more reliable than ones obtained from the much smaller data sets for each individual stress level. The Weibull distribution for a pooled fatigue data set for a carbon-fibre composite laminate is shown in Figure 6.7. The shape parameter for the pooled data is 1.1, and is similar to values obtained for other CFRP laminates by, for example, Whitney (1981) and Harris *et al.* (1997). The value of m obtained in this way can then be used, as shown above, to derive stress/life curves for desired failure probability levels.

6.3.4 CUMULATIVE DAMAGE LAWS

In most fatigue environments the mean and alternating stress amplitudes vary and may be presented as a spectrum of the frequency of occurrence of different stress levels. The concept of deterioration as a result of damage accumulation within the material with stress cycling is used in combination with the notion of residual strength, and the damage is quantified such that a parameter Δ represents the fraction of catastrophic damage sustained after n cycles (where $n < N_f$) so that at failure $\Delta = 1$ and $n = N_f$. Thus, Δ_i represents the fractional damage after cycling at a stress σ_i for a fraction of life n_i/N_i. Various proposals for damage laws have been derived, the most common of which is the

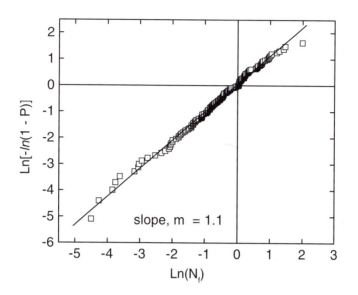

Fig. 6.7 Two-parameter Weibull plot for pooled, normalized fatigue data for a $[(\pm45,0_2)_2]_s$ T800/5245 CFRP laminate. The pooled data are for all stress levels at 5 different R ratios (Gathercole *et al.*, 1994).

Palmgren-Miner linear damage rule (Miner, 1945). The accumulation of damage is supposed to be linear with number of cycles, and is apparently independent of the value of the stress:

$$\Delta = n / N \qquad (6.2)$$

Thus, if the load spectrum is divided into blocks with n_i cycles at stress levels σ_i, then equation 6.2 applies incrementally to each block and the total damage is the sum of the individual 'block' damages, irrespective of the order of their application, so that:

$$\Delta = \Sigma \frac{n_i}{N_i} \qquad (=1 \text{ for failure}) \qquad (6.3)$$

This well-known, though rarely obeyed, rule was originally proposed for the prediction of the life of metallic components undergoing fatigue, and although it is a standard feature of text-book presentations of fatigue it has always been viewed with great suspicion by designers because it is often found to give non-conservative results, *i.e.* to predict lives greater than those observed experimentally. The simple concept of damage defined by equation 6.3 is not, of course, a mechanistic concept. It has been applied to many kinds of engineering materials, regardless of the actual nature of the damage mechanisms that contribute to the gradual deterioration and ultimate failure of an engineering component subject to cyclic loading.

The simplest step forward from the linear damage rule is to look for non-linear functions that still employ the damage parameter Δ, as defined by equation 6.3. In the

Marco-Starkey (1954) model, for example, a simple non-linear presentation suggests an equation of the form:

$$\Delta = (n / N)^{\beta} \tag{6.4}$$

where β is a function of stress amplitude. The condition for failure when $\Delta = 1.0$ is satisfied and the Miner rule is a special case of this law with $\beta = 1.0$.

6.4 FATIGUE OF REINFORCED PLASTICS

6.4.1 CONSTANT-STRESS FATIGUE CURVES

The dominant damage mechanisms in FRP vary both with the nature of the composite (the particular combination of fibres and matrices, reinforcement lay-up, *etc.*), and with the loading conditions (tension, bending, compression, *etc.*), and results for any one material under a given stress condition may appear to fit no general pattern. This is easily illustrated with reference to the gradual accumulation of results and materials development over the last two decades or so.

First, the forms of stress/life curves for various kinds of GRP have long been familiar to researchers, as was the fact that the fall of the curve with reducing stress was considerable. This fall, or average slope, is an indication of the fatigue resistance of the material, a more rapid fall being indicative of a greater susceptibility to fatigue damage. Mandell (1982) attempted to show that fatigue damage in GRP materials can be explained simply in terms of a gradual deterioration of the load-bearing fibres. He analyses a great range of experimental data to demonstrate that the behaviour of composites with long or short fibres, in any orientation, and with any matrix, can be explained in this way. He does this by effectively forcing $\sigma/\log N_f$ curves to fit a linear law of the form:

$$\sigma = \sigma_t - B \log N_f \tag{6.5}$$

where σ is the peak cyclic (tensile stress), σ_t is the monotonic tensile fracture stress, N_f is the fatigue life of the material at stress σ, and B is a constant, as illustrated in Figure 6.8. For a wide and disparate range of GRP materials, including moulded reinforced thermoplastics, Mandell obtained a constant value of about 10 for the ratio σ_t/B, with very little spread. In principle, the idea that the controlling mechanism in fatigue of composites is the gradual deterioration of the load-bearing fibres is logical and inevitable. What controls the actual life of any given sample or composite type is simply the manner in which other mechanisms, such as transverse-ply cracking in 0/90 laminates, or local resin cracking in woven-cloth composites, modify the rate of accumulation of damage in the load-bearing fibres. Nevertheless, a single mechanistic model which includes randomly-reinforced dough-moulding compounds and injection-moulded thermoplasts as well as woven and non-woven laminates is unexpected.

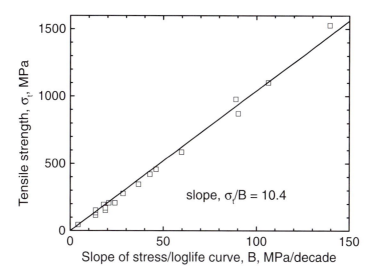

Fig. 6.8 Relationship between composite tensile strength and the slope of the fatigue $\sigma/\log N_f$ curve (after Mandell, 1982). Data points taken from the literature represent a wide range of GRP composites.

The $\sigma/\log N_f$ curves for many GRP materials are not linear, however, as illustrated by the selection of data for a group of glass-fibre composites shown in Figure 6.9. With GRP there are two factors which complicate the appearance of the stress/life curves. As a consequence of the high failure strain of the glass fibres and their sensitivity to moisture, the tensile strengths of GRP materials are sensitive to strain-rate and temperature, and during cycling at large strains there is usually a significant rise in temperature as a result of hysteretic heating which is not easily dissipated by the non-conducting constituents of the GRP.

The consequences of this have been demonstrated by Sims and Gladman (1978). First, when fatigue tests are carried out at constant frequency over the whole stress range of interest, the deformation rate is usually considerably greater than the rates normally used for measuring the monotonic strength, and as a result, since it is common for the strength to be included, as shown in Figure 6.9, as the point on the extreme left of the $\sigma/\log N_f$ curve (*i.e.* notionally at $N_f = 0.5$ for a repeated tension test, or $N_f = 0.25$ for a fully reversed tension/compression test) that measured strength is often actually lower than a value measured at the fatigue-test frequency. Second, if tests are run at constant frequency, the effective strain rate at each stress level will be different (higher at lower stresses), and the measured life values for each stress will not be associated with a common base-line of material behaviour: *i.e.* the fatigue curve will refer to a range of stresses, σ, which are proportions of a variable material property, say $\sigma_t(\dot{\varepsilon})$, instead of the material property that we normally define as the strength, σ_t. And third, as the peak stress level

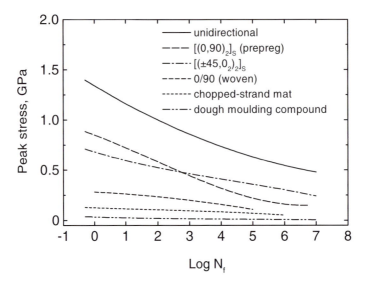

Fig. 6.9 Stress/loglife curves for a range of GRP materials at an R ratio, $\sigma_{min}/\sigma_{max}$, of 0.1. Data points are omitted for clarity, but may be found in the original diagram (Harris, 1994). The stress for a 'life' of 0.5 cycles is the monotonic tensile strength measured at a loading rate equivalent to that used for the fatigue tests.

falls and the life of the sample extends, the degree of hysteretic heating rises, and the effective baseline strength of the material therefore also falls, so reversing the effect of the higher effective deformation rate. As Sims and Gladman pointed out, the only way to avoid the effects of these interacting processes is to ensure that all $\sigma/\log N_f$ curves for GRP materials are determined at a fixed rate of load application (RLA), that the material strength is measured at the same RLA, and that the hysteretic effects are either eliminated or accounted for. These corrections can make substantial differences to the shapes of stress/life curves and to the validity of the design data obtained from them: many otherwise curved $\sigma/\log N_f$ graphs are also rendered more linear by these corrective measures.

By contrast with GRP, carbon-fibre composites are largely rate insensitive and because they deform less than GRP under working loads and are reasonably highly conducting, hysteretic heating effects are usually insignificant (Jones *et al.,* 1984). Some of the earliest CFRP, reinforced with the HM (high-modulus, or Type 1) species of carbon fibre had such low failure strains that fatigue experiments in tension were extremely difficult to carry out, and some of the first tension $\sigma/\log N_f$ curves published (Beaumont and Harris, 1972; Owen & Morris, 1972) were almost horizontal. The S/N line hardly fell outside the scatterband of measurements of the monotonic tensile strength, the fatigue response being effectively dominated by tensile failure of the fibres alone and the slight reduction in resistance after 10^6 or 10^7 cycles appeared to be a result of small viscoelastic (creep) strains rather than to cycling effects *per se* (Fuwa *et al.,* 1975). Only in torsion or bending

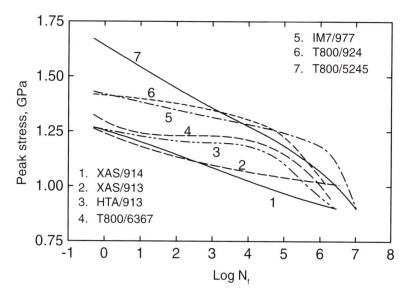

Fig. 6.10 Median stress/loglife curves at R = 0.1 for seven varieties of carbon-fibre composite, all with the lay-up $[(\pm45,0_2)_2]_S$ (Harris *et al.*, 1997).

tests was the slope of the $\sigma/\log N_f$ increased as these different stress systems allowed other damage mechanisms to occur.

A very small failure strain is often a disadvantage in engineering materials (some of the early CFRPs were extremely brittle) and developments in carbon-fibre manufacture led to materials that possessed higher failure strains (lower elastic moduli), although these improvements were often offset by concomitant reductions in strength. By the end of the 1970s, the most common polyacrylonitrile-based fibres, designated HMS, HTS and XAS, covered a range of moduli from about 400 GPa down to 200 GPa, and the tensile stress/loglife data for these materials fell approximately on straight lines with slopes which increased with decreasing fibre stiffness (Sturgeon, 1973). This is roughly in accord with the Mandell model except that Mandell's correlation was with composite strength rather than stiffness.

Increasing awareness on the part of designers of the qualities of fibre composites, and of carbon-fibre composites in particular, resulted in demands for fibres with combinations of strength and stiffness which were different from those which characterized the earlier materials. A particular call was for higher failure strains in association with high strength, and these demands led to developments in carbon-fibre manufacture and the availability of a much wider range of fibre characteristics. The fatigue performance of composites based on these newer fibres is clearly different from that of earlier CFRP, as illustrated by the range of results shown in Figure 6.10. These data are all for composites with the lay-up $[(\pm45,0_2)_2]_S$ and the curves are polynomial fits

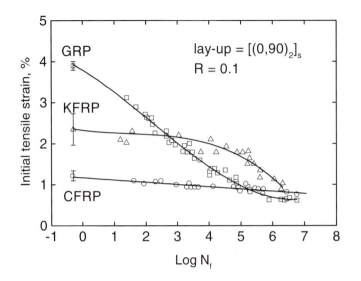

Fig. 6.11 Strain/life curves for $[(0,90)_2]_S$ laminates of composites reinforced with HTS carbon, Kevlar and E-glass fibres in a Code-69 epoxy resin (Fernando *et al.*, 1988: repeated tension fatigue, R = 0.1: RH = 65%).

to experimental data. Perhaps the most notable feature of this collection of data is that the monotonic strength characteristics of the reinforcing fibres are not directly translated into the fatigue response, *i.e.* the strongest fibres do not always generate the most fatigue-resistant composites. And whereas the earliest of CFRP composites had approximately linear, or even slightly upward turning $\sigma/\log N_f$ curves, the newer, increasingly 'high-performance' composites show downward curvature. We note that none of these curves for the newer materials has reached any kind of stable fatigue limit within the convenient laboratory test window extending to 10^7 cycles.

This downward curvature also occurs in composites reinforced with Kevlar-49 aromatic polyamide (aramid) fibres (Jones *et al.*, 1984; Fernando *et al.*, 1988). Typical behaviour of a cross-plied KFRP composite is compared with those of similar GRP and CFRP materials in Figure 6.11, plotted here in terms of initial peak strain, rather than stress. The composites are comparable in every respect except for the reinforcing fibres. The performance of the KFRP over the first three decades is similar to that of the CFRP on a normalized basis in that the slopes of the strain/life curves are similar. At maximum cyclic strain (or stress) levels only marginally below the tensile-failure scatter band, however, the fatigue resistance deteriorates rapidly, and a change in controlling failure mode occurs. At the highest stress levels (first three or four decades in Figure 6.11) failure of the KFRP occurs by the normal fibre-dominated failure process. At lower stresses, however, (much larger numbers of cycles) the internal structural weaknesses of the aramid filaments are exposed to the effects of the cyclic stress and the fibres themselves exhibit much lower fatigue resistance. This apparent weakness can be overcome to some extent by preventing complete unloading of the composite between cycles. Effectively, the down-turn of the $\sigma/\log N_f$ curve, which is not seen in the 0/90 CFRP in Figure 6.11, can be delayed by raising the minimum cyclic stress level.

These results for carbon, glass and aramid laminates illustrate very well a strain-control model of fatigue proposed by Talreja (1981). He suggests that the strain/loglife curves for polymer-matrix composites may be thought of in terms of three régimes within which separate mechanisms control fatigue failure. At high stress levels fibre breakage occurs (with or without interfacial debonding) which leads to failures within the normal tensile failure scatter band of the $0°$ plies. At lower cyclic stress levels, however, although statistical fibre breakage still occurs, it does not lead so rapidly to composite destruction and other competing mechanisms then have time to occur. These other mechanisms — matrix cracking, interfacial shear failure, *etc.* — can then influence the overall damage state provided the composite working strain level is sufficiently high, and the slope of the strain/loglife curve therefore begins to fall. There is, however, some notional fatigue-limit strain for the matrix itself and if working strains do not rise above this level the composite should not, in principle, fail in fatigue. The strain/loglife curve ought therefore to flatten out again and something in the nature of an endurance limit should be observed. Whether or not some or all of these stages are observed will clearly depend on the characteristics of the constituents and the lay-up geometry. In Figure 6.11, where the influence of the fibre deformation characteristics is more visible than in a stress/life plot, it can be seen that high working strains in the GRP prevent the establishment of stage 1 of the Talreja model, the curve moving almost immediately into stage 2, whereas the higher stiffness of the Kevlar fibre delays this transition in the aramid composites. Working strains in the CFRP are rarely sufficiently high as to exceed the matrix fatigue limit and so the strain/life curve (like the $\sigma/\log N_f$ curve) retains the low slope, characteristic of stage 1, for the entire stress range shown. It can be seen from the shapes of the $\sigma/\log N_f$ curves for various CFRP materials in Figure 6.10 that many of these high-performance composites exhibit similar features in tensile fatigue, regardless of the specific nature of the degradation mechanisms, although the extent to which some or all of the stages of Talreja's model are visible within a practical experimental test window depends on the nature of the constituents and of the fibre/matrix bond.

6.4.2 MATERIALS FACTORS AFFECTING FATIGUE BEHAVIOUR OF REINFORCED PLASTICS

Brittle fibres like glass, boron and carbon should not show the characteristic weakness of metals under fatigue loading conditions. Composites containing them should also, therefore, in principle, be fatigue resistant if the fibres carry the major part of the load and provided the reinforcing phase is not so extensible as to permit large elastic or viscoelastic deformations of the matrix as a function of time. These conditions are fulfilled when composites unidirectionally reinforced with high-modulus fibres like carbon or boron are tested in repeated tension, and their $\sigma/\log N_f$ curves are, in general, very flat.

As cycling continues, however, even small viscoelastic movements in the resin lead to local redistributions of stress which permit some random fibre damage to occur, and similar damage will occur in the neighbourhood of any stress concentrations. As a function of time, then, rather than as a function of the number of stress reversals, damage levels will build up to some critical level when the composite strength is no longer above the peak cyclic stress level, and failure will occur.

When the stress is not simply tensile, uniaxial, and aligned with the fibres more severe loads are placed on the matrix and the fatigue resistance of the composite is reduced. This occurs in simple flexural fatigue (Beaumont and Harris, 1972) and tension/compression loading of unidirectional composites, and also in fatigue tests on any kind of laminated materials. Ramani and Williams (1976), for example, tested 18-ply, (0, ± 30) carbon/epoxy laminates in various combinations of axial tension/compression cycling and showed an increasingly adverse effect on fatigue resistance as the compressive stress component was increased. Kunz and Beaumont (1975) and Berg and Salama (1972) showed that fatigue cracks propagated in CFRP during purely compressive cycling through the spreading of zones of fibre buckling failure. As we saw in chapter 4, compressive buckling is a characteristic mode of failure in fibre-reinforced materials of many kinds and it is usually initiated by local matrix shear failures, sometimes associated with debonding in imperfectly bonded composites. Ramani and Williams observed extensive shear damage resulting from matrix shear and from interlaminar and interfacial weakness.

Compression and shear stress systems will also be likely to be even more damaging to composites containing aramid fibres than to CFRP or GRP, as indicated by the poor compression response of KFRP described in chapter 4. The shear weakness is also clearly manifested in tensile fatigue of ±45 KFRP laminates, by comparison with similar samples of CFRP and GRP, as shown in Figure 6.12. In this mode of stressing the matrix and interface would normally be expected to dominate composite behaviour, the fibre characteristics playing little part. This is demonstrated in Figure 6.12 by the CFRP and GRP results, but the intrafibrillar weakness of the aramid fibre results in a considerable reduction in composite fatigue resistance, and microscopic observations of KFRP samples that have failed in this mode clearly show the kinking and splitting of the fibres resulting from this weakness (Figure 6.13). Similarly, when tested in torsion the weak parts of a composite (*i.e.* the resin and the interface) are loaded directly by the shear forces, the strength and rigidity of the fibres contributing relatively little to fatigue resistance, and the torsional rigidity of the material falls substantially during cycling. Some typical shear strain/loglife curves for two types of unidirectional carbon/epoxy composites are shown in Figure 6.14 (Phillips and Scott, 1976). The curves were obtained under constant angular (torsional) cycling conditions and the fatigue life is defined as the number of cycles to the occurrence of cracking.

There is much evidence from fatigue tests on reinforced plastics that the matrix and interface are the weak links as far as fatigue resistance is concerned. It appears that lower reactivity resins give better low-stress fatigue life in GRP, the optimum resin content being between 25 and 30 vol.%, although variations in resin content affect the fatigue strength at 10^6 cycles hardly at all by comparison with their effect on tensile strength

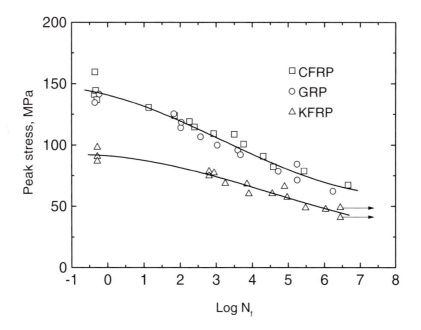

Fig. 6.12 Stress/loglife curves for the same materials as those of Figure 6.11 but tested at 45°to the main fibre directions: R = 0.1; RH = 65% (Jones *et al.*, 1984).

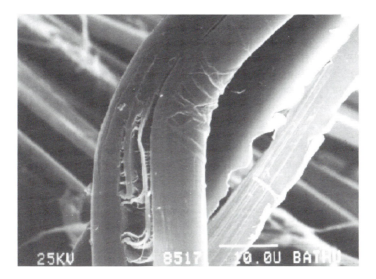

Fig. 6.13 Splitting and kinking damage in Kevlar-49 fibres after fatigue cycling (Fernando *et al.*, 1988)

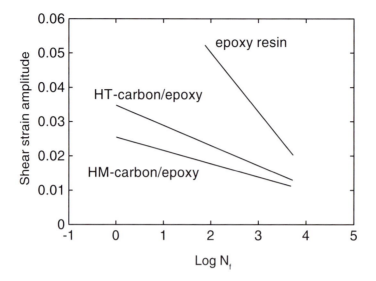

Fig. 6.14 Strain/loglife curves for unidirectional carbon/epoxy composites together with that of the unreinforced matrix resin. The reinforcements are type 1 (high-modulus) and type 2 (high-strength) carbon. The curves were obtained under constant angular (torsional) cycling conditions. The fatigue life is defined as the number of cycles to the occurrence of cracking. (Phillips and Scott, 1976)

(Owen, 1970). By contrast, variations in resin properties appear to have little effect on the fatigue strength of CFRP (Owen and Morris, 1970). Any materials treatment or processing that can improve the resistance of the matrix to crack propagation or the interfacial adhesion is likely to improve fatigue properties. Conversely, exposure of reinforced plastics to water often results in some degree of plasticisation of the matrix and weakening of the interfacial bond, and in GRP and aramid-fibre composites it may also affect the performance of the fibres themselves (Jones *et al.*, 1984). It follows that the fatigue of almost all FRP will be environment sensitive, the least sensitive being those consisting of moisture-resistant plastics like the polyaryls (PEEK, *etc.*) reinforced with carbon fibres which are themselves insensitive to degradation in moist atmospheres (Dickson *et al.*, 1985).

For manufacturing reasons (*e.g.* to secure good 'drape' characteristics when doubly curved surfaces are required) there is considerable advantage in using woven fabrics as reinforcements. Some of the effects of this on strength and toughness have already been seen in previous chapters, and it would be expected that these effects might follow through into the fatigue response. Curtis and Moore (1985) showed that both the monotonic tensile strength and the fatigue strengths of CFRP laminates reinforced with woven fabric were significantly poorer than the properties of similar materials made with non-woven fibres, largely as a consequence of the distortion in the load-carrying 0° fibres and the stresses induced at weave cross-over points. However, when the fabric was oriented at 45° to the loading direction both tensile and fatigue performances were slightly better than for non-woven material.

6.4.2.1 Hybrid Composites

The simplest expectation for the fatigue stress/life curve of a hybrid laminate would be that it would fall between those of the two single-fibre components, but published results do not offer an obvious picture on which to base predictive models. It would be expected that the overall fatigue response would depend upon the extent to which the particular fibre mix controlled strain levels in the composites, and, if aramid fibres were involved, on the extent to which the intra-fibrillar weakness of that fibre was exposed to the stress. Some early experiments by Phillips *et al.* (1976) on woven-cloth GRP/CFRP laminates suggested that adding CFRP to GRP improves its fatigue resistance more or less in proportion to the amount of CFRP added, *i.e.* in accordance with the rule of mixtures. Hofer *et al.* (1978) also found that the fatigue stress of unidirectional HTS-carbon/S-glass hybrids obeyed the rule of mixtures when in the as-manufactured state, but showed a positive deviation from linear when the composites were hygrothermally aged. Fernando *et al.* (1988) found that the failure stresses for lives of 10^5 and 10^6 cycles for unidirectional carbon/Kevlar-49/epoxy hybrids were linear functions of composition for both repeated tension and tension/compression loading. But since the tensile strengths of the same series of hybrids were given by the failure-strain model (Chapter 4, section 4.19), the fatigue ratio (fatigue stress for a given life divided by the tensile failure stress) showed a marked positive deviation from the linear weighting rule. The same workers (Dickson *et al.*, 1989) found that unidirectional CFRP/GRP fatigue strengths fell above a linear relationship, and the fatigue ratio therefore showed an even more marked positive synergistic effect. This suggests that factors controlling monotonic tensile (and compression) failure do not necessarily continue to determine failure under cyclic loading conditions, and that for fatigue applications there appear to be positive benefits in using hybrids in place of single-fibre composites.

6.4.2.2 Short-Fibre Composites

Although the fatigue resistance of high-V_f composites reinforced with continuous, aligned, rigid fibres is usually high, short-fibre composites of all kinds are much less resistant to fatigue damage because the weaker matrix is required to sustain a much greater proportion of the fluctuating load. Local failures are easily initiated in the matrix and these can destroy the integrity of the composite even though the fibres remain intact. The interface region is particularly susceptible to fatigue damage since the shear stresses at the interface may be reversing their direction at each cycle and there is always a high shear stress concentration at fibre ends. It is also possible that in both random and aligned short-fibre composites the ends of fibres and weak interfaces can become sites for fatigue-crack initiation.

The beneficial effects of short fibres in materials like injection-moulding compounds have been well-known for 40 years or more, and although increases in strength in such materials are modest, increases in toughness are much greater for the reasons discussed in Chapter 5. Since fatigue resistance depends in part on strength and in part on crack resistance, it would therefore be expected that these benefits would be advantageously

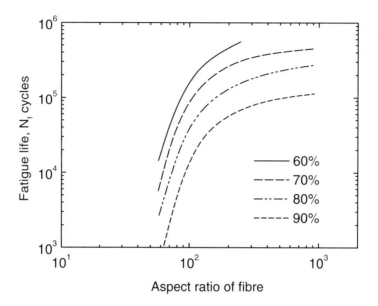

Fig. 6.15 Fatigue life versus aspect ratio for an applied stress of the indicated percentage of the composite tensile strength. The materials are short-fibre-reinforced boron/epoxy composites; repeated tension (Lavengood & Gulbransen, 1969).

translated into fatigue response. Lomax and O'Rourke (1966) reported that the endurance limit of polycarbonate at two million cycles is increased by a factor of seven when the polymer is filled with 40 vol.% of 6.4 mm long glass fibres. Lavengood and Gulbransen (1969) determined the number of cycles to failure of short-fibre boron/epoxy composites and found that for cycling at any given fraction of the failure stress the fatigue life increased rapidly with fibre aspect ratio, levelling off at l/d ratios of about 200. This therefore represents a critical aspect ratio above which the fatigue strength is a constant proportion of the flexural strength, as shown in Figure 6.15. These authors also studied the behaviour of aligned asbestos/epoxy composites and found them to behave in roughly the same way as boron/epoxy composites containing 25mm-long fibres. Owen *et al.* (1969) showed that the fatigue strength at a million cycles of polyester resin reinforced with random glass-fibre mat varied between 15 and 45% of the ultimate tensile strength. Huffine *et al.* (1973) studied the tension and flexural fatigue behaviour of glass/epoxy composites containing 44 vol.% of aligned E-glass fibres 12.5 mm long. Their results suggested that this material was potentially useful for dynamic structural applications since it appeared to have an endurance limit of 40% of the ultimate tensile strength.

    A comparison of the relative fatigue responses of composites containing continuous and discontinuous fibres is shown in Figure 6.16. The results are for unidirectional composites of the same carbon/epoxy system, XAS/914, in which the well-aligned prepreg

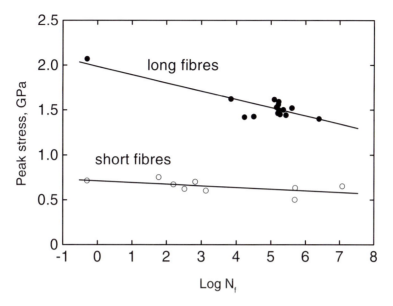

Fig. 6.16 Stress/loglife data for unidirectional composites of XAS/914 carbon/epoxy laminates reinforced with continuous and discontinuous fibres (R = 0.1) (Harris *et al.*, 1990).

for the short-fibre material, containing chopped fibres 3 mm long, was produced by the glycerol-alignment method described in Chapter 2: the fibre volume fraction is 0.35. The continuous-fibre laminate was produced from normal commercial prepreg. It can be seen that although the strength of the short-fibre composite is much lower than that of the continuous-fibre laminate, its fatigue response, as characterized by the slope of the $\sigma/\log N_f$ curve, is better. In fact, if the fatigue data are all normalized with respect to the monotonic tensile strengths of the two materials, the rate of fall-off is much greater for the conventional laminate. Working with similar aligned short-fibre composites, Moore (1982) also found that the fatigue behaviour of these materials compared favourably with those of continuous-fibre composites, although progressive cracking occurred from the fibre ends and fibre surface effects appeared to give more problems. It is assumed, therefore, that less fibre damage resulting from fatigue can occur if the fibres are already short, and the short-fibre ends apparently do not act as sources of excessive damage. It is not certain what the effect of fibre length would be in aligned composites such as these, but Hitchen *et al.* (1995) have shown that in random-short-fibre carbon/epoxy composites the fatigue life is independent of fibre length at any stress level. They also showed that the more flexible the matrix, the shorter the fatigue life. Friedrich *et al.* (1986) found that the incorporation of short glass or carbon fibres into PEEK to produce injection-moulding compounds produced only slight improvements in the fatigue and fracture toughness of this material.

Harris *et al.* (1977) studied the fatigue behaviour of commercial polyester DMC in both the dry and wet-aged conditions and for samples with drilled holes. Their results were characterized by a good deal of scatter, but indicated that in the low-stress (long-life) range the effects of moisture and stress concentrations were less marked than in the higher stress range. They showed that when their results were normalized by dividing by tensile strength, the $\sigma/\log N_f$ curve for the DMC fell only slightly more rapidly than the curves for other types of GRP (see Figure 6.9). Atzori *et al.* investigated the effects of a wide range of material and environmental variables on the fatigue behaviour of moulding compounds and developed a rule for estimating the fatigue strength from the monotonic tensile strength that was a modification of that of Mandell described in section 6.4.1.

In some cases the fatigue resistance of the matrix plastic has been found to be diminished by reinforcement with short fibres. For example, Shaver and Abrams (1971) found that randomly reinforced glass/Nylon composites had only half the endurance of the unfilled polymer. It should also be noted that the fatigue behaviour of most reinforced plastics will always be worse when the material is exposed to a wet environment.

6.4.3 OTHER FACTORS AFFECTING THE FATIGUE OF REINFORCED PLASTICS

6.4.3.1 Machines and Specimens

Testing machines of most kinds have been used for reinforced plastics, but testing rates must usually be carefully chosen to avoid excessive temperature increases caused by hysteretic heating. Heat dissipation by conduction is poor in many reinforced plastics, and even in CFRP it may be too low to prevent some temperature rise. In polymeric materials, quite small temperature increases may significantly change the mechanical properties, and excessive heating will also lead to thermal degradation. Recommendations of appropriate testing methods for reinforced plastics have been made by Owen (1970). Choice of appropriate specimen shape is strongly influenced by the characteristic nature of the composite being tested and has given rise to some difficulties. Early attempts to use test specimens similar to those used for metals led to unrepresentative modes of failure in composites that (a) were highly anisotropic and (b) had relatively low in-plane shear resistance, although techniques for testing most kinds of GRP had been early established by Boller (1959, 1964).

Much early work was carried out on waisted samples of the conventional kind, and designs were adapted for unidirectional CFRP and similar composites by using long samples with very large radii and relatively little change in cross-section in the gauge length. An alternative approach that has been used to prevent the familiar shear splitting in composites of low shear resistance (illustrated in Figure 6.3) was to reduce the sample thickness rather than the width, but this has now been almost universally superseded by the technique of using parallel-sided test coupons with end-tabs, either of GRP or soft aluminium, bonded onto the samples for gripping. Carefully done, this eliminates the risk of grip-damage (and resultant premature failure) without introducing significant stress concentrations at the ends of the test length and failures can usually be expected to occur in the test section.

Although sample size can have significant effects on the fracture toughness of metals it is not usually considered that fatigue tests on very small samples are likely to be unrepresentative of the material as a whole. In composite materials, however, the inhomogeneity resulting from the fibre distribution may be great enough to influence the fatigue response of a sample the size of which is comparable with the scale of the inhomogeneity. Thus, in woven-roving laminates, the width of a test piece should be sufficiently great to include several repeats of the weaving pattern. Similarly, a test piece for a hybrid laminate must be an order of magnitude or so greater than the scale of the hybridization.

6.4.3.2 Frequency and Related Effects

Testing has often been carried out at the highest testing rate that could be used without causing excessive heating of the sample. But even in CFRP, which are fairly good conductors, the occurrence of progressive damage, leading to lower stiffness and larger strains during a test, will give rise to hysteretic heating which may affect fatigue behaviour. In the presence of pre-existing design features like holes, or of growing cracks, high local temperature rises may occur at quite modest applied loads, as shown in Figure 6.17. The use of available $\sigma/\log N_f$ fatigue data for many reinforced plastics for design purposes is therefore complicated by uncertainty about the separate effects of frequency and autogenous heating of samples tested at different rates — an uncertainty that is exacerbated by the high degree of scatter normally found in fatigue-test results for these materials. Normalization of test data in terms of the monotonic tensile strength is also likely to lead to confusion because of the strong temperature- and rate-dependence of the monotonic mechanical properties of some reinforced plastics (particularly GRPs).

As mentioned earlier, Sims and Gladman (1978, 1980) have attempted to resolve the problem by studying the tensile and repeated tension (R = 0.1) behaviour of glass/epoxy laminates at constant rate of stress application instead of constant test frequency, a procedure which avoids the confusion of the varying loading rate in conventional constant-frequency sine-wave fatigue-test routines. By monitoring the specimen temperature rises during tests at different rates of loading, they were able to show that the temperature-dependence of the composite tensile strength transposed directly to the fatigue strength. They showed that fatigue tests at different rates could be normalized with respect to the tensile strength *measured at the same rate of stress application* to give a single, linear, master fatigue curve. The corrections make substantial differences to the interpretation of fatigue results, especially at longer lives. Their model is valid for materials of various constructions and is also able to take into account the effects of moisture in modifying the tensile properties.

6.4.3.3 Effects of Holes and Notches

There has been some disagreement about the effects of holes and notches on fatigue resistance of composites, perhaps because we are often tempted to generalize about the

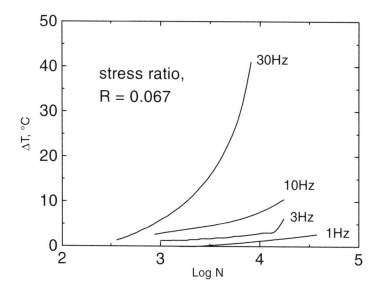

Fig. 6.17 Rise in sample temperature during cycling at different frequencies: 8-ply ±45 CFRP laminate (Sun and Chan, 1979)

class of materials we call 'composites' which, in reality, encompasses a wide range of materials of very different character. A brief survey of some of the available results will illustrate the danger in such generalisation.

Boller (1957, 1964) found that in GRP sharp notches were more detrimental than drilled holes, but that notches in general had little effect on fracture strength because of the large number of debonding sites present in the material to begin with. He also showed that small holes had no effect on long-life fatigue behaviour, presumably because progressive damage quickly removed the stress concentration due to the hole. Studies of the notch sensitivity of GRP by Prabhakaran and Sridhar (1975) confirm that values of the fatigue strength reduction factor, K_f, fell during cycling.

Owen and Bishop (1972), on the other hand, showed that holes are effective in initiating fatigue damage although they do not always affect final failure. They found that a circular hole acts as a stress concentrator at the onset of damage with reduction of the fatigue strength by a factor between 2 and 4 in many GRP materials and up to 10 in unidirectional GRP. They were also able to predict the fatigue stress for the onset of damage fairly accurately from a knowledge of the properties of plain samples and a stress analysis of the hole. Shütz and Gerharz (1977) showed that the notch sensitivity of $(0_2, \pm 45, 0_2, \pm 45, 90)_s$ carbon/epoxy laminates changed during cycling in such a way as to eliminate the stress concentration almost completely, whereas the opposite effect occurs during the fatigue cycling of metals. This progressive neutralisation of notches, which is often accompanied by an increase in residual strength, occurs by a variety of splitting mechanisms, especially longitudinal splits in 0° plies tangential to the hole. The accompanying reduction in notch sensitivity can even lead to fractures away from the cross-section containing the hole. This effect has also been observed in random, short-fibre composites such as polyester moulding compounds (DMC) (Harris *et al.*, 1977). More recent work on DMCs and SMCs by Atzori *et al.* (1994) also suggest that the fatigue properties of notched samples are similar to those of un-notched ones. Curtis and Moore (1985) showed that woven-fabric CFRP composites were slightly more notch-sensitive in fatigue than non-woven composites. Xiao and Bathias (1994) showed that the fatigue ratio of woven-fabric-reinforced glass/epoxy composites was unaffected by the presence of notches. And Kawai *et al.* (1996), comparing the behaviour of Nylon and epoxy composites reinforced with woven carbon fibres, found that un-notched samples of the carbon/Nylon had a higher fatigue strength that the carbon/epoxy material, whereas when notches were introduced, the relative performance of the two materials was reversed.

The crack-stopping ability of composites, which results from their inhomogeneity on a fine scale (the fibre/matrix interface) and on a gross scale (laminated structure) makes it difficult in many cases to apply a fracture mechanics approach to fatigue testing and design. The difficulty is most acute with unidirectional composites or laminates pressed from preimpregnated sheets of continuous and non-woven fibres. Stress concentrating effects of notches and holes may be almost completely eliminated by large-scale splitting in the $0°$ and $45°$ plies and by delaminations between the plies, the net result often being disintegration of the composite. In laminates containing woven cloth or CSM, however, or in mouldings containing random chopped fibres, the scale of this damage is limited by the structure of the composite and cracks and notches will often propagate in a more normal fashion, especially in wide plates. Owen and Bishop (1974) have shown that it is possible to use a fracture-mechanics approach, based on the Paris power law:

$$\frac{da}{dN} = A(\Delta K)^p \qquad (6.6)$$

relating crack growth rate, da/dN, to the range of stress intensity factor, ΔK, for some GRP laminates. The exponent, p, in this relationship was 12.75 for CSM/polyester composites and 6.4 for plain-weave fabric GRP, a substantial difference not accounted for by the different tensile strengths of the two materials. To overcome the problem of accurately defining crack length, a compliance calibration was used to determine an *equivalent* crack length, a procedure somewhat similar to that of using a plastic- or process-zone size correction when applying fracture mechanics to other materials. For constant load-range cycling Owen and Bishop constructed log (ΔK) versus N curves which were initially flat but turned up rapidly at certain critical levels. This instability defined a 'fatigue life' when the cyclic ΔK level reached the value of K_c for ordinary monotonic crack growth.

Underwood and Kendall (1975) also used a fracture-mechanics approach to study the growth of a damage zone during the fatigue of notched GRP laminates. They obtained much lower values of the Paris-law exponent, p, — 2.5 for 0/90 laminates and 4 for (0, ±60) laminates — much closer to the values normally obtained for metallic materials. They found that their results were best described by linear plots of ΔK versus cycles-to-failure which extrapolated back to the monotonic fracture toughness values. They observed that, depending on the type of material, large diffuse damage zones could increase the residual strength, while small, sharply defined zones would reduce it.

Vancon *et al.* (1984) have shown that for the very much simpler case of propagation of delamination cracks in multi-ply laminates, fracture mechanics methods and the Paris power law are straightforwardly applicable.

6.4.3.4 Effect of Stress System and the Stress Ratio, R

Although interpretation of the results of tests carried out under axial loading conditions (*i.e.* repeated tension, repeated compression, and mixed tension/compression) are perhaps the most straightforward to interpret, a good deal of testing has also been done in other modes, including shear (torsion) and flexure. What happens in any kind of test, however, depends sensitively on the structure of the composite and manner in which the applied load develops stresses within the material. In a unidirectional rod subjected to torsion, for example, it is the matrix and interface, both sources of weakness in reinforced plastics, that determine the fatigue response: the fibres play little part (Beaumont and Harris, 1972; Phillips and Scott, 1976). The situation is similar when a ±45 laminate is tested in repeated tension, as we saw earlier. In flexural tests, the stress distribution varies across the thickness of the sample and it would be expected that the performance of the material in this mode of testing would be controlled by the maximum tension and compression stresses on the sample faces. In early tests on high-stiffness materials (see Beaumont and Harris, 1972, for example) it was often observed that the flexural fatigue behaviour was markedly poorer than that obtained in repeated tension. This was presumably because in some of these tests, especially those conducted on small test pieces, significant levels of interlaminar shear loading could be generated, and the final fatigue failure of the sample was initiated prematurely by interface or matrix damage. In larger samples of more modern composites, the flexural fatigue response may be very similar to that obtained in tension, as illustrated in Figure 6.18, a further indication of the fact, mentioned in chapter 4, that compression failures rarely occur in the flexural loading of thin laminates, despite their known weakness under direct compression loading conditions.

The effects of direct compression loading are best illustrated by the use of stress/loglife curves at different R ratio, such as the family of curves in Figure 6.19. These median-life curves are plotted in terms of peak stress as a function of $\log N_f$ and compression stresses are represented as negative. Note that an R ratio of 0 represents cycling between 0 and some positive stress level (*i.e.* ideal repeated-tension, or T/T, cycling) but it is usual to retain a small positive load at the bottom of the cycle in order to avoid the possibility of any overshoot into compression: hence the use of an R ratio of 0.1 to represent T/T cycling. Similarly, an ideal repeated-compression cycle would have an R ratio of ∞, which complicates the mathematical handing of R ratios, so a small negative load is usually retained at the top of a C/C cycle: hence the R ratio of 10 in Figure 6.19. An R ratio of −1 represents a fully reversed T/C cycle.

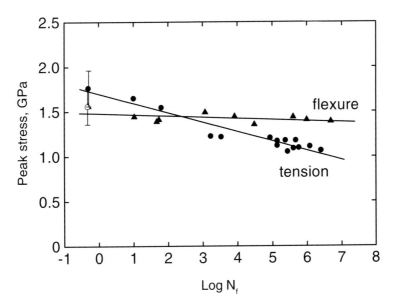

Fig. 6.18 Comparison of stress/loglife data for a unidirectional XAS/913 laminate tested in tension and flexure (Charrière, 1985).

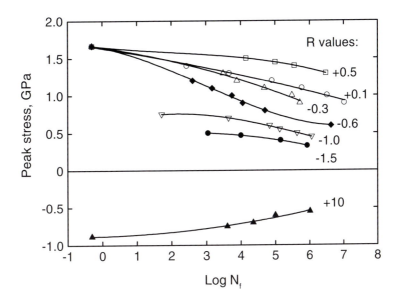

Fig. 6.19 Median stress/loglife curves for a $[(\pm45,0_2)_2]_S$ T800/5245 laminate at various R ratios (Gathercole *et al.*, 1994). The plotted curves are best-fit polynomial curves of second or third order.

It is interesting to note that the median s/logN$_f$ curves for repeated tension (T/T) and repeated compression (C/C) extrapolate back to the related monotonic tension and compression strength values, and when the two sets of median-life data are normalized with respect to those strength values there is a high degree of correspondence between them. As the compression component of cycling increases (R increasingly negative) the stress range, $2\sigma_{alt}$, to which the sample is subjected must at first increase, so that if the data were plotted as stress range versus loglife the data points for R = –0.3 would then apparently lie above those for R = +0.1. This is a familiar feature of fatigue in fibre composites, and it leads to the notion that some element of compression load in the cycle can apparently improve the fatigue response. It also results in a well-known aspect of composites fatigue, *viz.* that master diagrams of the constant-life, or Goodman, variety are displaced from the symmetrical location about R = –1 (*i.e.* the alternating-stress axis) (Owen & Morris, 1972; Schütz and Gerharz, 1977; Kim, 1988). It has recently been shown (Harris *et al.*, 1997) that the effects of R ratio can be illustrated by presenting the fatigue data on a normalized constant-life diagram by means of the function:

$$a = f(1 - m)^u (c + m)^v \tag{6.7}$$

where $a = \sigma_{alt}/\sigma_t$, $m = \sigma_m/\sigma_t$ and $c = \sigma_c/\sigma_t$ (definitions of σ_{alt} and σ_m were given in section 6.3.2, and σ_t and σ_c are the monotonic tensile and compressive strengths, respectively). The stress function, f, depends on the test material, and in particular on the value of the ratio σ_c/σ_t, and the exponents u and v, usually quite close to each other in value, are functions of logN$_f$. Since this is a bell-shaped function, it is more akin, as a failure criterion, to the old parabolic Gerber relationship (1874) than to the more frequently used linear Goodman law (1899).

The extreme values of *m* for a = 0 are 1 on the tension side of the ordinate and –c on the compression side: the mean stress range is thus (1 + c). The setting up of this parametric form of the constant life curve is in fact a double normalisation with respect to the monotonic tensile strength of the hybrid material. A family of curves representing different constant values of life, plotted as a three-dimensional diagram, is shown in Figure 6.20. There are naturally differences between the shapes of constant-life curves for different kinds of composite. Figure 6.21, for example, shows curves for CFRP (HTA/913) and GRP (E-glass/913) laminates of identical lay-up and matrix resin for a life of 10^5 cycles. It can be seen that the GRP curve is much more asymmetric than that for the carbon composite, and that its maximum, which is lower than that for the CFRP, lies to the left rather than to the right of the alternating-stress axis. These differences may all be ascribed to the difference in the reinforcing fibres.

6.4.4 DAMAGE ACCUMULATION AND RESIDUAL STRENGTH

The progressive damage which occurs in a composite sample during fatigue will affect the mechanical properties of the material to an extent which depends on the

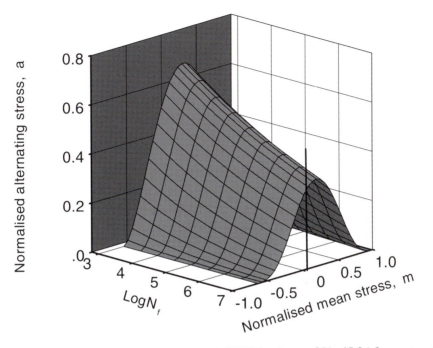

Fig. 6.20 Constant-life curves for a T800/924 CFRP laminate of $[(\pm 45,0_2)_2]_S$ construction (Gathercole *et al.*, 1994).

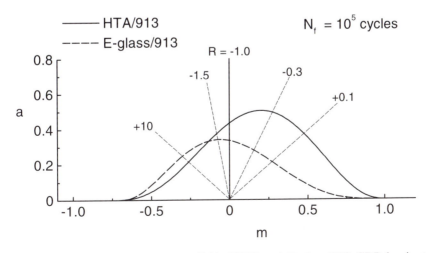

Fig. 6.21 Constant-life plots for HTA/913 CFRP and E-glass/913 GRP laminates of $[(\pm45,0_2)_2]_S$ construction for a life of 10^5 cycles (Beheshty *et al.*, 1999).

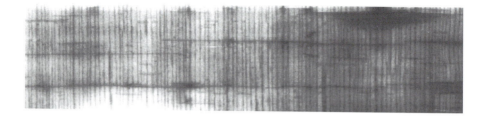

Fig. 6.22 X-ray photograph of a 0/90 carbon-fibre/PEEK composite after 8.5×10^5 fatigue cycles in repeated tension at a peak stress of 200 MPa (Dickson *et al.*, 1985).

composition and lay-up of the composite and on the mode of testing. In a unidirectional, high-modulus CFRP, for example, there may be no significant changes in composite stiffness prior to failure (Beaumont & Harris, 1972). By contrast, in cross-plied (0/90) GRP and CFRP laminates transverse-ply cracking occurs early in the life of the sample, causing a significant stiffness reduction, perhaps of the order of 10%. Thereafter, for the greater part of the life following this initial deterioration there may be little further change in stiffness (although longitudinal cracks may appear as a result of the lateral constraint imposed by the transverse plies) until close to failure when the elastic modulus falls rapidly (Reifsnider and Jamison, 1982: Poursartip and Beaumont, 1982). An illustration of the extent of this progressive damage is shown in the x-ray photograph of Figure 6.22, showing both transverse and longitudinal cracking in a 0/90 laminate of XAS-carbon/ PEEK after cycling in repeated tension. We note that even the use of a high-toughness matrix like PEEK does not eliminate this cracking. Boniface and Ogin (1989) have shown that this kind of transverse-crack growth can be satisfactorily modelled by the Paris relationship (equation 6.6) with exponents, p, between 2 and 6. When the reinforcement is in the form of a woven fabric, the occurrence of transverse-ply cracking is hindered, and the changes in stiffness accompanying cycling are less marked. And even in laminates containing only ±45 and 0° plies, there may be no fall in stiffness until near to failure, as illustrated by some typical experimental results for a $[(\pm45,0_2)_2]_S$ CFRP laminate shown in Figure 6.23. The microstructural damage events which lead to these changes are readily monitored by means of such methods as acoustic emission monitoring (Fuwa *et al.*, 1975; Chen and Harris, 1993), as shown in this figure.

We referred earlier to the 'wear-out' model of fatigue damage. It is a matter of common experience that both the rigidities and strengths of composite materials are affected adversely by repeated loading as damage accumulates in the material. In some circumstances, both strength and stiffness may initially increase slightly as the fibres in off-axis plies or slightly misaligned fibres in 0° plies re-orient themselves in the visco-elastic matrix under the influence of tensile loads, but the dominating effect is that due to the accumulating damage in the material. The deterioration during cycling usually depends on the stress as shown in Figure 6.24: the higher the peak stress, the earlier the

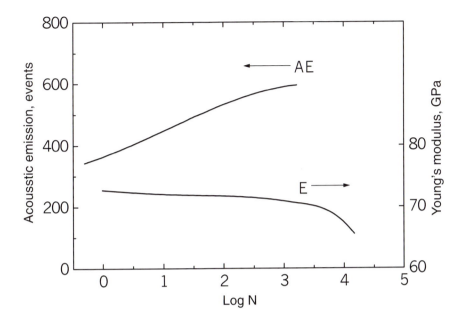

Fig. 6.23 Reduction in stiffness and accompanying acoustic emission in the early stages of fatigue cycling of a $[(\pm45,0_2)_2]_S$ XAS/914 CFRP laminate at a stress ratio, R = 0.1 (Chen & Harris, 1993)

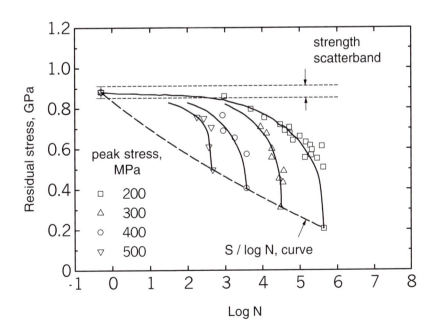

Fig. 6.24 Residual strength curves for samples of 0/90 GRP laminate subjected to fatigue cycling at an R ratio of 0.1 and various stress levels (Adam *et al.*, 1986).

deterioration starts. In accordance with the wear-out model, the end-point of each residual-strength curve lies on the normal $\sigma/\log N_f$ curve for the material.

In their work on the comparable group of CFRP, GRP and KFRP cross-plied laminates already referred to, Adam *et al.* (1986) found that the shapes of the residual strength curves for all three materials were similar and could be represented by an interaction curve of the form:

$$t^a + s^b = 1 \qquad\qquad (6.8)$$

where t is a function of the number of cycles, n, sustained at a given stress, σ_{max}, for which the expected fatigue life is N_f:

$$t = \frac{\log n - \alpha}{\log N_f - \alpha} \qquad\qquad (6.9)$$

The constant α is equal to $\log (0.5)$ and simply accounts for the fact that the normal strength of the material corresponds to the lower limit of the cycles scale at 1/2 cycle. The normalized residual strength ratio, s, is defined as:

$$s = \frac{\sigma_R - \sigma_{max}}{\sigma_t - \sigma_{max}} \qquad\qquad (6.10)$$

where σ_t is again the monotonic tensile strength of the material and σ_R is the residual strength after cycling at a peak fatigue stress σ_{max}. The exponents *a* and *b* are materials- and environment-dependent parameters which may have physical significance although they must be obtained by curve fitting. For rate-sensitive materials like GRP and KFRP, σ_t and σ_R must be determined at the same loading rate as that used for the fatigue cycling. The loss in strength due to cycling, $(\sigma_t - \sigma_R)$, may then be regarded as a damage function:

$$\Delta = 1 - s = \left(1 - t^a\right)^{\frac{1}{b}} \qquad\qquad (6.11)$$

The shape of the curve can, in principle, vary from a linear form (with a = b = 1), through a circular quadrant for powers of two, to an extremely angular variation for very high powers. Since both the gradual wear-out and 'sudden-death' type of behaviour can be accommodated within a single general model, a treatment of this kind has advantages over non-normalising procedures. Despite apparent mechanistic dissimilarities in the fatigue behaviour of the three types of composite studied in this work, the residual strength results suggested that a common mechanism of damage accumulation led to final failure since the damage law of equation 6.8, with suitably chosen values of *a* and *b*, fits all of the data for all three materials, as shown in Figure 6.25. Having established the values of *a* and *b* for a given material, the residual strength for any fatigue stress can then be evaluated:

$$\sigma_R = \left(\sigma_t - \sigma_{max}\right)\left(1 - t^a\right)^{\frac{1}{b}} + \sigma_{max} \qquad\qquad (6.12)$$

or, alternatively, the remaining life, $(N_f - n)$ can be estimated. It seems unlikely that a single mathematical model could cope with the whole range of composite types and lay-ups

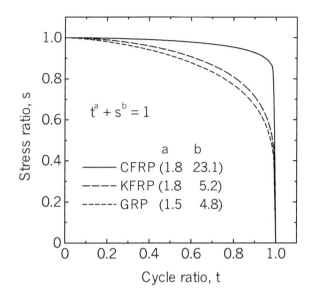

Fig. 6.25 Normalized residual strength curves for three comparable 0/90 laminates consisting of HTS carbon, Kevlar-49 and E-glass fibres in a common epoxy resin (Adam *et al.*, 1986). The values of the exponents for the fitted curves are shown in brackets.

available, and although this model requires validation for each system of interest — materials, lay-ups, stress systems — this validation would be reasonably economical of test time and material, given that the interaction curve could be deduced from relatively few data.

The foregoing discussion suggests that the changes in mechanical properties of fatigued composites may be caused by a single damage mechanism, but this is unlikely to be the case. Detailed studies of the damage processes occurring during cycling of CFRP materials (Chen & Harris, 1993) show that sequences of damage occur throughout life, and these sequences can be mapped, as shown in Figure 6.26, on a conventional $\sigma/\log N_f$ diagram. The supposition is that each damage-mechanism curve, including the final failure curve ($\sigma/\log N_f$ curve), represents part of an S-shaped decay curve of the kind postulated by Talreja (1981), although in the experimental window we see only a part of each curve. It is also supposed that at sufficiently low stresses, corresponding to the notional endurance limit, all curves will flatten out and converge at large numbers of cycles. One of the problems of developing damage-growth models for life-prediction purposes (see section 6.4.6) is that the damage development laws are likely to be different for each specific damage mechanism and will be structure-dependent, as illustrated by the results of crack-density measurements shown in Figure 6.27. The measurements were made on a T800/5245 CFRP laminate of $[(\pm45,0_2)_2]_s$ lay-up, and it can be seen that the development of cracks in the outer (unconstrained) and inner (constrained) 45° plies

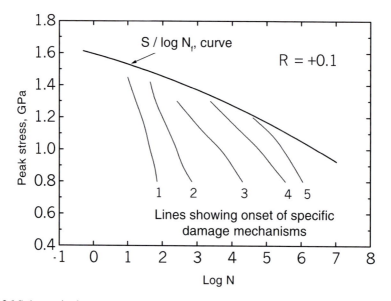

Fig. 6.26 Schematic damage mechanism maps for a T800/5245 $[(\pm45,0_2)_2]_S$ CFRP laminate tested in repeated tension fatigue (R = 0.1) (Chen & Harris, 1993). Key: 1 = fibre fracture in 0° plies; 2 = matrix cracks in outer 45° plies; 3 = fibre drop-out; 4 = matrix cracks in inner 45° plies; 5 = delamination at 0/45 interfaces.

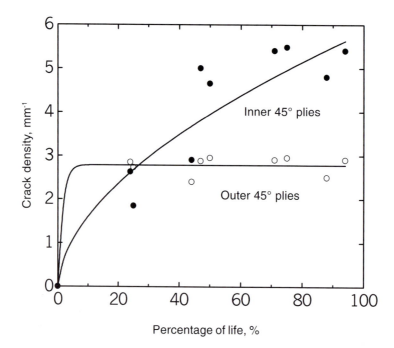

Fig. 6.27 Density of cracks in 45° plies in a $[(\pm45,0_2)_2]_S$ T800/5245 CFRP laminate during cycling at a peak stress of 1GPa and an R ratio of 0.1 (Grimm, 1996).

proceeds at quite different rates and that the saturation levels for the two types of crack are also quite different.

Howe and Owen (1972) studied the accumulation of damage during cyclic loading with the object of obtaining useful working relationships of the Miner-rule type that might be used in design. They used optical microscopy to study the development of debonding sites and resin cracks in CSM/polyester composites. They suggested that although debonding did not itself cause reductions in strength it served to initiate resin cracks which did weaken the material. For resin cracking they proposed a non-linear damage law, independent of stress level, which gives the damage, Δ, as

$$\Delta = \Sigma \left[A\left(\frac{n}{N}\right) + B\left(\frac{n}{N}\right)^2 \right] \tag{6.13}$$

where n is the number of cycles sustained by the composite at a stress level which would normally cause failure after N cycles, and A and B are constants. B is negative and Δ is equal to unity at failure. They used a modification of this law to predict residual strength after cycling of CSM/polyester laminates, this strength being dependent upon the growth of resin cracks.

6.4.5 FATIGUE UNDER CONDITIONS OF VARIABLE STRESS: BLOCK LOADING

Although studies of damage accumulation processes have been correlated moderately satisfactorily with fatigue-testing régimes involving a limited number of stress levels, it is even more necessary to make good correlations when variable-amplitude or flight-loading sequences are used for practical design purposes. Shütz and Gerharz (1977) have reported results of life-prediction experiments with HT- carbon/epoxy laminates of $0.4V_f$ and $(0_2, \pm 45, 0_2, \pm 45, 90)_s$ construction. They used flight-spectrum tests typical of the conditions at an aircraft wing root, and they also obtained conventional fatigue data over a range of different R values from which they made Miner-rule life predictions at two mean stress levels for comparison with the variable-amplitude test results. The linear-damage rule predicted lives up to three times those actually measured — far too wide a margin for comfortable design work — and they concluded that this was probably due to damage contributed by low-load cycles in the compressive region which is not accounted for in the Miner estimate.

In some more recent experiments, Adam *et al.* (1994) carried out some simple four-unit block loading sequences on samples of a $[(\pm 45, 0_2)_2]_s$ T800/5245 CFRP laminate. They first carried out a series of all-tension (TTTT) experiments for which the numbers of cycles at each of the stress levels were chosen so that the block should account for 20% of the lifetime of the specimen if Miner's rule were to be obeyed. Each unit in the sequence should therefore have contributed 5% of the total damage to failure. Since the contribution to the total life of the specimen by each of the units within the block programme is 5% of the median life, each block of four units therefore had a notional Miner number, M, of 0.2, *i.e.*

$$\frac{n_1}{m(N_1)} + \frac{n_2}{m(N_2)} + \frac{n_3}{m(N_3)} + \frac{n_4}{m(N_4)} = M = 0.2 \tag{6.14}$$

where n_i is the number of cycles in the unit and $m(N_i)$ is the median life at that particular stress level. If Miner's rule is followed, the block should repeat itself 5 times before failure occurs. Following the first all-tension (TTTT) series, the next stage was to introduce a compressive element into the loading sequence so as to construct TTCT blocks.

It is logical to assume that if the life data for a given constant stress level are randomly distributed then the Miner sum, M, defined by equation 6.14, should be similarly distributed, regardless of whether the mean value or median value of M is equal to unity. The Miner sums from all six groups for each sequence were therefore pooled and analysed in terms of the Weibull model, discussed earlier, which showed that there was a marked difference between the mean values of the Miner number for the two groups of tests, 1.1 for the TTTT sequence, and 0.36 for the TTCT sequence. To all intents and purposes, then, the Miner rule is valid for all-tension block-loading sequences for this material, and the effect of introducing a single compression block into an otherwise all-tension group was clearly highly damaging, Miner's law no longer being obeyed and by a substantial margin, the mean life having been reduced by some 60% as a result of the substitution. Similarly, the substitution of a repeated tension unit into an all-compression sequence also resulted in a marked reduction in the mean Miner sum. It therefore appears to be the stress reversal as such which is damaging rather than the actual sequence or the actual stress levels. This has serious consequences for designers since most practical cycling régimes, are likely to include stress excursions into both tension and compression. This is certainly true of the familiar 'FALSTAFF' and other load-spectrum sequences used in the laboratory to represent actual life conditions for aircraft components.

6.4.6 LIFE PREDICTION

There have been many attempts to develop life-prediction methods for reinforced-plastics, but the complexity of the response of these materials to stress makes it very difficult to develop universally applicable models. Some of the established methods are based on structural models, while others are less specific. Most of the structurally based procedures depend on modelling the degradation of material properties as a consequence of microstructural damage (e.g. Lee et al., 1989; Hwang and Han, 1989) and many are based on the concept of a 'characteristic damage state' (CDS) developed by Reifsnider and co-workers (see, for example, Reifsnider et al., 1979). Many of these methods involve both statistical and laminate-theory analyses and give reasonably good predictions. But many require a good deal of prior data determination and are structure-specific. It is beyond the scope of this book to treat this subject in any detail, but we shall refer to two recent approaches which appear to have some generality.

The first is based on the constant-life analysis referred to in section 6.4.3.4. Equation 6.7 is a parametric description of the stress/R-ratio/life 'environment' for composite fatigue which appears to be valid for a wide variety of material types and structures (Harris *et al.*, 1997). In this equation:

$$a = f(1 - m)^u (c + m)^v$$

of which the terms have already been defined, the shape and location of the fatigue 'surface' are determined by four parameters, the compression and tensile strengths (which define c), the exponents u and v, and the factor f. Since c is a material property and the alternating and mean components of stress (a,m) are normalized with respect to the tensile strength, u, v and f must be obtained by a parametric analysis of fatigue data sets. The exponents u and v (both functions of fatigue life) are usually very similar in value, unless the constant-life curves are very asymmetric, and for a wide range of CFRP materials are close to 2. And the parameter f is an inverse-power function of the ratio of the compression strength to the tensile strength, *i.e.* of c, of the form (Beheshty and Harris, 1999):

$$f = Ac^{-p} \tag{6.15}$$

where A and p are functions of life. The (a,m) data may be obtained from median-life fatigue results, or, by pooling fatigue data sets as described earlier to obtain valid statistical information, they may be chosen to represent any required failure probability. If no information other than the tensile and compression strengths is available for a given composite, a preliminary attempt can still be made to predict stress/life curves on the basis of accumulated data for other related materials. A limited fatigue-test programme can then be carried out to provide stress/life data at three or four stresses at, say, three R ratios and the predictions of the model will become gradually more refined and reliable. The virtue of this method is that it can provide designers with useful design information at very early stages in the development of new materials, before detailed descriptions of the fatigue response have become available, as shown by Harris *et al.* (1997). While this may well be all that is necessary for non-critical applications, in the case of critical applications such as aircraft structures it would of course always be necessary to confirm predictions from any model before a design could be validated. The procedure described above requires data specific to a given type of reinforcing fibre. It appears from these results that stress/life curves for materials damaged by low-velocity impact can also be predicted from data for the undamaged laminate provided the changes in the tensile and compression strengths caused by the impact damage are known (Beheshty & Harris, 1999).

Although a high level of understanding of the mechanisms of fatigue damage accumulation and the effects of this damage on the ability of a material or structure to sustain the service loads for which it was originally designed is of importance to materials developers, it must be said that the designer himself has little interest in the actual physical mechanisms of degradation. What he needs is simply a way of analysing available data to predict the likelihood of failure under a specified set of conditions. In recent years artificial neural networks (ANNs) have emerged as a new branch of computing suitable for applications in a wide range of fields. They offer a means of dealing with many multi-variate properties for which there is no exact analytical model and fatigue seems to be the kind of materials property that is suitable for ANN analysis. Neural networks provide a compact way of coping with the large amounts of characterisation data generated in the study of a multi-variate dependent property such as fatigue. They also provide a very simple means of assessing the likely outcome of the application of a specified set of conditions, precisely what is required by a designer who needs to make safe use of complex fatigue data for complex materials like composites.

An ANN is a type of computer programme that can be 'taught' to emulate the relationships between sets of input data and corresponding output data and the results of a characterisation programme can therefore be 'learnt' by the network and patterns in the data can be established. Once the network has been 'trained' it can be used to predict the outcome of a particular set of input parameters. Unlike other types of analysis there is no need to know the actual relationship between inputs and outputs and the question of determining physical laws and relationships need not be addressed. The network simply compares patterns of inputs with those it has been 'taught' and provides an output which takes full account of its accumulated 'experience'. It is a highly developed interpolation system and its outputs provide a good indication of the certainty of an outcome, a property which may be a useful indicator of where additional experimental information is required. The ability to signal the 'certainty' of the outcome is particularly important for materials like composites for which the fatigue response is often highly variable even when the monotonic strength properties of the same material may be quite closely definable.

As we have seen, fatigue in composites involves many variables, including a variety of materials characteristics (*e.g.* fibre/resin mix, lay-up, moisture content) as well as the specific fatigue variables (mean and alternating stresses, variable stress characteristics, *etc.*), all of which may be treated as inputs to an ANN. One of the first questions that arises relates to the issue of what constitutes 'failure'. It is more likely to be the minimum life in a data set than the median or mean life, and it is important to decide whether the input should simply be the lowest value of life recorded in a data set, or some value estimated from the application of extreme-value theory (Castillo, 1988), as discussed earlier. Some recent results obtained by Lee *et al.* (1995, 1999) suggest that the ANN approach may provide a useful method of fatigue data analysis and life prediction.

6.5 FATIGUE OF METAL-MATRIX COMPOSITES

Much early work on MMCs related to metals reinforced with high-strength metallic wires like tungsten and cold-drawn steel. These were essentially model composites, and were rapidly supplanted as experimental materials by modern aluminium and titanium alloys mostly reinforced with boron, carbon and SiC fibres of various kinds. Nevertheless,

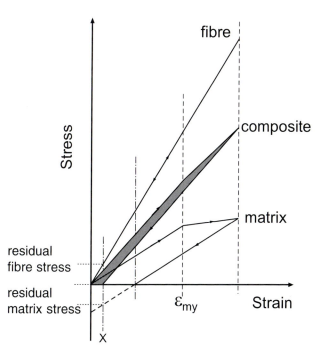

Fig. 6.28 Schematic illustration of the first tensile stress/strain cycle of a metal-matrix composite loaded beyond the matrix yield stress, *i.e.* into the elastic-plastic range.

some of the early results provide useful insights into the fatigue response of this kind of material. It will be appreciated that in a metal/metal system, both components will be susceptible to conventional fatigue failure, whereas in a metal/ceramic system, only the matrix will be susceptible. However, since metallic fatigue is essentially a brittle-failure process, the difference may not be great.

When a composite with a low-yield-stress matrix is cycled in repeated tension, the matrix will be subjected to a more complex stress régime than the level of external applied stress would suggest. During the first cycle, shown schematically in Figure 6.28, in the régime where both fibres and matrix deform elastically, the initial elastic modulus of the composite, usually designated E_I, is given by the normal rule of mixtures. At the yield point of the matrix, at the strain ε_{my}, the composite deforms elasto-plastically with a somewhat lower stiffness (a constant E_{II} if we assume linear work-hardening behaviour in the matrix) while the fibres continue in their elastic state and the matrix deforms irreversibly. On reversing the sense of the applied deformation, the composite describes a hysteresis loop, as shown, while the matrix and fibres unload elastically. The matrix reaches a zero-stress state at a finite level of strain, and the stresses in the constituents will reach an equilibrium state at the point X, the fibres being in tension and the matrix in compression, as shown, and the composite is left with a permanent strain.

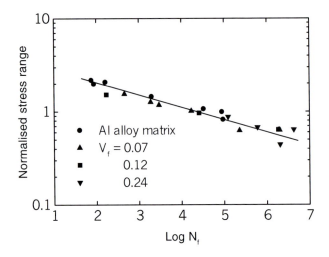

Fig. 6.29 Normalized stress/loglife data for an Al/4%Cu alloy and three composites containing tungsten wire reinforcement (McGuire and Harris (1974). The data are for reversed axial tension/compression cycling (R = -1) and the stress ranges for the data points are normalized with respect to the tensile strength of each material.

Further cycling in repeated tension will result in entirely elastic loading and unloading of both fibres and matrix from this new stress state (Dvorak and Tarn, 1975). However, if the stress cycle is fully reversed (*i.e.* for R = –1), with the matrix stress exceeding the matrix yield stress in each direction, energy will continue to be dissipated in the material until the plastic strain range has been reduced as a result of cyclic work-hardening. The effect of the elasto-plastic stress cycle in the matrix may be substantial and could lead to over-ageing in a peak-aged aluminium matrix, for example, perhaps resulting in more rapid work-hardening and crack growth than would occur in purely elastic cycling. Thus, although it might at first sight be expected that in a metal/metal system the fatigue response should be predictable by a mixtures-rule model for the fatigue stress for a given number of cycles:

$$\sigma_c(N_f) = \sigma_f(N_f)V_f + \sigma_m(N_f)(1 - V_f) \qquad (6.16)$$

the low-cycle fatigue of the matrix may prevent this. McGuire and Harris (1974) studied the fatigue response of unidirectional composites consisting of a tungsten-wire-reinforced aluminium alloy (Al/4%Cu) produced by vacuum diffusion bonding, and their results show the effects of heat treating the matrix alloy to peak hardness or over-ageing it to produce a brittle matrix state. Their σ/logN$_f$ curves showed that increasing V$_f$ increased the fatigue resistance of the composite, and this increase in fatigue resistance seemed to be a direct result of the increased strength of the composite since if their stress/life data for each composite composition are normalized by dividing the cyclic stress-range values by the appropriate sample tensile strength the points for all of the composites and the

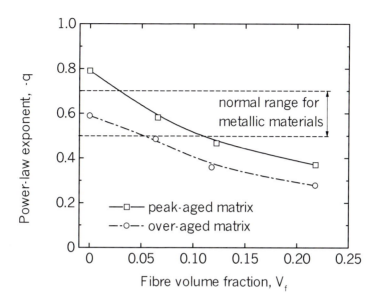

Fig. 6.30 Variation of the Coffin-Manson exponent, q, with fibre content for W/Al-Cu composites in low-cycle fatigue (McGuire & Harris, 1974).

unreinforced matrix fall on a single line which appears to obey Basquin's Law very well (Figure 6.29). Both matrix and fibres exhibit distinctive signs of fatigue failure, as shown in Figure 6.2, in clear contrast with the normal tensile fracture mode of both constituents. The composites with the higher cyclic work-hardening rate (peak aged) possessed the better fatigue resistance, but the fatigue stresses (for $N_f = 10^7$ cycles) for both heat-treatment conditions fell well below the mixture-rule prediction of equation 6.16. In other metal/metal systems including steel/silver and tungsten/copper (Courtney & Wulff, 1966) and tungsten/silver (Morris & Steigerwald, 1967) it was shown that the fatigue ratio actually decreased with increasing fibre content.

McGuire and Harris also carried out low-cycle fatigue experiments at constant total strain range, and in these tests the composites showed work-hardening and work-softening characteristics that were affected by matrix plastic properties. The observed changes did not indicate any difference in the nature of deformation, however, and the exponent, q, in the Manson-Coffin law relating plastic strain range to the number of cycles to failure,

$$\Delta\varepsilon_p N_f^q = \text{constant} \tag{6.17}$$

fell smoothly with increasing V_f from values characteristic of conventional metallic materials, of the order of 0.5, to something like half that value at $V_f \approx 0.2$, as shown in Figure 6.30. This indicates that the improvement in fatigue properties results from the effect of the reinforcing fibres in limiting the composite plastic strain range of the matrix.

The matrix plastic strain range has been identified by several workers as being the key parameter determining the fatigue characteristics not only of metal/metal systems, but also of metals reinforced with brittle ceramic fibres like carbon and silica. The work of Baker and co-workers, for example (Baker, 1966; Baker *et al.*, 1966), suggested that in both types of system, fatigue failure occurred mainly by the growth of matrix or interface cracks, linking *via* fractured fibres in the case of brittle-fibre systems, and that the main effect of the reinforcing fibres in improving fatigue resistance was to reduce the matrix plastic-strain range. Courtney & Wulff (1966) invoked a fatigue model due to Orowan (1939) which also ascribed the fatigue failure (of W/Cu and steel/Ag) to matrix cracking resulting from cyclic work-hardening as a result of the elastic-plastic deformation described above. Lee and Harris (1974), working with tungsten/copper composites, commented on an associated effect, recalling the work of Kelly & Lilholt (1969) who showed that during the deformation of an MMC, the stress levels in the matrix were often considerably greater than would be predicted by the rule of mixtures, probably because of the highly perturbed and constrained state of stress in the matrix adjacent to the fibres. The difference between the resulting enhanced matrix flow stress and the unconstrained matrix yield stress increases as V_f increases and as the fibre diameter decreases. The result during fatigue is to reduce both the matrix ductility and the matrix plastic strain-range, and thereby increase the composite life. By contrast with some of the other results referred to above, they showed positive deviations from the mixtures rule between 4 and 20%, the greater deviations being associated with finer reinforcing filaments.

Leddet and Bunsell (1979), studying the fatigue of boron/aluminium-alloy composites in rotating bending, have shown that fibre/matrix interactions caused by heat treatment change the mechanisms which lead to progressive deterioration during cycling. They suggested that with a soft matrix the predominant damage mechanism is fatigue cracking in the matrix parallel to the fibres and originating from initial fibre fractures. In a slightly harder matrix cracks propagate through the matrix and around the fibres without causing fibre failure, but in a highly work-hardening matrix successive fibre failure can occur. The fatigue behaviour of B/Al composites depends on the intrinsic matrix mechanical properties but is controlled by the fibre strength.

By contrast with reinforced plastics, crack growth from pre-existing notches often occurs normally in composites MMCs. Harris *et al.* (1977) showed this to be the case with composites of V_f up to 0.22, but in samples with 37 vol.% of reinforcement the closely-spaced fibres led to the development of a highly localized plastic zone parallel with the fibres that inhibited planar crack growth and encouraged the same kind of longitudinal splitting that occurs in unidirectional reinforced plastics. For their W/Al materials, the crack growth rate, da/dN, correlated well with a stress-intensity-range parameter, similar to ΔK, of the form $\sigma\sqrt{a}$, following the Paris Law (equation 6.6), as illustrated in Figure 6.31. The slopes of the curves, p, for different composites were identical with that for the unreinforced matrix, about 5. These results again indicate that fatigue response was determined simply by the strength and work-hardening ability of the matrix, without significant differences in the controlling damage mechanism. It is clear that the matrix alloy work-hardens sufficiently to cause fibre failure, but by a conventional fatigue mechanism rather than brittle (tensile) failure. Similar effects of work-hardening behaviour would not necessarily have occurred had the fibres been non-

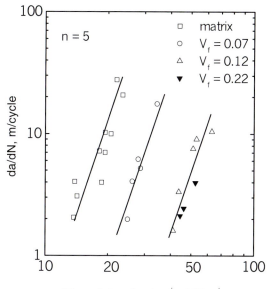

Fig. 6.31 Paris law plots for fatigue crack growth in W/Al-Cu composites (Harris *et al.*, 1976). The average slope of the curves is about 5.

metallic. A comparison between the results shown in Figure 6.31 and the stress-intensity factor measurements for crack propagation in the same materials under monotonic loading shows the extent to which cyclic loading conditions are more damaging (Figure 6.32).

Crack propagation in some unidirectional short-fibre Cu/W composites shows slightly different features (Figure 6.33). In this case, the slope for the matrix is much higher than that for the Al/4%Cu alloy in Figure 6.31, about 10 instead of 5, and the intrinsic fatigue resistance of the copper is thus much lower. Addition of a small volume fraction of tungsten filaments nevertheless reduces the slope to the same level as that for the Al/W composites. The curious feature of this graph, however, is that for initial cracks below some threshold level, the short fibres actually cause the growing cracks to slow down as their lengths increase before conventional Paris-law behaviour is established. Similar results have been reported by Bowen (1993) and Barney *et al.* (1993) for titanium-alloy MMCs and by Pegoretti & Ricco (1999) for composites of polypropylene reinforced with short glass fibres. Bowen suggests an approach to the prediction of this crack-arrest mechanism.

6.6 FATIGUE OF CERAMIC-MATRIX COMPOSITES

Although it has traditionally been considered that ceramics do not suffer from fatigue, this view may have arisen partly because of the difficulty of carrying out fatigue tests on

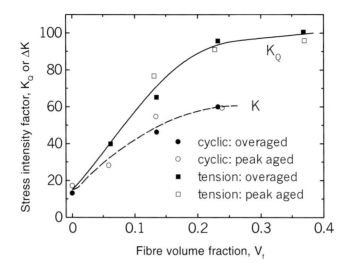

Fig. 6.32 Variation of effective stress intensity factor with fibre content for monotonic and cyclic loading conditions (McGuire and Harris, 1974 and 1976).

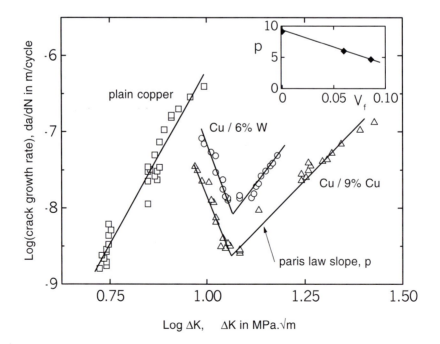

Fig. 6.33 Paris-law plots for fatigue crack propagation in composites consisting of short, aligned tungsten fibres in copper (Harris and Ramani, 1975).

brittle materials and partly because fatigue phenomena may easily be obscured by the high level of variability associated with brittle failure. Certainly, fatigue crack growth phenomena have been observed in a variety of brittle materials and a range of mechanisms, including microcracking, small-scale plasticity resulting from dislocation movements and stress-induced phase transformations, have been invoked to explain the observations (Suresh *et al.*, 1988). Nevertheless, although fine-scale effects such as these will certainly occur in ceramic-matrix composites, especially those reinforced with particulate fillers or fine whiskers, the results of fatigue in continuous-fibre CMCs are on a much larger scale and are due largely to cracking. Ritchie *et al.* (1993) have suggested that cyclic fatigue in low-ductility materials such as these is conceptually distinct from metallic fatigue, and is largely driven by cycle-dependent suppression of some of the crack shielding mechanisms that normally operate in monolithic ceramics.

In the reinforcement of metals and polymers, the modulus ratio, E_f/E_m, is usually high — 100 or more in high-performance reinforced plastics, perhaps less than 10 in MMCs — whereas in CMCs there may be little or no difference in the stiffnesses of the components. Fibres are added to ceramics to improve the toughness and reduce the variability, rather than to improve the rigidity. Thus, although, as we saw in chapter 4, CMCs can be made to support high stresses and many have considerably greater toughness than monolithic ceramics, they achieve this at a price, namely, that extensive cracking occurs in the very early stages of loading. In a 0/90 laminate of a GRP or CFRP laminate, the transverse plies are sources of weakness, and these crack easily at low loads when the stress exceeds their notional transverse strength which may be even lower than the normal tensile strength of the resin matrix. But since, in CMCs, the modulus ratio may be close to unity, so that there is no stiffening effect, the matrix even in a unidirectional composite must crack at composite stresses which correspond to very low loads in the matrix compared to the load-bearing ability of the fibres, as shown by the Aveston-Cooper-Kelly (ACK) model described in chapter 4. For the SiC/CAS composites used as an illustration of the ACK model, it was shown that the model predicts that matrix cracking should start at about 135 MPa (about one third of the composite tensile strength). However, the monitoring of tensile tests by means of acoustic emission (AE) sensors shows that microfailure events can be detected almost as soon as loading begins, certainly at axial strain levels less than 0.05% (Harris *et al.*, 1992). Two further comparisons with a 0/90 GRP laminate now invite themselves. The cracking of the 90° plies in a 0/90 GRP composite reaches its peak rate (as shown by AE activity) at a strain of about 0.7% (Harris *et al.*, 1979), and if the stress is removed and the composite reloaded no new cracking activity occurs until the previous stress level has been reached (Fuwa *et al.*, 1975). By contrast, if a *unidirectional* SiC/CAS composite is reloaded a number of times, the microcracking activity continues to occur even at the lowest load levels. The potential problems of this early matrix cracking were apparent from the earliest days of research into fibre-reinforced ceramics and glasses (Phillips, 1988).

Thus, although a cycled 0/90 plastic laminate loses a little of its initial stiffness early in its fatigue life, this transverse ply cracking — shown in the x-ray photograph of Figure 6.22 — does not seriously impair the fatigue life of the laminate, the damage

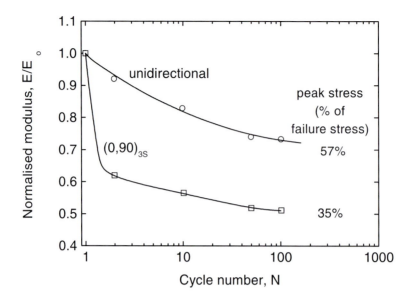

Fig. 6.34 Effect of repeated tension cycling on the elastic moduli of unidirectional and cross-plied SiC/CAS composites. The composite tensile strengths were 334 MPa (ud) and 210 MPa (0,90). (Habib *et al.*, 1993)

sustained in the CMC leads to continuing reductions in stiffness and residual strength. The changes in the stiffness during a few hundred repeated tension cycles of a unidirectional and a $(0/90)_{3s}$ SiC/CAS laminate are shown in Figure 6.34. The elastic modulus of the unidirectional composite falls to about 75% of its initial value by the 100th cycle, while that of the $(0,90)_{3s}$ material falls by 40% on the initial cycle, and thereafter falls more slowly, and at about the same rate as that of the unidirectional composite, to about 50% of its initial value. Prewo (1986) reported similar stiffness losses during repeated loading of SiC/LAS composites, and Sorensen *et al.* (1993) show how frictional effects related to interfacial sliding result in tangent moduli that are different on loading and unloading. It appears that the interfacial shear strength decreases during cycling and that this results in a reduction in the efficiency of stress transfer into the bridging fibres, with consequent deterioration of fatigue resistance.

Since it is easier to carry out flexural fatigue experiments on ceramics than axial tests, much of the available data for CMCs are of this kind. For example, strain/loglife results for a $(0,90)_{3s}$ material are shown in Figure 6.35 and associated residual strength and stiffness curves are shown in Figure 6.36. The residual stiffness curves of Figure 6.36 show that for strain levels between 0.45% and 0.84%, there is an initial reduction in stiffness, at a reducing rate over the first 100 or so cycles, which is almost the same whatever the strain level. Thereafter, the curves become increasingly

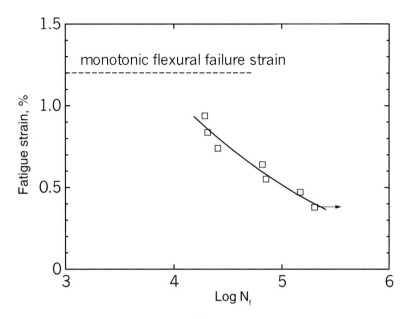

Fig. 6.35 Strain/life curve for a $(0/90)_{3S}$ SiC/CAS composite obtained in constant-moment flexural loading (R = 0), (Habib *et al.*, 1993)

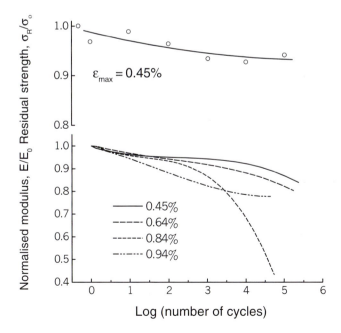

Fig. 6.36 Residual strength and stiffness of $(0/90)_{3S}$ SiC/CAS composites during flexural cycling (Habib *et al.*, 1993).

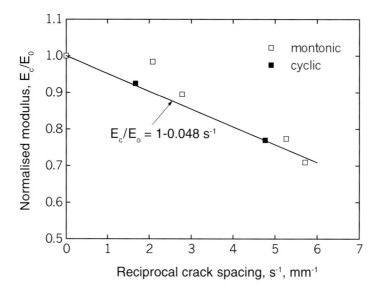

Fig. 6.37 Relationship between crack spacing and elastic modulus for unidirectional SiC/ CAS composites loaded monotonically and cyclically in tension (Habib *et al.*, 1993).

separated, all taking on a marked downward curvature the extent of which is now greater the higher the fatigue strain. However, at the lower two strain levels, the observed reductions are of the order of only 15% by 10^5 cycles, but at 0.84% strain, the curve is dropping very rapidly after only about 1000 reversals. At the higher strain of 0.94%, about 80% of the monotonic failure strain, the initially more rapid reduction in stiffness is followed by a gradual reduction in the rate of loss of stiffness and as a result of the onset of delamination the curve then crosses over the rapidly falling line for cycling at 0.84% strain. The effect of this delamination would be to reduce the stiffness of the compression side of the sample and shift the neutral axis, thereby lowering the effective tensile stress on the tension face and prolonging the life. The loss of flexural strength after cycling a sample to a peak strain of 0.45%, equivalent to a peak tension-face stress of about 200 MPa, for different periods of time was only of the order of 6%, by comparison with a loss of stiffness of 15%, and the residual strength curve appears to have flattened out after about 1000 cycles whereas the stiffness loss begins to accelerate after this point. This pattern of behaviour is somewhat different from that which is usually found in structural composites in which there is often a proportional change in strength and stiffness implying a constant failure strain.

The damage mechanisms contributing to these changes appear to be of the same kind as those which occur during monotonic loading to failure. This has been shown by the extensive AE studies of Harris *et al.* (1992) and also by measurements of the spacing of matrix cracks in samples deformed in both monotonic and cyclic loading (Figure 6.37). A simple relationship between the stiffness of a unidirectional composite and the reciprocal crack spacing may be expected, as can be demonstrated by an argument similar to that used to obtain the strength of a composite reinforced with aligned, short fibres. When a resin-based composite containing strong, brittle fibres is loaded, the fibres break into shorter and shorter lengths until, when they reach the critical length, they are just too short to be broken down any further and the maximum tensile load that can be reached in the middle of the fibre is just below the fibre strength. The *mean* load supported by the broken fibres is then only half the fibre breaking load, therefore. In using the rule of mixtures to determine the strength of the composite we apply the efficiency factor, $(1 - \ell_c/2\ell)$, to the fibre stress component, as shown in section 4.1.8 of chapter 4. This ensures that when the fibre is unbroken the efficiency factor is unity, and when the fibre is at the critical length, its value is 0.5. For a brittle-matrix/brittle-fibre composite in which the matrix has a failure strain lower than that of the fibres, the inverse situation occurs and the matrix breaks into increasingly short blocks until the critical spacing is reached, at which point the blocks are then too short to allow the transferred stress to break them down any further. We have seen that this critical spacing is close to 0.2 mm, for both monotonic and cyclic loading, in SiC/CAS composites. The short-fibre argument is not, of course, restricted to fibres: the model makes no assumptions about the composite geometry. We may, therefore, write the modified rule of mixtures:

$$E_c = E_f\,V_f + E_m\left(1 - \frac{s_c}{2s}\right) \tag{6.18}$$

where s_c is the limiting value of the matrix crack spacing, s. Inserting known data for this SiC/CAS material ($E_f V_f = 67$ GPa; $E_m V_m = 63$ GPa; and $s_c = 0.2$) and dividing through by the expected value of E_c, 130 GPa (given by the unmodified rule of mixtures) in order to normalize the model, as in Figure 6.37, we have:

$$E_c/E_o = 1 - 0.048\ s^{-1}\ (\text{for } s > s_c) \tag{6.19}$$

and when s reaches s_c, no further reductions in stiffness occur. This relationship coincides almost exactly with the best-fit line through the actual measurements. The two sets of data included in Figure 6.37 are sufficiently similar to suggest that the damage mechanisms occurring during repeated loading and incremental loading are the same, a suggestion which is corroborated by the similarity of the acoustic emission patterns obtained from the two kinds of experiment.

The results shown above and those for SiC/LAS composites reported by Prewo (1987) show some interesting similarities and differences. For the repeated tension cycling of unidirectional composites of Nicalon-reinforced LAS-I glass ceramic, for example, for peak stresses close to the failure stress the $\sigma/\log N_f$ curve was very flat, and there was no loss in residual strength or stiffness after 10^5 cycles, even for cycling at about 80% of the tensile strength. For Nicalon/LAS-II composites fatigued in repeated tension, below the knee of the stress/strain curve there was no loss in either stiffness or residual strength up to 10^4 cycles and the stress/strain loops showed no hysteresis. Above the knee (or, more accurately reporting Prewo, above the 'proportional limit' of his stress/strain curves) no fatigue failures occurred, but there were marked changes in the shapes of the hysteresis loops. The so-called proportional limit appeared to be reduced as cycling proceeded, from about 300 MPa on the first cycle to about 120 MPa by the third, and the initially wide damping loop rapidly became narrower. However, the initial tangent modulus on each cycle remained the same as that of the uncycled material. When SiC/LAS-II samples were cycled in bending, the $\sigma/\log N_f$ was again very flat, almost coinciding with the scatter band of monotonic strength measurements, by contrast with Figure 6.35, and there was again no loss in residual strength. Furthermore, the fatigue failures were all initiated on the compression side of the test piece. It is very difficult to reconcile these marked differences in behaviour of two relatively similar types of composite. The basic properties and damage mechanisms of the two types of material do not differ appreciably, and even though in some cases we are comparing unidirectional and cross-plied materials our own results suggest that this difference in lay-up does not greatly influence the overall response of the composite to mechanical deformation. It does appear from Prewo's published fractographs, however, that the scale of fibre pull-out is significantly greater than that observed by Harris et al. (1992) and it may well be that it is the nature of the fibre/matrix bond which is responsible for the observed differences in fatigue response.

Solti et al. (1996) have recently developed a simplified model for the fatigue life of unidirectional CMCs. The model is a modified shear-lag analysis in which the microstructural damage is estimated from simple damage criteria. It takes into account a number of damage mechanisms, including matrix cracking, interfacial debonding and slip, fibre fracture and fibre pull-out. The model estimates the average matrix crack density, discussed above, and predicts the stress/strain hysteresis and racheting described by several authors, as well as the fatigue response. The predictions of the model are in good agreement with experimental results.

6.7 CONCLUSION

We have attempted to demonstrate the extent to which conventional fatigue testing procedures have been applied to composite materials and to indicate the limitations, arising from the peculiar nature of composites and the manner in which damage accumulates in them, that must be imposed on any interpretation of results. It is important to emphasise that no general conclusions, relating to composites as a class, can be drawn from any set of fatigue results on one type of composite and that the results referred to in this chapter are intended to served as illustrations rather than to provide rules.

6.8 REFERENCES

T. Adam, R.F. Dickson, C.J. Jones, H. Reiter and B. Harris: *Proceedings of the Institute Mechanical Engineers: Mechanical Engineering Science*, **200(C3)**,1986, 155-166.

T. Adam, N. Gathercole, H. Reiter and B. Harris: *International Journal of Fatigue*, **16**, 1994, 533-548.

M.F. Ashby and D.R.H. Jones: *Engineering Materials*, Pergamon Press, Oxford, **Vol.1**, 1980.

A.A. Baker and J.E. Mason, D. Cratchley: *Journal of Materials Science,* **1**, 1966, 229-237.

A.A. Baker: *Applied Materials Research*, **3**, 1966, 143-153,

C. Barney, D.C. Cardona and P. Bowen: Proceedings of 3rd Japan International SAMPE Symposium, *Advanced Materials — New Processes and Reliability*,T Kishi, N Takeda and Y Kagawa eds., 1993 Chiba, Japan, Japan Chapter of SAMPE, Tokyo, Japan, **2,** 1993, 1830-1835.

P.W.R. Beaumont and B. Harris: Proceedings of an International Conference on *Carbon Fibres: Their Composites and Applications*, Plastics Institute, London, 1972, 283-291.

M.H. Beheshty and B. Harris: Proceedings of International Conference on *Fatigue of Composites (ICFC), Paris*, Société Française de Métallurgie et de Matériaux, Paris, 1997, 355-362.

M.H. Beheshty, B. Harris and T. Adam: Composites A, 1999, *in press*.

C.A. Berg and N. Salama: *Journal of Materials*, **7**, 1972, 216-230.

K.H. Boller: Proceedings of 14th Annual Technical Conference of the Reinforced Plastics/Composites Institute of SPI, Society for Plastics Industry, New York, 1959, paper 6C.

K.H. Boller: *Modern Plastics*, **41**, 1964, 145 *et seq.*

L. Boniface and S.L. Ogin: *Journal of Composite Materials*, **23**, 1989, 735-754.

P. Bowen: *Proceedings 3rd Japan International Symposium, Advanced Materials — New Processes and Reliability*, T Kishi, N Takeda and Y Kagawa, eds.,1993, Chiba, Japan, Japan Chapter of SAMPE, Tokyo, **2,**1993**,** 2117-2124.

K.V. Bury: *Statistical Models in Applied Science*, J Wiley and Sons, London,1975.

E. Castillo: *Extreme Value Theory In Engineering*, Academic Press, Boston/London,1988.

B. Charrière: unpublished results, School of Materials Science, University of Bath,1985.

A.S. Chen and B. Harris: *Journal of Materials Science*, **28**, 1993, 2013-2027.

T.H. Courtney and J. Wulff: *Journal of Materials Science*, **1**, 1966, 383-388.

P.T. Curtis and B.B. Moore: Proceedings of Fifth International Conference on *Composite Materials (ICCM5)* San Diego, W.C Harrigan, J Strife and A.K Dhingra, eds., Metallurgical Society of AIME, Warrendale, PA, U.S.A., 1985, 293-314.

R.F. Dickson, G. Fernando, T. Adam, H. Reiter and B. Harris: *Journal of Materials Science*, **24**, 1989, 227-233.

R.F. Dickson, C.J. Jones, B. Harris, D.C. Leach and D.R. Moore: *Journal of Materials Science*, **20**, 1985 60-70.

G.J. Dvorak, J.Q. Tarn: *in Fatigue of Composite Materials STP 569*, J.R. Hancock, ed., American Society for Testing and Materials, Philadelphia, U.S.A., 1975, 145-168.

G. Fernando, R.F. Dickson, T. Adam, H. Reiter and B. Harris: *Journal of Materials Science*, **23**, 1988, 3732-3743.

A.M. Freudenthal and E.J. Gumbel: *Proceedings of Royal Society, London*, **A216**, 1953, 309-332.

K. Friedrich, R. Walter, H. Voss and J. Karger-Kocsis: *Composites*, **17**, 1986, 205-216.

M. Fuwa, B. Harris and A.R. Bunsell: *Journal of Physics D: Applied Physics*, **8**, 1975, 1460-1471.

N. Gathercole, H. Reiter, T. Adam and B. Harris: *International Journal of Fatigue*, **16**, 1994, 523-532.

W. Gerber: *Z Bayer Archit Ing Ver*, **6**, 1874, 101.

J. Goodman: *Mechanics Applied to Engineering,* Longman Green, Harlow, U.K, 1899.

B. Grimm: unpublished results, Department of Materials Science & Engineering, University of Bath, Bath, U.K, 1996.

F.A. Habib, R.A.J. Taylor, R.G. Cooke and B. Harris: *Composites*, **24**, 1993, 157-165.

B. Harris and S.V. Ramani: *Journal of Materials Science*, **10**, 1975, 83-93.

B. Harris, A.O. Ankara, D. Cawthorne and S.M.T. Bye: *Composites*, **8**, 1977, 185-189.

B. Harris, F.J. Guild and C.R. Brown: *Journal of Physics D: Applied Physics*, **12**, 1979, 1385-1407.

B. Harris, H. Reiter, T. Adam, R.F. Dickson and G. Fernando: *Composites*, **21**, 1990, 232-242.

B. Harris, F.A. Habib and R.G. Cooke: *Proceedings of Royal Society, London*, **A437**, 1992, 109-131.

B. Harris: *Fatigue of Glass-Fibre Composites, in Handbook of Polymer-Fibre Composites,* FR Jones ed., Longman Green, Harlow, U.K, 1994, 309-316.

B. Harris, N. Gathercole, J.A. Lee, H. Reiter and T. Adam: *Philosophical Transactions of the Royal Society, London*, **A355**, 1997, 1259-1294.

S.A. Hitchen, S.L. Ogin and P.A. Smith: *Composites*, **26**, 1995, 303-308.

K.E. Hofer, M. Stander and L.C. Bennett: *Polymer Engineering Science*, **18**, 1978, 120-127.

R.J. Howe and M.J. Owen: *Proceedings of Eighth International Reinforced Plastics Congress,* British Plastics Federation, London, 1972, 137-148.

C.L. Huffine, D.F. Solum and Wachtler: *Failure Modes In Composites,* I Toth, ed., Metallurgical Society of AIME, New York, 1973, 455-471.

W.B. Hwang and K.S. Han: *Composite Materials: Fatigue & Fracture (2) STP* 1012 P.A Lagace, ed., 1989, 87-102.

C.J. Jones, R.F. Dickson, T. Adam, H. Reiter and B. Harris: *Proceedings of Royal Society, London*, **A396**, 1984, 315-338.

M. Kawai, M. Morishita, K. Fuzi, T. Sakurai and K. Kemmochi: *Composites*, **A27**, 1996, 492-502.

A. Kelly and H. Lilholt: *Philosophical Magazine*, **20**, 1969, 311-328.

R.Y. Kim: *Composites Design 4th Edition:* SW Tsai, ed., Think Composites, Dayton, Ohio, U.S.A., 1988, Chapter 19.

S.C. Kunz and P.W.R. Beaumont: *Fatigue of Composite Materials STP 569*, American Society for Testing and Materials, Philadelphia, U.S.A., 1975, 71-91.

R.E. Lavengood and L.D. Gulbransen: *Polymer Engineering Science*, **19**, 1969, 365.

I. Leddet and A.R. Bunsell: *Composite Materials: Testing and Design (5) STP 674*, SW Tsai, ed., American Society for Testing and Materials, Philadelphia, U.S.A., 1979, 581-596,

J.A. Lee, B. Harris and D.P. Almond: *Proceedings of 2nd International Conference on Composites Engineering ICCE/2, August 1995*, D Hui, ed., International Community for Composites Engineering New Orleans, U.S.A, 1995, 443-444.

J.A. Lee, B. Harris, D.P. Almond and F. Hammett: *Composites*, **28A**, 1997, 5-15.

J.A. Lee, D.P. Almond and B. Harris: *to be published*, 1999.

J.W. Lee, I.M. Daniel and G. Yaniv: *Composite Materials: Fatigue & Fracture (2) STP 1012*, PA Lagace, ed., American Society for Testing and Materials, Philadelphia, U.S.A, 1989, 19-28.

R.E. Lee and S.J. Harris: *Journal of Materials Science*, **9**, 1974, 359-368.

J.W. Lomax and J.T. O'Rourke: *Proceedings 21st Annual Technical Conference of Reinforced Plastics/Composites Institute of SPI, Society for Plastics Industry, New York*, 1966, paper X-5.

J.F. Mandell: *Developments in Reinforced Plastics 2*, G Pritchard, ed., Applied Science Publishers London, 1982, 67-108.

S.M. Marco and W.L. Starkey: *ASME Transactions*, **76**, 1954, 627.

M.A. McGuire and B. Harris: *Journal of Physics D: Applied Physics*, **7**, 1974, 1788-1802

M.A. Miner: *Journal of Applied Mechanics*, **12**, 1945, A159-164.

B.B. Moore: *Proceedings of 3rd Risø Symposium Fatigue & Creep of Composite Materials*, 1982, H Lilholt and R Talreja, eds., Risø National Laboratory, Roskilde, Denmark, 1982, 245-257.

A.W.H. Morris and E.A. Steigerwald: *Transactions of AIME,* **239**, 1967, 730-739.

E. Orowan: *Proceedings of Royal Society*, *London*, **A171**, 1939, 79-106.

M.J. Owen, R. Dukes and T.R. Smith: *Proceedings Annual Technical Conference of Reinforced Plastics/Composites Institute of SPI, Society for Plastics Industry,* New York, 1969, paper 14A.

M.J. Owen: *in Glass Reinforced Plastics*, B. Parkyn, ed., Iliffe, London, 1970, 251-267.

M.J. Owen and S. Morris: *Proceedings of 25th Annual Technical Conference of Reinforced Plastics/Composites Institute of SPI, Society for Plastics Industry, New York*, 1970, paper 8-E.

M.J. Owen and P.T. Bishop: *Journal of Physics D: Applied Physics*, **6**, 1972, 2057-2069.

M.J. Owen and P.T. Bishop: *Journal of Physics D: Applied Physics*, **7**, 1974, 1214-1224.

M.J. Owen and S. Morris: *Carbon Fibres: Their Composites and Applications,* Plastics Institute, London, 1972, 292-302.

A. Pegoretti and T. Ricco: *Composites Science & Technology,* 1999, in press.

D.C. Phillips and J.M. Scott: *AERE Technical Report AERE G645, Atomic Energy Authority, Harwell, U.K,* 1976.

D.C. Phillips: *Proceedings 9th Risø Symposium on Mechanical and Physical Behaviour of Metallic and Ceramic Composites 1988* SI Andersen H Lilholt & OB Pedersen, eds., Risø National Laboratory, Roskilde, Denmark, 1988, 183-199

L.N. Phillips, J.S. Bradley and J.B.Sturgeon: *RAE, Farnborough, Technical Memorandum MAT241,* Ministry of Defence Procurement Executive, U.K, 1976.

A. Poursartip and P.W.R. Beaumont: *Proceedings IUTAM Symposium Mechanics of Composite Materials, VPI,* Blacksburg, Virginia, U.S.A., 1982.

K.M. Prewo: *Journal of Material Science,* **21**, 1986, 3590-3600.

K.M. Prewo: *Journal of Material Science,* **22**, 1987, 2695-2701.

S.V. Ramani and D.P. Williams: *Failure Modes in Composites,* Metallurgical Society of AIME, New York, 1976, 115-140.

K.L Reifsnider and R.D. Jamison: *International Journal of Fatigue,* **4**, 1982, 187-198.

K.L. Reifsnider, E.G. Henneke and W.W. Stinchcome: *Defect-Property Relationships in Composite Materials, AFML-TR-81 part 4,* U.S Air Force Wright Laboratories, Dayton, Ohio, 1979, 1979.

R.O. Ritchie, R.H. Dauskardt and K.T.Venkateswara Rao: *Proceedings 3rd Japan Internatl SAMPE Symposium, Advanced Materials — New Processes and Reliability,* T Kishi, N Takeda and Y Kagawa, eds., 1993, Chiba, Japan, Japan Chapter of SAMPE, Tokyo, Japan, **2** 1966-1976, 1993.

D. Schütz and J.J. Gerharz: *Composites,* **8**, 1977, 245-250.

R.G. Shaver and E.F. Abrams: *Proceedings of. Annual Technical Conference of SPE,* **17**, 1971, 378-381.

G.D. Sims and D.G. Gladman: *Plastics and Rubber: Material and Application,* **1**, 41-48, and 1980, **3**, 1978, 122-128.

T.R. Smith and M.J. Owen: *Proceedings 6th International Reinforced Plastics Congress,* British Plastics Federation, London, 1968, paper 27.

J.P. Solti, S. Mall and D.D. Robertson: *Journal of Composites Technology and Research,* **18**, 1996, 167-178.

B.F. Sorensen, R. Talreja and O.T. Sorensen: *Composites,* **24**, 1993, 129-140.

J.B. Sturgeon: *Proceedings 28th Annual Technical Conference of Reinforced Plastics/ Composites Institute of SPI, Society for Plastics Industry, New York,* 1973.

J.B. Sturgeon: *RAE, Farnborough Technical Memorandum MAT 228,* Ministry of Defence Procurement Executive U.K., 1975.

C.T. Sun and B.W.S.Chan: *Composite Materials; Testing and Design STP 675,* S.W. Tsai, ed., American Society for Testing and Materials, Philadelphia, U.S.A, 1979, 418-430.

S. Suresh, L.X. Han and J.J. Petrovic: *Journal of American Ceramic Society,* **71**,

1988, C158-161.

R. Talreja: *Proceedings of Royal Society, London*, **A378**, 1981, 461-475.

M. Vancon, J. Odorico and C. Bathias: *Comptes Rendus des Quatrièmes Journées Nationales sur les Composites, JNC4,* Editions Pluralis, Paris, 1984, 93-120.

W. Weibull: *Fatigue Testing and Analysis of Results,* Pergamon, Oxford, for AGARD, NATO, Paris, 1961.

J.M. Whitney: *Fatigue of Fibrous Composite Materials STP 723*, American Society for Testing and Materials, Philadelphia, U.S.A, 1981, 133-151.

J.Y. Xiao and C.Bathias: *Composites Science and Technology*, **50**, 1994, 141-148.

7. Environmental Effects

7.1 INTRODUCTION

The foregoing chapters deal, for the most part, with the mechanical response of composite materials under what might be called 'normal' conditions, that is, under a variety of régimes of loading, but with no consideration of possible interactions between the applied stress system and environments that differ appreciably from ordinary laboratory conditions. This is naturally limiting, and yet this 'normal' performance must always be established before the effects of other more aggressive environments can be distinguished. By and large, it is the effects of temperature, ultra-violet radiation, and chemical environments, including oxidising atmospheres, that first come to mind when considering practical applications of composites in general. But, ironically, it is moist environments that are likely to be the major source of trouble to designers in most applications.

There are some obvious problems of which the designer must be aware. As far as temperature is concerned, a number of effects might be mentioned. The first is that at elevated temperatures creep may occur in one or both of the components. No measurable creep will occur in glass, carbon, or boron fibres over the temperature ranges within which most resins, even the more thermally-stable ones, remain undamaged. In continuously-reinforced plastics, therefore, time-dependent deformation will be limited by the extent to which the continuous fibres support the major part of the load. In short-fibre composite mouldings, however, the upper usable temperature limit may be much lower than that at which resin degradation begins to occur.

Leaving aside chemical degradation of matrix resins which may begin, in the less stable polymers, at relatively modest temperatures, a second effect of temperature is in the material response associated with changes in temperature, rather than the effects of exposure at constant temperature. We saw in Chapter 2 that substantial levels of residual stress can be developed as a result of thermal expansion mismatch. Under such circumstances these can be so great as to cause fibre damage, inter-ply cracks and transverse ply cracks. Of particular concern in aerospace structures is the deleterious effect of rapid excursions to high temperature, a phenomenon known as 'thermal spiking', which can result in significant reduction of the residual strength of the spiked laminate. Collings and Stone (1985), for example, observed that the thermal spiking of dry, cross-plied XAS/914 CFRP resulted in a 7% reduction in the ILSS, although no visible signs of degradation occurred. On the other hand, they found that similar spiking of laminates containing moisture produced large-scale damage in the form of microscopic interlaminar cracks which reduced the ILSS to some 75% of the dry room-temperature value, and almost doubled the equilibrium moisture content of the laminates. The combined effect

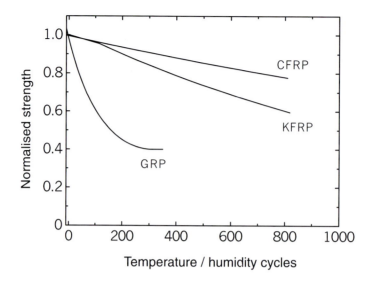

Fig. 7.1 Reductions in the strengths of 0/90 laminates of CFRP, GRP and KFRP as a consequence of temperature/humidity cycling. The cycle consisted of 12 hours at 2°C and 0% RH, followed by 12 hours at 85°C and 100%RH. (Dickson *et al.*, 1984).

of temperature and moisture is frequently observed to be more damaging than the effect of either separately. The differing responses of cross-plied epoxy-based composites reinforced with glass, aramid, and carbon fibres to combined temperature/humidity cycling are illustrated by some results of Dickson *et al.* (1984) shown in Figure 7.1.

Water diffuses readily through many of the thermoset and thermoplastic polymers that are commonly used as matrix materials. And where the environment has access to the fibre/matrix interface, at the cut edges of a component, for example, water may be 'wicked in' by capillarity along the interface. Moisture therefore has ready access to those parts of the composite from whence it derives its load-bearing properties, to a greater or lesser extent, depending on the nature of the matrix polymer concerned, and in the long term this can result in deterioration of composite properties.

Most forms of high-energy radiation are damaging to polymers because of the relatively low energies required to cause chemical damage. The most obvious effect in long-chain polymers is the degradation associated with scission of the main chains, which results in a reduction in the average molecular weight, and additional cross-linking between chains, resulting in network formation. Scission often causes a lowering of the polymer viscosity and softening temperature and reduction of mechanical strength, and it may also lead, in some cases, to an increase in the degree of crystallinity. Cross-linking, on the other hand, leads to an increase in strength and ductility up to some limiting cross-link density beyond which the strength is again reduced and the polymer embrittled. These are therefore opposing effects which nevertheless occur simultaneously. But the long-term effect of irradiation is almost always serious embrittlement of the polymer. Internal stresses are also developed which, in the presence of an external stress and an aggressive environment, may result in rapid disintegration. Exposure of thermoset resin composites to ultra-violet radiation can promote additional cross-linking in the surface resin which may result in surface microcracks. Most practical reinforcing fibres, with the exception of the aromatic polyamides and high-stretch polyolefins, are unlikely to be affected by uv radiation, and since the bulk of the reinforcing fibres in a composite will largely be screened from the uv by the surface layers of resin and fibre or by pigmented paint coatings, the response of a reinforced plastic laminate to uv will be largely dominated by what happens to the resin. In artificial exposure experiments lasting some 2500 hours in an ultra-violet cabinet, Dickson *et al.* (1984) found the properties of cross-plied laminates of glass and Kevlar-49 epoxy-based composites to be substantially unaffected even though long-term exposure caused considerable darkening of the surface resin. Carbon fibre appears to have the same effect as lamp-black which is sometimes added to polymers as a means of stabilising the material against ultra-violet degradation: the fibres, like the particles, efficiently filter the radiation. It is nevertheless often observed that outdoor exposure results in serious damage to the surface layers of laminates, sometimes to such an extent that the resin-rich surface layer or gel-coat is completely eroded, leaving the underlying fibre exposed. This kind of damage, which is perhaps likely to be a result of the combined effects of ultra-violet, regular wetting and drying, and diurnal temperature cycling, is particularly serious in the case of glass-fibre- or polymer-fibre-reinforced resins since it leaves the reinforcing elements totally bare of protection against both mechanical damage and further environmental attack.

A useful illustration of the difficulties facing a designer in relation to environmental effects is that of the application of composites in space vehicles. Many of the requirements are the same as those for aeronautical structures, since there is a need to have low weight and high stiffness in order to minimize loads and avoid the occurrence of buckling frequencies. But the environmental effects are special. Service temperatures range from cryogenic to many hundreds of degrees, and the temperature changes are therefore very great: consequently, severe thermal stresses may arise which may cause serious distortion. Dimensional stability is at a premium, for stable antennae and optical platforms, for example, and materials need to be transparent to radio-frequency waves and stable towards both uv radiation and moisture. Many of these effects lead to residual strains, microcracking, and thermal distortion. Porosity and internal debonds also give problems with moisture, so a high degree of perfection and high-quality adhesive bonding procedures are needed to ensure dimensional stability. A particular problem for the space vehicle designer is that reinforced plastics are degraded by atomic oxygen and by radiation. Paillous and Pailler (1994) showed that the synergistic action of electrons and thermal cycling degrades the matrices of carbon/epoxy composites by chain scission, cross-linking and microcrack damage, while the erosion caused by atomic oxygen led to decreases in flexural stiffness. These effects are exacerbated by impact events in low earth orbit.

The designer will usually be looking to use metal-matrix composites for applications at temperatures significantly higher than ambient (and higher than temperatures for which there are already plenty of good reinforced plastics available). In such cases, creep of both components may occur, and a conventional creep-design approach may be needed. A side effect of high-temperature exposure in MMCs is that at temperatures sufficiently high to permit creep there will also be a possibility of diffusion-controlled interfacial reactions which may, at best, modify the load transfer conditions (often favourably) or, at worst, may cause dramatic reduction in the load-bearing capability of the reinforcement. McLean (1985) has given a recent review of the high-temperature behaviour of metal-matrix systems.

In metal-matrix composites an aqueous environment may cause severe corrosion if the matrix/reinforcement combination forms an appropriate galvanic couple. Carbon-fibre-reinforced aluminium alloys were identified in the early seventies as being particularly prone to attack (see, for example, the review by Pfeifer (1977). A wide range of corrosion resistance was observed in various systems studied at that time, much of the work referred to in this review relating to the effects of marine, salt-spray and relative-humidity exposures. It was reported that the principal mode of corrosion was pitting commencing at the interface at points where the fibres were exposed, at cut or machined surfaces for example, the pitting being often followed by severe exfoliation leading to complete destruction of test panels. The composite corrosion resistance was strongly affected by the composition and structure of the matrix alloy, and was highest in composites in which a good fibre/matrix bond was achieved. Similar corrosion problems may be expected when carbon-fibre-reinforced plastics are used in conjunction with metallic components in the presence of the almost inevitable aqueous environment.

7.2 HYGROTHERMAL SENSITIVITY OF REINFORCED PLASTICS

7.2.1 FIBRE EFFECTS

Although reinforcing fibres like carbon, boron, and other ceramic reinforcements are insensitive to the effects of moisture, others, particularly glass and aramid fibres like Kevlar-49, are affected by moisture even at low exposure levels. Glass and Kevlar-49 fibres are known to have mechanical properties that are dependent on time and temperature, giving rise to the familiar 'static fatigue' effect illustrated in Figure 7.2 (Aveston et al., 1980) which is characterized by the fact that the time to failure under a constant load depends on the load level.

In the case of glass, this is a form of *environmental stress cracking*, or *stress-corrosion*, resulting from the leaching out of the network modifier, NaO_2, from the glass structure. The resulting alkaline environment then attacks the normally-stable SiO_2 network and reduces the strength of the glass fibre. Stress-corrosion essentially results from the conjoint effects of the corrosive environment and the applied stress. The effect occurs more rapidly in acid or alkaline environments than in water alone, but even in pure water the rate of loss of strength is high and the extent of the weakening is substantial. The loss of strength

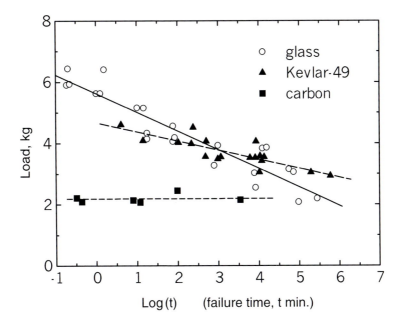

Fig. 7.2 Time-dependent failure of carbon, glass and Kevlar-49 fibres (Aveston *et al.*, 1980).

of ordinary E-glass fibres in the alkaline environment of a damp concrete matrix seriously limited the effective use of glass-fibre-reinforced cement products (GRC) in the building industry. This problem led to development at the U.K. Building Research Establishment, and subsequent exploitation by Pilkington, of a zirconia-containing glass known as 'CemFil', which is more alkali-resistant than other glass compositions (Majumdar, 1970). Results on the stress-corrosion of this CemFil glass fibre in aqueous alkaline environments (Proctor and Yale, 1980) suggested that the material offered a far-from-perfect solution, but it has been shown that under the more realistic conditions of service exposure of thin-walled CemFil-reinforced cement pipe the practical life of such materials is greater than the laboratory results of Proctor and Yale suggested (ARC, 1979).

As the results in Figure 7.2 show, the stress-rupture curve for carbon filaments is horizontal, the fibre therefore being completely insensitive to the environment, while Kevlar-49 shows a time-dependent response somewhere between that of carbon (nil) and that of glass (high). The aramid fibre, like glass, is therefore also environmentally sensitive although the stress-corrosion mechanism is different. The yarn may absorb some 6% by weight of water in about 30 hours at a relative humidity of 96% (Smith, 1979). This moisture absorption is accompanied by dimensional changes which are of importance for the manufacturers of composites. Any absorbed moisture must be removed from the yarn before prepregging or other laminating procedures since drying out after cure may lead to fibre/matrix decohesion and consequent loss of composite properties. Moisture absorption at ambient temperatures does not impair the short-term strength of

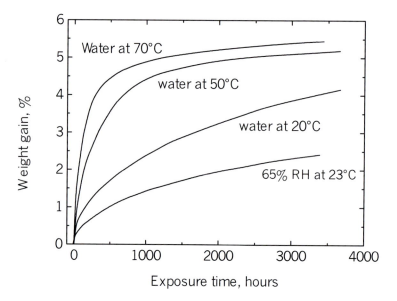

Fig. 7.3 Moisture absorption in Elf 37SA epoxy resin during exposure to wet or moist environments at various temperatures (Charrière, 1985).

the fibre, but Smith (1979) has shown that high-temperature wet-ageing seriously reduces its strength.

7.2.2 RESIN EFFECTS

The open molecular structure of glassy thermoplastic and thermoset polymers permits relatively rapid diffusion of moisture. Depending on the nature of the polymer, considerable quantities of moisture may be absorbed into the material, and both the rate of absorption and the saturation absorption level will depend on the conditions of exposure (*i.e.* temperature and relative humidity). This is illustrated in Figure 7.3 by some results for an experimental epoxy resin manufactured by Elf-Aquitaine.

Moisture ingress in polymers usually follows Fickian kinetics, the amount of moisture absorbed being proportional to $\sqrt{(time)}$ in the early stages of exposure, as predicted by the one-dimensional Fickian model of Shen and Springer (1981), for example. A condition for the applicability of Fick's Law is that the diffusion is concentration-independent, but the diffusion behaviour of many polymers, both glassy and cross-linked, under particular conditions, may not be adequately described by a law which predicates fixed boundary conditions, especially when the movement of the diffusing species is linked with molecular relaxations or morphological effects in the polymer. Deviations from Fickian behaviour are then observed which may be associated with structural changes, resin degradation, or even simple relaxation in response to the diffusing species. Frequently, the early

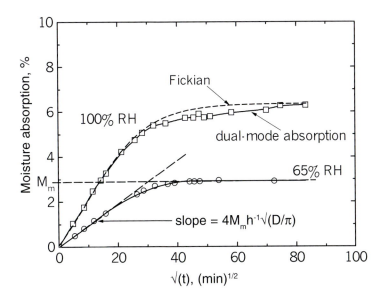

Fig. 7.4 Moisture absorption curves for 914 epoxy resin at 23°C and two different levels of relative humidity (RH) (Fernando, 1986).

response is Fickian and this may be followed by a quasi-equilibrium stage with a much slower approach to final true equilibrium (dual-mode sorption). True Fickian and dual-mode sorption effects are illustrated in Figure 7.4 by some results obtained at two different humidities for Ciba-Geigy 914 epoxide resin (Fernando, 1986).

The equilibrium moisture content depends on the chemical and physical nature of the polymer. The absorption of water strongly reduces the mechanical properties of hydrogen-bonded polymers like conventional polyamides, but in polyester and epoxide resins it acts as a diluent, increasing the free volume and facilitating molecular motion under stress. This plasticising effect reduces the stiffness of the resin, lowers the glass-transition temperature, T_g, and increases the magnitude of time-dependent effects like stress relaxation. The reduction in T_g by moisture absorption may be by as much as 100°C in some cases, as Figure 7.5 from the work of de Iasi and Whiteside (1978) shows. Crystalline thermoplasts are much more resistant to moisture penetration than epoxide resins. The semi-crystalline high-performance polymer poly(ether ether ketone) (PEEK), for example, appears to be largely immune from environmental effects. Mensitieri *et al.* (1996) report only a 2°C depression of T_g in samples equilibrated in water at 60°C. Although PEEK contains oxygen atoms that may potentially form hydrogen bonds, it is not moisture sensitive because these atoms are sterically shielded by the aromatic rings.

The wetting and drying of a resin are accompanied by dimensional changes which generate residual stresses in a resin containing rigid reinforcement. Solvent attack on resins can cause both physical and chemical effects, the latter usually being either

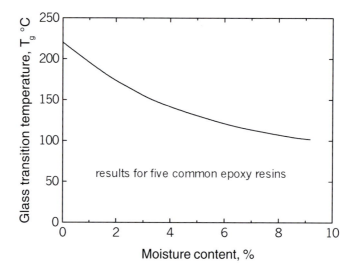

Fig. 7.5 Effect of moisture content on the glass transition temperature of common epoxide resins (de Iasi & Whiteside, 1978).

hydrolysis (of the ester linkages in a polyester, for example) or oxidation, although these processes are slow at ambient temperatures. Chemical attack naturally reduces the mechanical properties of the common resins.

7.2.3 COMPOSITE EFFECTS

In a normal composite material, moisture will enter by diffusion through the bulk resin, by capillary flow through pores in the resin, and also along the fibre/resin interface. Fickian kinetics again often appear to apply, but deviations from Fickian behaviour are frequently observed, such as where accelerated moisture pick-up occurs as a result of the presence of micro-voids or cracks in the resin. The moisture diffusion in an anisotropic composite lamina is of course also likely to be anisotropic, the rate of diffusion being more rapid in the direction of fibre alignment because of the continuity of the resin diffusion path. The kinetics of moisture uptake are therefore also markedly influenced by the lay-up of a laminate (Collings & Stone, 1985).

For the case of Fickian diffusion (*i.e.* assuming that moisture transport is through the resin only) the diffusion coefficients for a single unidirectional lamina can be approximated by:

$$D_{11} = (1 - V_f)D_r \tag{7.1}$$

$$D_{22} = [1 - \sqrt{(V_f/\pi)}]D_r \tag{7.2}$$

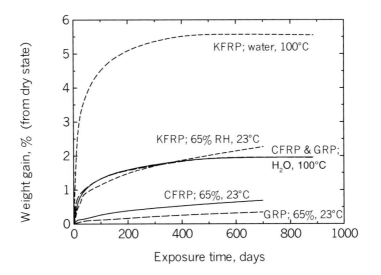

Fig. 7.6 Absorption of water during exposure of composite laminates at ambient temperature (65% relative humidity) and at 100°C (in water). The laminates are 0/90 laminates of HTS carbon, E-glass and Kevlar-49 in Code-69 epoxy resin.

for diffusion parallel and perpendicular to the fibres, respectively, D_r being the diffusion coefficient for the matrix resin ($D_r \gg D_f$). The diffusivity at any angle, θ, to the fibre direction is then

$$D_\theta = D_{11} \cos^2 \theta + D_{22} \sin^2 \theta \tag{7.3}$$

If the response of a single lamina is known (or can be calculated from models such as those of Springer) the overall moisture absorption behaviour of a multi-ply laminate can therefore be calculated. Computer programs for this purpose have also been developed (Curtis, 1981).

Bueche and Kelly (1960) developed a theory for the prediction of the T_g of plasticized resin composites which gives:

$$T_g = \frac{\alpha_r (1 - V_f) T_{gr} + \alpha_f V_f T_{gf}}{\alpha_r (1 - V_f) + \alpha_f V_f} \tag{7.4}$$

where α is a thermal expansion coefficient and V_f is the fibre (or filler) volume fraction, the subscripts r and f referring to the resin and fibre (or filler) components. The theory predicts the trend in T_g reasonably well. The nature of the fibre can also exert an influence on the process of moisture uptake, as illustrated by the graphs in Figure 7.6. The materials are of the same 0/90 lay-up and contain the same epoxide matrix resin, and it can be seen that at both ambient temperature in moist air and in boiling water there are only slight differences between the absorption in composites containing carbon and glass fibres,

Fig. 7.7 Tensile fracture surface of an E-glass/epoxy GRP laminate after boiling in water (photograph by Dr RF Dickson).

whereas the absorption in the aramid-fibre composite is substantially greater as a consequence of the hydrophilic nature of the aromatic polyamide.

Both reversible and irreversible mechanisms of degradation may occur as a result of moisture ingress. For example, if water travels down the interface or is able to accumulate at the interface after diffusion through the bulk resin it may cause hydrolytic breakdown of any chemical bonding between the fibre and the resin and bonds may be disrupted by the swelling of the resin. Either of these mechanisms of bond failure will impair the efficiency of stress transfer between matrix and reinforcement. The strength and stiffness of the composite will then be recovered on drying out only if new chemical bonds can be re-established. If the bonding is simply mechanical, however, and the absorption and desorption of water are purely physical processes related to the resin or the interfacial bond (*i.e.* supposing the fibre strength to be also unaffected), the composite would be expected to recover its properties on drying out. Thus, while the tensile strengths of carbon and Kevlar laminates are little affected by hygrothermal treatments, even as severe as boiling in water, a GRP laminate may lose some 50% of its strength. However, much of this damage is apparently reversible since even boiled GRP laminates may regain all or most of their original strengths on re-drying (Dickson *et al.*, 1984). This suggests that in carbon and Kevlar composites there is only a physical bond between fibres and matrix, and that this is not affected by hot/wet ageing, whereas in the GRP there is a chemical bond (presumably resulting from the presence of a silane coupling agent on the fibres) which is destroyed by the hygrothermal treatment but apparently re-established on removal of the moisture. The marked change in the fracture surface of a GRP laminate after exposure to boiling water is shown in Figure 7.7: a failure surface of this kind, which is in marked contrast to the normal bushy fractures shown in Figure 5.7 of chapter 5, is inevitably associated with brittle behaviour.

The reversibility of this moisture effect, and the fact that the laminate behaviour goes from brittle in the dry state to ductile when wet and back to brittle on re-drying,

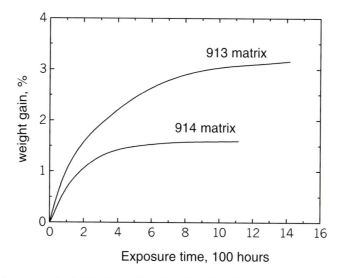

Fig. 7.8 Moisture uptake in 16-ply undirectional carbon-fibre composites with two different epoxy resin matrices exposed to water at 23°C (Charrière, 1985).

supports the general belief in the plasticising effect of moisture on the resin. It should be pointed out, however, that other experimenters, working with other resins and different combinations of environmental and stress conditions, have observed irreversible damage, such as resin cracking and permanent bond breakdown, with the inevitable irreversible degradation of composite properties. The chemical stability and resistance to moisture penetration of the matrix polymer, which depend sensitively on chemical and molecular structure and state of cure, will usually determine the extent of both reversible and irreversible composite damage, although resin void-content and composite fibre-content will also exert an effect. It is generally recognized that laminates based on polyester resins are less moisture resistant than epoxy-based composites, and that phenolic laminates are more water-resistant than either of these. But within each group of resins there is in fact a wide range of individual types and compositions, from which the designer may select according to his requirements, with a correspondingly wide range of chemical stability and moisture resistance. In the polyester range, for example, there is increasing chemical resistance through the sequence orthophthalic, isophthalic, neopentyl-glycol, and bisphenol polyester resins. Much of the research on matrix resins has as a general goal the improvement of the polymer tolerance to hot/wet conditions. One of the difficulties here is that improvements in hygrothermal tolerance can usually only be achieved at the cost of a raised processing temperature. In Figure 7.8, for example, we see that the saturation uptake of water into a carbon-fibre composite based on Ciba-Geigy 913 resin is twice that in a similar laminate made with the same company's 914, and as we saw in chapter 1, the former is processed at 120°C while the latter requires a processing temperature of 170°C.

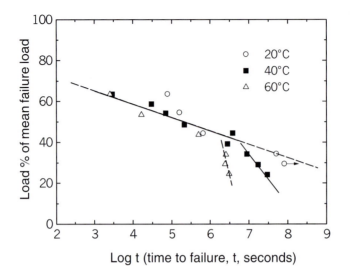

Fig. 7.9 Time-dependent failure of GRP materials in water at various temperatures (Phillips *et al.*, 1983). The mean ultimate tensile failure load of the material is 256 N/mm.

7.3 TIME-DEPENDENT EFFECTS

The results of Aveston *et al.* in Figure 7.2 showed that the load to cause failure in 10^5 minutes of E-glass strands tested in distilled water was less than 50% of the normal breaking load. They also found, to their surprise, that the situation was not improved when the strands were impregnated with resin. The apparent strengths of the impregnated strands fell to one sixth of the normal value, regardless of whether the impregnating resin was epoxy or polyester (although the epoxy resin appeared to delay the onset of the strength reduction somewhat) presumably because of the lower moisture permeability of the epoxy resin.

Phillips *et al.* (1983) have reported the results of an extensive series of stress-rupture experiments on a range of GRP, with a variety of different resins and reinforcement types, in distilled water and in sea-water at different temperatures. Their results confirm the general idea that the strength of moist GRP is controlled by the plain moisture sensitivity of the fibre strands. They were unable to find any effect of variation in immersion conditions, but found that different resin types produced detectable differences in life at a given stress level. Vinyl-ester resin showed considerable advantage over the less expensive isophthalic polyester, for example. It has often been supposed that stress-rupture results for GRP materials fitted a linear stress vs. log time relationship (Zhurkov, 1965), and attempts at time-temperature superposition based on this assumption have often been proposed for design purposes. The work of Phillips *et al.* however, suggests that the situation may in reality be much more complex, and their results were frequently best fitted by a two-slope curve, as shown in Figure 7.9. Below a certain well-defined load level, results were strongly temperature-dependent, the effect being greater than could be accounted for solely on the basis of a thermal-activation model, while above this critical load level, temperature had little effect on sample life.

Jones *et al.* (1983) have studied the more dramatic effects of acid environments on glass filaments and glass/epoxy laminates. They confirm that the prime cause of stress-corrosion in GRP is the susceptibility of the glass fibres themselves to stress-corrosion. Thus, while the main function of the resin is to protect the fibres, they found that the glass/resin interface itself was also subject to attack, with the result that the environment — sulphuric acid in their experiments — penetrates the laminate. At applied strain levels greater than 0.15%, failure times for laminates were the same as those for the fibres, while at lower strain levels the resin protected the fibres and the laminate was more resistant to stress-corrosion. However, they also found that crystallizable calcium-rich glass corrosion products accumulating in unexposed areas of the composite could result in more rapid failures than normal at these low strain levels. Stress-corrosion effects of this kind often markedly change the fracture mode of a GRP composite. Instead of the kind of brushy fracture that is commonly observed in high-toughness composites, cracks appear to be able to propagate easily from fibre to fibre without being diverted along the interface, as shown by Figure 5.4 in chapter 5.

7.3.1 GRP IN CORROSIVE ENVIRONMENTS

This brief survey of environmental effects suggests that reinforced plastics components and structures may not be an obvious first choice for applications where stability against chemical attack is a pre-requisite. Nevertheless, a great deal of GRP is used for the containment of corrosive and other hostile liquids and designers specify them with increasing confidence despite the fact that relevant long-term stress-corrosion data are not always available and extrapolation from laboratory-scale data is always dangerous. This is usually possible because the manufacturer is able to use expensive, chemically-resistant and moisture-resistant resins, either for the whole structure if the extra cost can be justified, or as a thin, renewable surface gel-coat which protects an underlying GRP based on a cheaper (and less-resistant) matrix resin. Such a gel-coat could be additionally stabilized against ultra-violet if necessary, by pigmenting with TiO_2, for example. A useful review of corrosion and environmental degradation of GRP composites of various kinds has been published by Hogg and Hull (1983).

7.4 HIGH-TEMPERATURE STRENGTH AND CREEP OF COMPOSITES

7.4.1 INTRODUCTION

The strengthening achieved through the operation of conventional solid-state mechanisms in metals and plastics is easily destroyed by thermal activation as the

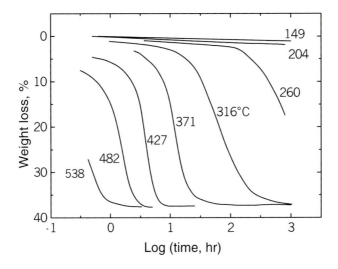

Fig. 7.10 Weight loss during thermal degradation of glass-fibre/phenolic laminate at elevated temperatures (after Boller, 1960). The temperatures are all in degrees Celsius,

temperature is raised and there are few alloy or polymer systems that can continue to give satisfactory service when atomic or molecular mobility is enhanced by heating. But since many of the strong solids used for reinforcement are inherently resistant to this form of instability, up to relatively high temperatures, an important advantage of fibre composites is that they may be expected to retain their strengths to much higher temperatures than would normally be possible. This would be true for composites reinforced with inorganic synthetic fibres like carbon, boron, SiC or Al_2O_3, although metallic fibres are naturally subject to the same limitations as metallic matrices and glassy fibres lose their strengths at the glass-transition temperature when molecular networks become freely mobile.

In reinforced plastics there are few matrix materials that can withstand temperatures above 300°C (see chapter 1), and the limiting factor for the most stable of matrices is likely to be chemical degradation. On exposure to air at elevated temperatures, thermosetting materials gradually become degraded through chemical changes which result in loss of material accompanied by loss of mechanical integrity. Early work by Boller (1960) illustrated this for glass/phenolic laminates: Figure 7.10, redrawn from his article, shows that the phenolic matrix is not stable above 200°C for more than two hours or so. Families of curves of this kind can be shifted along the abscissa to form a continuous master curve having an exponential form where the weight loss is proportional to exp(–kt), k being a constant. The deterioration in strength and modulus follows a similar exponential decay. The same kind of time/temperature superposition is used in the study of polymers to produce master curves of creep compliance or relaxation modulus over a wide range of times on a logarithmic scale (see McCrum *et al.*, 1988, for example).

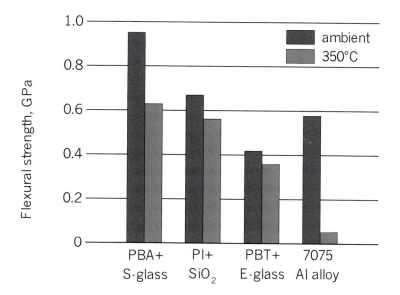

Fig. 7.11 Flexural strengths at ambient temperature and 350°C of some laminates based on thermally stable polymers polybenzoxazole (PBA), polyimide (PI) and polybenzothiazole (PBT), compared with the behaviour of an age-hardened 7075 aluminium alloy (Aponyi, 1967)

Most thermoplastic polymers are naturally unsuitable for high-temperature composites, although some of the more sophisticated polymers like the polyaryls (*e.g.* PEEK) have continuous working temperatures that are quite close to 300°C. Some of the more complex thermoset polymers, too, possess remarkable resistance to degradation — polymers like the polyimides. Figure 7.11 compares the room-temperature and elevated-temperature strengths of composites based on some of these newer polymers. It can be seen that, in common with many other reinforced plastics, the strengths of these materials at ambient temperature are comparable with those of heat-treated aluminium alloys, but they have a clear advantage over aluminium alloys at temperatures where the metallic materials over-age as a consequence of solid-state diffusion.

Although aluminium alloys and thermoplastics rapidly lose their load-bearing ability and rigidity as the temperature rises above ambient, the consequences of this for users of composites based on these materials will depend to an extent on how the load is distributed relative to the fibres. Consider a unidirectional composite containing 60 vol% of rigid ceramic fibres, for example, in which the matrix strength is one tenth that of the fibres. A simple application of the mixtures rule shows that if, as a result of a rise in temperature of, say, 300°C, the matrix loses half its strength, the drop in the uniaxial composite strength will be only about 3.5%.

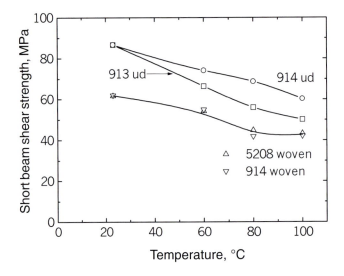

Fig. 7.12 Effect of temperature on the short-beam shear strengths of some unidirectional and woven-fabric carbon-fibre laminates (Charrière, 1985).

But while such composites may perform very well by comparison with aluminium alloys when loaded in tension or flexure, even at 300°C, they will creep readily under the action of interlaminar shear forces when the visco-elastic behaviour of the matrix and interface govern the composite response. As an illustration, the results of short-beam shear tests on some carbon-fibre composites are shown as a function of temperature in Figure 7.12. Two types of comparison may be made. First, the importance of the thermal stability of the resin depends on the composite lay-up: the composite based on the high-temperature-cure 914 resin retains a higher proportion of its room-temperature strength than the low-temperature-cure 913 composite as the temperature rises. Second, when the reinforcement is woven, resistance to shear loading is still lost as the temperature rises, but less rapidly than in the case of a unidirectional laminate, and the nature of the matrix resin is less important.

In metallic composites the high-temperature performance will often be severely limited by fibre/matrix interaction. Interfaces in most artificially produced composites with metallic or ceramic matrices are unstable at elevated temperatures, and solid-state diffusion often occurs unless a thermal/diffusion barrier coating has been deposited on the fibre. This often results, at first, in slightly improved properties as the adhesion improves, but this is usually followed by deterioration of composite strength as the fibre surface is attacked and chemical compounds begin to form. Work on the development of diffusion barrier coatings to overcome these problems currently has a high priority in the composites research field, and some of the ceramic fibres now commercially available are already protected by graded-property coatings. Chemical reactions will also occur in CMCs if the exposure temperature is sufficiently high, and functionally graded diffusion-

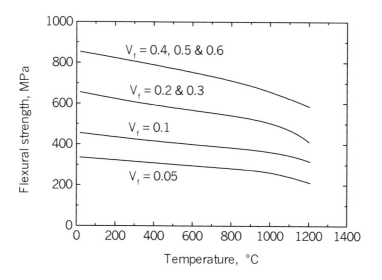

Fig. 7.13 The effect of temperature on the flexural strength of SiC-whisker-reinforced alumina (Tiegs and Becher, 1987).

barrier coatings are also necessary. But although advantageous short-term performance at elevated temperatures can be obtained in CMCs, as illustrated by the results in Figure 7.13, since brittle-matrix systems inevitably crack under modest loads, exposing the reinforcing filaments to the working environment, a CMC in service at high temperature will lose strength over time unless the fibres are protected by a diffusion barrier. This can be seen from the results of some short-term flexural creep-rupture tests on $(0/90)_{3S}$ Nicalon/CAS composites tested in air at 600°C. Figure 7.14 shows that under loads as low as 60% of the instantaneous breaking load the life of the material is only three or four hours.

7.4.2 CREEP

7.4.2.1 Introduction: Creep Phenomena

Although the mechanisms by which metals, plastics and ceramics deform in creep are different, the phenomena which are observed in these materials are similar. The basic mathematical laws of creep often seem to apply across the material classes, even though the physico-chemical phenomena are different. Studies of creep behaviour of rod-shaped samples in tension show that the creep response, characterized in terms of the time-dependent strain at constant stress and temperature, may be represented by a general law of the kind:

$$\varepsilon(t) = \varepsilon_0 + \alpha t^{1/3} + \beta t + \kappa t^3 \tag{7.5}$$

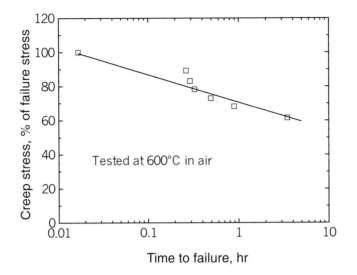

Fig. 7.14 Flexural creep-rupture data for a $(0,90)_{3S}$ SiC/CAS ceramic-matrix composite at 600°C in air.

in which the first term represents the initial elastic strain on loading, and the remaining terms are referred to as *primary*, *secondary* and *tertiary creep*, α, β and κ being constants. An equation of a simplified form, known as Findley's equation (Findley and Kholsla, 1955), is also frequently used to describe the creep behaviour of polymeric materials. Primary creep is sometimes referred to as *transient creep*, and secondary creep is usually referred to as *steady-state* creep. Which of the time-dependent components are actually observed in a given creep test depends on the stress level and the temperature. At temperatures above half the absolute melting point ($\frac{1}{2}T_m$) and fairly high stress levels, for example, a metal creep curve may appear to be curving continuously upwards, showing little, if any, steady-state behaviour and no transient creep. By contrast, at low temperatures and modest stresses, many metallic materials show only the transient component, deformation ceasing after a given time or appearing to become asymptotic to some maximum strain level. These features are illustrated schematically in Figure 7.15. In many creep investigations, it is the steady state that attracts most attention because the temperature- and stress-dependence of the constant deformation rate, $d\varepsilon/dt$, usually denoted $\dot{\varepsilon}_{II}$, can be interpreted in terms of the atomic or molecular mechanisms of deformation and the rate theory of chemical reactions (Glasstone *et al.*, 1941). The stress-dependence of the secondary creep rate of many materials can be written in terms of the power-law equation:

$$\dot{\varepsilon}_{II} = A\sigma^n \qquad (T = \text{constant}) \tag{7.6}$$

so that a log/log plot of creep rate versus stress is linear, and this is often combined with an Arrhenius expression for the temperature dependence in a general creep law:

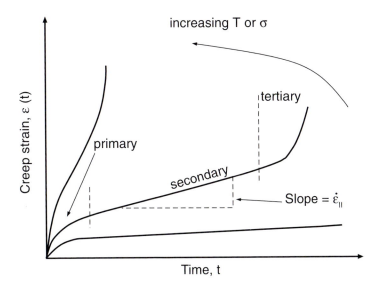

increasing T or σ

primary

secondary

tertiary

Slope = $\dot{\varepsilon}_{\mathrm{II}}$

Creep strain, ε (t)

Time, t

Fig. 7.15 Schematic illustration of the appearance of metallic creep curves.

$$\dot{\varepsilon}_{11} = B\sigma^n e^{-(Q/kT)} \qquad (7.7)$$

In these equations, A, B and *n* are materials constants, *k* is Boltzmann's constant, T is in degrees Kelvin, and Q is loosely described as being an 'activation energy' for whatever deformation process is controlling the rate of the mechanism causing creep. If a single mechanism is responsible for the observed strain, then in principle Q may be calculated from physical knowledge of the mechanism. For example, at temperatures above $\frac{1}{2}T_m$, solid-state diffusion occurs very rapidly in metals, and creep is mostly a result of dislocation climb as a result of vacancy movement through the lattice. The observed value of Q obtained by plotting log($\dot{\varepsilon}_{\mathrm{II}}$) versus 1/T is therefore usually equal to the activation energy for self-diffusion (Dorn, 1957). At high homologous temperatures and low stresses the value of *n* may fall to near unity and the deformation approximates Newtonian flow (strain rate proportional to stress), while at low temperatures and high stresses the deformation may be controlled by dislocation movements other than climb (*i.e.* glide, cross-slip *etc.*), in which case the value of *n* may be very large. These changes are shown schematically in Figure 7.16, and are illustrated in Figure 7.17 by some results for the creep of cold-drawn tungsten filaments, 76 μm in diameter, at various temperatures. It can be seen that the creep process at the highest temperature represented, above 1600 K — just about $\frac{1}{2}T_m$ — the deformation process appears to be controlled by self diffusion, whereas at the lowest temperature, about 900 K, the value of *n* is characteristic of non-diffusion processes.

The modelling of the time-dependent deformation of polymers is done in a slightly different fashion. Various combinations of elastic springs and viscous dashpots are often

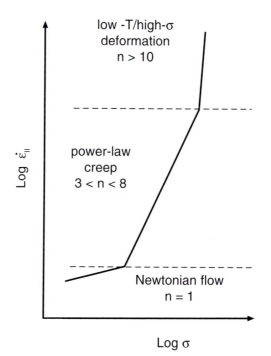

Fig. 7.16 Schematic illustration of the variation of the secondary creep with stress for metallic materials.

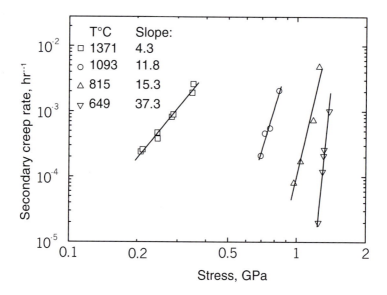

Fig. 7.17 Secondary creep rate of 76 μm diameter tungsten wires as functions of stress and temperature (Harris & Ellison, 1966).

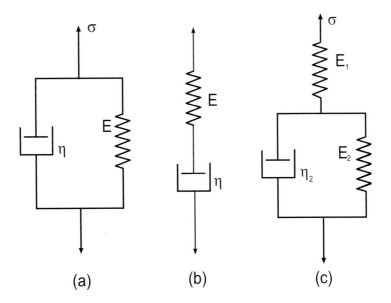

Fig. 7.18 Simple spring and dashpot models that are used to represent time-dependent behaviour of polymers. The parallel model, (a), is known as a Voigt element, the series model, (b), is known as a Maxwell element, and the combination of the two represented by (c) is known as a standard linear solid (SLS).

used to represent their visco-elastic response, and there is a highly developed mathematical theory of visco-elasticity (Ferry, 1980). The three simplest spring/dashpot models are those shown in Figure 7.18. The Hookean spring is supposed to have an elastic modulus E, and the Newtonian dashpot has a viscosity η. The parallel element (a) responds like a damped automotive shock absorber, its maximum strain level being limited by the extension of the spring. The series model, (b), however, has no strain limit and deforms in a viscous manner after an initial elastic extension. Neither of these represents perfectly the real behaviour of a polymer, and more complex combinations such as that labelled (c), known as a standard linear solid (or Zener solid), provides a somewhat better model. The derivation of the creep curve for the Voigt model (model (a) in Figure 7.18) will illustrate the modelling procedure.

The responses of the separate elements are the normal definitions of elastic and viscous deformation:

$\sigma = E\varepsilon$ (Hookean), and

$\sigma = \eta \dfrac{d\varepsilon}{dt}$ (Newtonian)

The stress acting on the element is shared between the two components: the total stress on the element is the sum of the stresses in the two components:

$$\sigma = \sigma_{spring} + \sigma_{dashpot}$$

and hence we obtain the constitutive equation for the Voigt element:

$$\sigma = E\varepsilon + \eta \frac{d\varepsilon}{dt} \tag{7.8}$$

The ratio η/E is called the time constant, τ, and this is referred to as a *retardation time* or a *relaxation time*, depending on the deformation conditions. Rewriting the constitutive equation to incorporate this definition, we obtain the differential equation:

$$\frac{d\varepsilon}{dt} + \frac{\varepsilon}{\tau} = \frac{\sigma}{E\tau} \tag{7.9}$$

which on integration yields the creep equation:

$$\varepsilon(t) = \frac{\sigma}{E}\left(1 - e^{-t/\tau}\right) \tag{7.10}$$

This strain, which is fully recoverable, is shown as a time function in Figure 7.19 together with the responses of the simple elastic and viscous components and the series Maxwell element. The exponential form is not unlike that of the bottom curve in Figure 7.15, and it is generally considered that the Voigt model does represent the creep behaviour of polymers to a first approximation (Ward, 1971). It is more common in discussing these models to present them in terms of a *creep compliance*, J(t):

$$J(t) = E^{-1}[1 - \exp(-t/\tau)] \tag{7.11}$$

which has the normal units of compliance, $mm^2.N^{-1}$.

The constitutive equation for the standard linear solid (SLS) is more complicated because there are two time constants instead of one:

$$\sigma + \tau_1 \frac{d\sigma}{dt} = E\left(\varepsilon + \tau_2 \frac{d\varepsilon}{dt}\right) \tag{7.12}$$

where E is now a combination of the stiffnesses of the two springs, $(1/E_1 + 1/E_2)^{-1}$, and the two time constants in the equation are $\tau_1 = \eta_2/(E_1 + E_2)$ and $\tau_2 = \eta_2/E_2$. The deformation of the SLS, like that of the Voigt element, is also reversible since the overall control of extension is by the two spring elements.

The foregoing comments about spring/dashpot models relate only to linear visco-elastic behaviour. The concept of a linear visco-elastic solid differs from that of a linear elastic material in that it involves the time-dependence characterized by the derivatives, as in equation 7.12. In a linear visco-elastic material, if a particular stress causes a given strain in a given time, a stress which is twice as great will cause twice the strain in the same amount of time. The assumption of visco-elasticity, which in the Voigt model (but not in the Maxwell model) implies reversibility, is often an invalid assumption for the behaviour of polymers if the strain is not small. A fuller treatment of the behaviour of non-linear viscoelastic behaviour in polymers and polymer composites has been given by Schapery (1996) and Emri (1996).

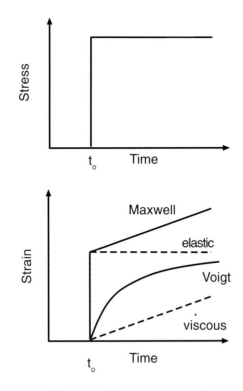

Fig. 7.19 Strain responses of simple solids to a stress applied at time t_o.

The time-dependent and frequency-dependent behaviour of polymers are closely linked, *via* the theory of linear visco-elasticity (Ferry, 1980), and are frequently studied by the experimental method known as dynamic mechanical and thermal analysis (DMTA). Because of its visco-elastic nature, energy is dissipated when a polymer undergoes load cycling and this dissipation, characterized by the hysteresis or energy loss per cycle, reaches a maximum whenever the frequency of molecular motion is close to the cycling frequency. The energy loss is usually measured as the ratio of the loss modulus to the storage modulus, E''/E', which is approximately equal to the phase angle, $\tan\delta$. Thus, measurements of $\tan\delta$ as a function of temperature will identify the occurrence of specific mechanisms of molecular activity. DMTA is thus a sensitive means of detecting changes in the mobility of molecules and for investigating phase structure and morphology. For polymeric materials, it is able to identify a range of relaxations and transitions, and is thus a potentially powerful tool for assessing the effects of such features of composite materials as matrix polymer modification (*e.g.* 'flexiblization'), matrix crystallinity, transcrystallinity at a fibre/thermoplastic-matrix interface, and interphase behaviour or other effects of fibre surface treatments.

An example of the use of DMTA for the study of fibre composites is shown in Figure 7.20. The materials are a plain epoxy resin and unidirectional composites of this

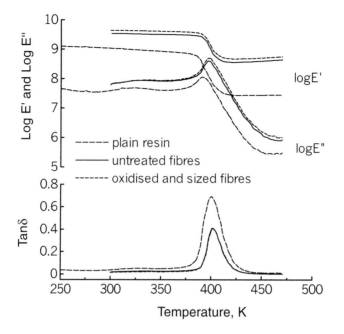

Fig. 7.20 Loss and storage moduli and the loss tangent for epoxy resin and also samples of *unidirectional* composites containing carbon fibres with no surface treatment and in the oxidized and sized conditions. The resin is Ciba Geigy LY1927GB epoxy resin and the unidirectional composites contain 10 vol% of ENKA HTA carbon fibres. (Harris *et al.*, 1993)

resin reinforced with 10 vol% of carbon fibres, both without surface treatment and with a normal, commercial, oxidized and sized surface. The upper graph shows the storage and loss moduli, E' and E". The storage moduli for the three materials all show the same features — a relatively flat section at low temperatures with a marked drop at about 400 K to a second flat region at high temperature. The fall coincides with a peak in the loss modulus which identifies the glass-transition temperature, T_g. It can be seen that the stiffness of the composites, as indicated by the real component of the dynamic flexural modulus, E', is significantly higher than that of the resin, as would be expected, over the whole temperature range, and the effect of the commercial surface treatment is to increase the composite stiffness slightly above that for the composite with untreated fibres. The curves for the loss modulus, E", show that the hysteretic loss in the composites is also higher than that in the plain resin, and we assume that this is because of something like frictional losses at fibre/resin interfaces. There is no clear separation between the E" curves for the two surface treatments, however, and this is confirmed by the tanδ curves in the lower graph. But there is some suggestion that the temperature of the loss peak (often referred to as the α peak) associated with the glass transition is raised slightly by the presence of the fibres. The composite thus reflects the visco-elastic features of the matrix resin. Although all of the graphs exhibit a single, major α peak at about 130°C, the resin curves also show a broad, weakly defined β transition in the region of 50°C.

This very much simplified sketch can serve as only an introduction to a very extensive subject, and for further details specialist books on creep should be consulted (*e.g.* for metals: Garofalo, 1965, Evans & Wilshire, 1993; for plastics: Ward, 1971, McCrum *et al.*, 1988).

7.4.2.2 Creep of Composites

Creep, like high-temperature strength, depends upon the stability of the fibre and the extent to which it can relieve the matrix of direct load. Thus, in composites reinforced with continuous, well-aligned carbon or boron fibres no creep strains should be observed apart from those which accompany small amounts of straightening or realignment of the fibres under stress. But even though the fibres themselves do not exhibit any creep deformation, small redistributions of load due to matrix creep will still lead to time-dependent fibre breakage because of the statistical distribution of defects in the brittle fibres. Lifshitz and Rotem (1972) have developed a statistical model for the time-dependent failure of reinforced plastics which predicts the reduction of strength of composites under constant tensile loads. In GRP this loss in strength with time can be substantial. Lifshitz and Rotem found that failure occurred at room temperature in 0.60 V_f unidirectional glass/polyester composites after 10^5 minutes at 50% of the ordinary tensile strength, while the value for glass/epoxy composites was about 70%. Fuwa *et al.* (1975) showed by means of acoustic emission studies that in unidirectional carbon/epoxy composites the random fibre damage ceased after a few minutes even at stresses greater than 90% of the tensile strength. On the other hand, creep in laminates under interlaminar shear stresses or in short-fibre composites under any kind of load will be rapid if the matrix has no inherent creep resistance. Creep will also occur in laminates with fibres at some angle to the stress axis in a manner characteristic of the constrained matrix material.

Early creep models were largely developed for all-metal systems in which the creep behaviour depends on how the phenomena described in the previous section affect the components of the composite. There are a number of separate possible cases:

i.   Continuous fibres; matrix creeping, fibres elastic; stress too low to cause damage to fibres
ii. Continuous fibres; matrix creeping, fibres elastic; stress high enough to cause breakdown of fibres
iii. Continuous fibres; matrix and fibres creeping; stress too low to cause damage to fibres
iv. Continuous fibres; matrix and fibres creeping; stress high enough to cause breakdown of fibres
v.   Short fibres; matrix creeping, fibres elastic
vi. Short fibres; matrix and fibres creeping.

Consider, for example, a metal-matrix composite in which the matrix creeps but the fibre remains elastic and is undamaged by the stress. The creep strain of the matrix would be constrained by the capacity of the fibres for elastic deformation, so that the maximum possible creep strain of the system would be equal to $\sigma_c/E_f V_f$, where σ_c is the stress acting on the composite. The deformation of the material would approach this strain level asymptotically, much like the response of the Voigt element shown in Figure 7.19. Following the spring/dashpot arguments given earlier, the stress on the composite is divided between the parallel components according to the mixture rule:

$$\sigma_c = \sigma_m V_m + \sigma_f V_f \qquad (7.13)$$

The strain in each of the two components is the same as that of the composite (the *iso-strain model*), but since we are dealing with time-dependent effects we can only write:

$$\dot{\varepsilon}_c = \dot{\varepsilon}_m = \dot{\varepsilon}_f$$

At a given temperature, then, the fibre deformation equation is:

$$\dot{\varepsilon}_f = \frac{1}{E_f}\left(\frac{d\sigma_f}{dt}\right) \qquad (7.14)$$

and the matrix deformation equation is:

$$\dot{\varepsilon}_m = \frac{1}{E_m}\left(\frac{d\sigma_m}{dt}\right) + B_m \sigma_m^P \qquad (7.15)$$

the first term being the elastic strain rate and the second the plastic creep rate, assuming power-law creep (equation 7.6) with an exponent *p*. Combining equations 7.14 and 7.15 in the differentiated form of equation 7.13, we obtain an equation for the creep rate of the composite (McLean, 1982):

$$\dot{\varepsilon}_c = \alpha B_m \sigma_m^P \left(1 - \frac{\varepsilon}{\varepsilon_c}\right)^p \qquad (7.16)$$

where α is given by:

$$\alpha^{-1} = \left[1 + \left(\frac{E_f V_f}{E_m V_m}\right)\right](1 - V_f)^p$$

McLean showed that this model was able to predict reasonably accurately the creep behaviour of the *in situ* eutectic composite $\gamma/\gamma'/Cr_3C_2$.

If the temperature is sufficiently high for both fibres and matrix to creep, the deformation equations for the two components, ignoring the elastic strains now, are:

$$\begin{aligned} \dot{\varepsilon}_m &= B_m \sigma_m^P \\ \dot{\varepsilon}_m &= B_m \sigma_m^P \end{aligned} \quad T = \text{constant}$$

where B_f and *q* are the power-law constants for the fibres. The stress on the composite is then given by:

$$\sigma_c = V_f \left(\frac{\dot{\varepsilon}_c}{B_f}\right)^{1/p} + (1 - V_f)\left(\frac{\dot{\varepsilon}_c}{B_m}\right)^{1/q} \qquad (7.17)$$

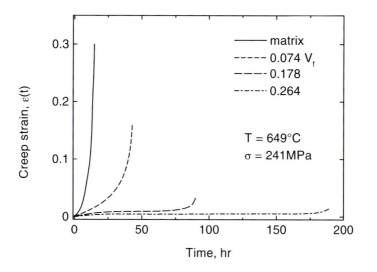

Fig. 7.21 Curves of creep strain versus time for unidirectional composites consisting of Inconel reinforced with tungsten wires (Ellison and Harris, 1966). The stress level was 241 MPa and the temperature 649°C.

This equation cannot be solved explicitly for the composite creep rate, but it can be used to derive graphical relationships between creep rate and stress for particular materials if experimental data for B_m, B_f, p and q are available. As an example, we take some results of Ellison and Harris (1966) for the creep of an MMC consisting of an Inconel-600 nickel alloy reinforced with the cold-drawn tungsten wires for which some creep data were shown in Figure 7.17. A set of creep curves for the plain matrix and composites of three different volume fractions at 649°C and a stress of 241 MPa are shown in Figure 7.21 and the effect of the fibre additions on the rupture time and the strain to failure can be seen from Figure 7.22. If the experimental values of B_f, B_m, p and q for the matrix and for the reinforcing wires are extracted from the publications of Harris and Ellison (1996), curves of predicted behaviour of composites based on the mixture-rule model of equation 7.17 can easily be derived for a given temperature, as in Figure 7.23. As shown by Street (1971) the predicted result for the lowest V_f composite ($V_f = 0.074$) corresponds reasonably well with the experimental result, but as the fibre volume fraction increased, the predictions underestimate the measured creep rates by increasingly large amounts. There are a number of reasons why the model of equation 7.17 fails to describe the experimental data. The first relates to the validity of the equation. We have assumed power-law behaviour for both matrix and fibres. In other words, we are dealing in equation 7.17 with *steady-state creep* only. Yet it is clear from Figure 7.21 that the secondary creep regions for fibres and matrix cannot coincide and the basic premise on which the equation is founded must therefore be invalid. Since we do not have a simple model for the stress-dependence of transient creep, however, we cannot re-derive the model with more appropriate stress-dependent functions.

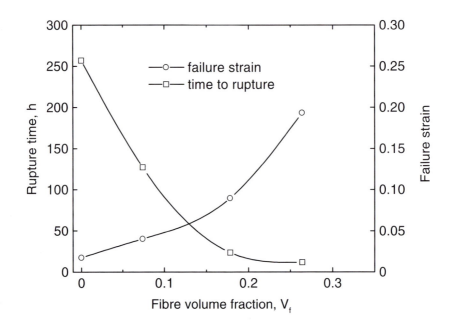

Fig. 7.22 Creep rupture time and failure strain as functions of fibre volume fraction for unidirectional composites of Inconel-600 reinforced with tungsten wires creep-tested at 649°C and 241 MPa (Ellison & Harris, 1966).

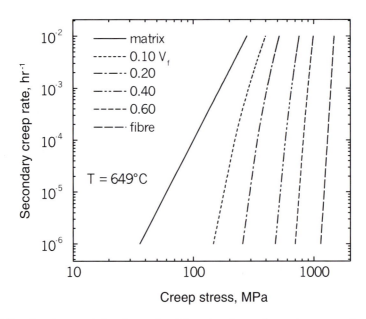

Fig. 7.23 Predicted creep behaviour of unidirectional metal-matrix composites with both fibre and matrix creeping. The data for fibre and matrix are based on results of Ellison and Harris (1966) for composites of Inconel reinforced with tungsten wires.

A second possible cause for the discrepancy is that we assume a homogeneous state of stress within the composite. It has been shown, however, that in MMCs inhomogeneous deformation resulting from the differences in elastic properties of the two components leads to greater levels of work hardening in the matrix close to the fibre than in material more remote from the interface (Kelly and Lilholt, 1969). It may be, then, that the matrix is creeping at a higher true stress level than is assumed in the model of equation 7.17. Kelly and Tyson (1966) had also pointed out that although the addition of fibres substantially reduced the creep rate of a short-fibre composite, $\dot{\varepsilon}$ was nevertheless governed by the creep of the matrix under the action of the shear stresses close to the interface.

In practice the creep resistance of a matrix will usually be much lower than that of a reinforcing fibre, and we might expect that at high stress and high temperature the creep of the composite would be controlled by that of the fibres alone under the fibre stress, $\sigma_c V_f$. Street (1971) pointed out that at low V_f neglect of the matrix contribution could result in significant *over-estimates* of the composite creep rate, but Harris (1972) showed that the times-to-failure of these Inconel/tungsten composites at the equivalent fibre stress $\sigma_c V_f$ agreed with the creep behaviour of the isolated wires for test times of 100 hours or so at 650°C. When the time to failure was increased well beyond this by increasing V_f, however, deterioration of the wires by intermetallic diffusion led to a reduction in creep resistance below that predicted from the isolated fibre properties.

Street also points out that, on the basis of equation 7.17, the composite will not exhibit a unique value of the power-law exponent, n_c (*i.e. n* in equations 7.6 and 7.7), although the matrix and fibres do (*p* and *q* in equation 7.17). As implied by Figure 7.23, for most V_f the value of n_c is determined by the fibres at low creep rates, but falls at higher $\dot{\varepsilon}$ as the effect of the matrix increases. The measured values of n_c thus depend on both volume fraction and deformation rate.

The creep behaviour of short-fibre composites is governed by the matrix rather than the fibres. A simplistic approach would be to suggest that the stress component in equation 7.17 is replaced by a term describing the average stress in the short fibres, similar to that discussed in section 4.18 of chapter 4. For a non-creeping fibre, the revised equation then becomes:

$$\sigma_c = V_f \sigma_{fu} \left(1 - \frac{\ell_c}{2\ell}\right) + \left(1 - V_f\right) \left(\frac{\dot{\varepsilon}_c}{B_m}\right)^{1/q} \tag{7.18}$$

where ℓ_c is the fibre critical length, and σ_{fu} is the fibre tensile strength. Assuming that the fibres are longer than or equal to the critical length, the effect is then for the fibres simply to shield the matrix from part of the load. The creep strain rate can then be obtained explicitly as:

$$\dot{\varepsilon}_c = B_m \left[\frac{\sigma_c - V_f \sigma_{fu} \left(1 - \ell_c / 2\ell\right)}{1 - V_f}\right]^p \tag{7.19}$$

This can only be a lower bound to the composite creep rate, however, because other processes will occur during the creep of a short-fibre composite. Shear stresses at fibre ends will be relaxed, causing load redistribution, stress relaxation will occur in the matrix, and load will be transferred by shear into the fibres. Creep of the fibres themselves may also occur if the temperature and stress are sufficiently high. Starting from a more rigorous evaluation of the average stress in the fibre than that assumed in equation 7.18, Kelly and Street (1972b) derived a model which took most of these factors into account. They found that, although the fibres reduce the effective load on the matrix, the presence of the fibres actually increases the rate of matrix shear, and this increase is inversely proportional to the fibre spacing. They showed that for rigid, perfectly bonded fibres of aspect ratio ℓ/d:

$$\dot{\varepsilon}_c = B_m \sigma_p^c \left[V_f \, \Phi(\ell/d)^{(1+1/p)} + V_m \right]^{-p} \tag{7.20}$$

where $\Phi(\ell/d)$ is a load-transfer parameter dependent on V_f and on the power-law exponent, p. For $p \approx 6$ and $V_f = 0.44$, Φ is approximately 0.5: at $V_f = 0$, equation 7.20 reduces to the creep response of the plain matrix, as required. This model agrees well with experimental measurements by the same authors (1972a) on composites of lead reinforced with phosphor-bronze wires.

Mileiko (1970) also developed a model for the creep of short-fibre composites in an attempt to take these mechanisms into account. His was a simple shear model of steady-state creep which assumed that the creep 'character' of the interface was the same as that of the matrix and that the shear deformation obeyed the same power-law creep model as the tensile deformation. It treated hexagonal and plate-like reinforcement arrays, and examined the rôles of V_f, aspect ratio, and fibre distribution. With the terminology modified to conform with the foregoing discussion, Mileiko's model gives the composite creep rate as:

$$\dot{\varepsilon}_c = B_m \sigma_p^c \frac{2^{2p+1}}{(\ell/d)^{p+1}} \left(\frac{1-V}{V_f^{p+1}} \right) \tag{7.21}$$

Although Mileiko assumed that the matrix load-carrying capacity was negligible, his model nevertheless gives good agreement with the results of Kelly and Tyson (1966) for silver/tungsten composites mentioned earlier. The model only agrees with the predictions of the Kelly and Street (1972b) model if a matrix term is added.

These early models ignored the possibility that fibres may fracture as deformation proceeds. McLean (1986), however, has modelled the creep of aligned composites containing brittle fibres that fracture as a result of the interaction between the fibre flaw distributions and the local stress distributions in the creeping material. A recent analysis by Durodola et al. (1994) modifies McLean's model to take into account the effects of inherent residual thermal stresses, matrix primary creep, and the effect of matrix yielding. Good agreement is obtained between the predictions of this model and experimental results for SiC/Ti-6Al-4V MMCs. Some success has also been obtained in the application of these power-law creep models to CMCs: a recent summary is given by Chawla (1993).

Results are not always what might be expected, however. For example, Nixon *et al.* (1990), studying the steady-state creep of hot-pressed composites of Si_3N_4 reinforced with 20 and 30 vol% of SiC whiskers, found that under some circumstances the composites crept more rapidly than the unreinforced matrix ceramic. The cause of this appears to be that grain-boundary sliding by viscous flow was the controlling creep mechanism in these composites, and this viscous flow (with power-law exponents of the order of unity) was limited by crystallisation of a glassy phase at the grain boundaries. The SiC contained some silica which was contributed to the overall grain-boundary glass composition with a resultant decrease in the degree of crystallisation and hence an increase in the creep rate. It is imperative, when dealing with systems required for elevated-temperature exposure, therefore, that all compositional and constitutional aspects of composite behaviour be considered in addition to straightforward micro-structural aspects.

Much of the above discussion is based on the simple mixture-rule assumption that the composite creep behaviour is a weighted average of the behaviour of the separate constituents and ignores detailed effects of microstructure. And while this may be satisfactory for many of the synthetic composites manufactured by combining processes, it may not be valid for materials such as *in situ* composites which may contain reinforcements of sub-micron dimensions which provide extra creep resistance, perhaps as a consequence of dispersion strengthening. In directionally solidified alloys there is evidence that the creep-resistance of material with fine, closely-spaced fibres is much greater than that of the same alloy grown with coarse, widely-spaced fibres, even though there is not much difference between their room-temperature strengths. It is also well known that aspect ratio plays an important rôle: the creep rates of an aluminium alloy reinforced with SiC whiskers, for example, are some two orders of magnitude lower than that of an identical alloy reinforced with the same volume fraction of SiC particles (Nieh, 1984).

The creep behaviour of polymer composites largely concerns materials in which the matrix is visco-elastic and the fibres are inert at most use temperatures. The 'creeping-matrix/elastic-fibres' models are therefore appropriate for unidirectional or 0/90 laminates, but not for the majority of practical composites where fibres may be random, or at 45° to the stress. In such cases, composite deformation may often be modelled by the same kinds of procedures as those used for unreinforced linear visco-elastic polymers (Emri, 1996) as illustrated by the room-temperature creep curves for some unreinforced and glass-reinforced Nylon manufactured by injection moulding shown in Figure 7.24. On the strain *vs.* log(time) basis the curves have the normal concave-upwards appearance, and the effect of the glass-fibre reinforcement is readily seen. Many examples have been published of the use of the Findlay equation, referred to earlier, and time/temperature superposition to obtain master curves of creep strain (or creep compliance) *vs.* temperature or the related curve of relaxation modulus *vs.* temperature (or time). For glass-fabric-reinforced thermoplastics, Brüller (1996), for example, shows that these materials behave in a linear visco-elastic fashion when loaded in the warp and weft directions, but are highly non-linear when loaded at 45° to the fibre directions. Ma *et al.* (1997), working on $(\pm 45)_{4S}$ carbon-fibre/PEEK composites, obtained results at temperatures up to 200°C

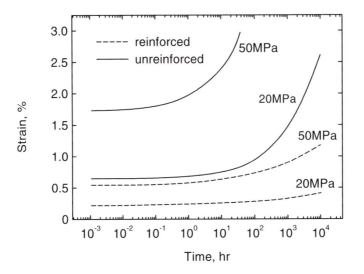

Fig. 7.24 Creep curves for unreinforced and glass-filled Nylon in air at 20°C (after Crawford, 1981)

for which the Findlay relationship gave a good description, and they determined a master curve by time/temperature superposition which gave good predictions of creep behaviour. Miranda Guedes *et al.* (1996) also determined master creep curves in flexure for carbon/epoxy laminates, despite the fact that bending tests do not produce uniform strains.

7.5 REFERENCES

T.J. Aponyi: *Modern Plastics*, 1967, 1967, 151-154.

ARC Concrete Slimline: *The Specification and Performance of Slimline Glass Fibre Reinforced Concrete Pipes*, R.M. Farahar, *et al.*, Paper Presented at a technical seminar held at ARC Concrete Research Centre, 1979.

K.H.G. Ashbee, F.C. Frank and R.C. Wyatt: *Proceedings of Royal Society*, *London*, **A300**, 1967, 415-419.

J. Aveston, A Kelly and J.M. Sillwood: *Proceedings ICCM3 (Paris) Advances in Composite Materials*, Pergamon, Oxford, **1**, 1980, 556-568.

F. Beuche and F.N. Kelly: *Journal of Polymer Science*, **45**, 1960, 267-269.

K.H. Boller: *Strength Properties of Reinforced Plastics at Elevated Temperatures* WADC Tech. Report 59-569, Wright Air Development Centre, Dayton, Ohio, 1960.

O.S. Brüller: *Progress in Durability Analysis of Composite Systems* A.H. Cardon, H. Fukuda and K. Reifsnider, eds.: A.A. Balkema, Rotterdam, The Netherlands, 1996, 39-44.

B. Charrière: Unpublished results, School of Materials Science, University of Bath, 1985.

K.K. Chawla: 1993, *Ceramic Matrix Composites*, Chapman and Hall, London.

T. Collings and D.E.W. Stone: *Composites Structures*, **3**, 1985, 341-378.

R.J. Crawford: *Plastics Engineering,* Pergamon Press, Oxford, 1981.

P.T. Curtis: *A Basic Computer Program to Calculate Moisture Content in Resin and Fibre Resin Composites*, RAE, Farnborough, Technical Memo MAT 375,1981,1981.

R. Delasi and J.B. Whiteside: *Advanced Composite Materials -Environmental Effects*, J.R. Vinson, ed., ASTM STP 658, 1978, 2-20.

R.F. Dickson, C.J. Jones, B. Harris, H. Reiter and T. Adam: *Proceedings of Fibre-Reinforced Composites '84*, Plastics and Rubber Institute, London, 1984, paper 30 1-11.

J.E. Dorn: *Creep and Recovery*, American Society for Metals, Cleveland, Ohio, U.S.A, 1957, 255 *et seq.*

J.F. Durodola, C. Ruiz and B. Derby: *Composites Engineering*, **4**, 1994, 1241-1255.

E.G. Ellison and B. Harris: *Applied Materials Research*, **5**, 1966, 33-40.

I. Emri: *Durability Analysis of Structural Composite Systems,* A.H. Cardon, ed., A.A. Balkema, Rotterdam, The Netherlands, 1996, 85-122.

R.W. Evans and B. Wilshire: *Introduction to Creep*, Institute of Materials, London, 1993.

G. Fernando: PhD thesis, School of Materials Science, University of Bath,1986.

J.D. Ferry: *Viscoelastic Properties of Polymers*, John Wiley, New York,1980.

W.N. Findley and G. Khosla: *Journal of Applied Physics*, **26**, 1955, 821-832.

M. Fuwa, B. Harris and A.R. Bunsell: *Journal of Physics D: Applied Physics*, **8**, 1975, 1460-1471.

F. Garofalo: *Fundamentals of Creep and Creep Rupture in Metals*, Macmillan Company, New York, 1965.

S. Glasstone, K.J. Laidler and H. Eyring: *Theory of Rate Processes: the Kinetics of Chemical Reactions, Viscosity, Diffusion and Electrochemical Phenomena*, McGraw-Hill London,1941.

B. Harris and E.G. Ellison: *Transactions of ASM,* **59**, 1966, 744-754.

B. Harris: *Composites*, **8**, 1972, 152-167.

B. Harris: *Metal Science*, **14**, 1980, 351-362.

B. Harris, O.G. Braddell, D.P. Almond, C. Lefèbvre and J. Verbist: *Journal of Materials Science*, **28**, 1993, 3353-3366.

P.J. Hogg and D. Hull: *Developments in GRP Technology*, B. Harris, ed., Elsevier Applied Sciences, London, 1983, 37-90.

F.R. Jones, J.W. Rock and J.E. Bailey: *Journal of Materials Science*, **18**, 1983, 1059-1071.

A. Kelly and H. Lilholt: *Philosophical Magazine*, **20**, 1969, 311-328.

A. Kelly and K.N. Street: *Proceedings of Royal Society, London*, **A328**, 1972a, 276-282,

A. Kelly and K.N. Street: *Proceedings of Royal Society, London*, **A328**, 1972b, 283-293.

A. Kelly and W.R. Tyson: *Journal of Mechanics and Physics of Solids*, **14**, 1966, 177-186.

J.M. Lifshitz and A. Rotem: *Journal of Materials Science*, **7**, 1972, 861-869.

C.C.M. Ma, N.H. Tai, S.H. Wu, S.H. Lin, J.F. Wu and J.M. Lin: *Composites*, **28B**, 1997, 407-417.

A.J. Majumdar: *Proceedings of Royal Society, London*, **A319**, 1970, 69-79.

N.G. McCrum, C.P. Buckley and C.B. Bucknall: *Principles of Polymer Engineering*, Oxford University Press Oxford, 1988.

M. McLean: *Proceedings 3rd Risø Symposium Fatigue & Creep of Composite Materials*, 1982, H. Lilholt and R Talreja, eds., Risø National Laboratory, Roskilde, Denmark, 1982, 77-88.

M. McLean: *Composites Science and Technology*, **23**, 1985, 37-52.

M. McLean: *Proceedings of 4th Conference of the Irish Durability and Fracture Committee*, F.R Montgomery, ed., Queen's University, Belfast, 1986, 202-213.

G. Mensitieri, A. Apicella, L. Nicolais and M.A. Del Nobile: in *Progress in Durability Analysis of Composite Systems*, A.H. Cardon , H. Fukuda and K. Reifsnider, A.A. Balkema, Rotterdam, The Netherlands, 1996, 249-258.

S.T. Mileiko: *Journal of Materials Science*, **5**, 1970, 254-261.

R. Miranda Guedes, A.T. Marques and A.H. Cardon: *Progress in Durability Analysis of Composite Systems*, A.H. Cardon, H. Fukuda and K. Reifsnider, eds., A.A. Balkema, Rotterdam, The Netherlands, 1996, 211-214.

T.G. Nieh: *Metallic Transactions*, **15A**, 1984, 139-146.

R.D. Nixon, D.A. Koester, S. Chevacharoenkul and R.F. Davis: *Composites Science and Technology*, **37**, 1990, 313-328.

A. Paillous and C. Pailler: *Composites*, **25**, 1994, 287-295.

W.H. Pfeifer: *Hybrid and Select Metal-Matrix Composites*, W.J. Renton, ed., AIAA, New York, 1977, 159-255.

D.C. Phillips: *Proceedings of 9th Risø International Symposium - Mechanical and Physical Behaviour of Metallic and Ceramic Composites*, S.I. Andersen, H. Lilholt and O.B. Pedersen, eds., Risø National Laboratory Roskilde Denmark, 1988, 183-199.

M.G. Phillips, N. Heppel, R.C. Wyatt and Norwood: *Proceedings of 38th Technical Conference of Reinforced Plastics/Composites Institute of SPI*, 1983, paper 2-D.

M.G. Phillips, R.C. Wyatt and L.S. Norwood: *Proceedings of First Paisley Conference on Composites Structures*, I.H. Marshall, ed., Applied Science Publishers, London, 1981, 79-91.

B.A. Proctor and B. Yale: in *New Fibres and their Composites*, W. Watt, B. Harris and A.C. Ham, eds., Royal Society, London, 1980, 19-28.

R.A. Schapery: *Progress in Durability Analysis of Composite Systems*, A.H. Cardon, H. Fukuda and K. Reifsnider, eds., A.A. Balkema, Rotterdam, The Netherlands, 1996, 21-38.

C.H. Shen and G.S. Springer: *Journal of Composite Materials*, **10**, 1976, 2-20.

W.S. Smith: *Environmental Effects on Aramid Composites, Proceedings of SPE Meeting, 1979 Los Angeles*, 1979, paper A-34.

K.N. Street: *Proceedings of NPL Conference on The Properties of Fibre Composites*, Teddington, U.K., IPC Science and Technology Press, Guildford, Surrey, 1971, 36-46.

T.N. Tiegs and P.F. Becher: *Materials Science Research*, **20**, in *Proceedings of a Conference on Tailoring Multiphase and Composite Ceramics,* R.T. Tessler, G.L. Messing, C.G. Pantano and R.E. Newham, eds., 1985, Plenum Press, New York, 1986.

I.M. Ward: *Mechanical Properties of Solid Polymers*, Wiley-Interscience, London, 1971.

S.N. Zhurkov: *International Journal of Fracture Mechanics*, **1**, 1965, 311-322.

8. Problems

1. Show that the theoretical maximum volume fraction for identical continuous fibres of circular cross section in a composite material is 0.91.
   How does this figure relate to the volume fractions of practical composites?

2. Why is it important for a multi-ply fibre composite laminate to have a balanced structure?
   Use Krenchel's method to determine the stiffness of a 32-ply carbon-fibre-reinforced plastic laminate whose structure is represented by the formula $[(\pm 30, 0_2, \pm 60, 90_2)_2]_s$ if the fibre Young's modulus is 280 GPa, the resin modulus is 3.2 GPa and the volume fraction of the individual plies is 0.65.

3. A uniaxially reinforced glass-fibre/epoxy composite with $V_f = 0.65$ is loaded by a tensile stress of 145 MPa acting at an angle of 30° to the fibres. If the tensile modulus of glass fibres is 70 GPa and the Poisson ratio of glass is 0.25, and the same two properties for the epoxy resin are 3 GPa and 0.33, respectively, calculate the tensile strain in the composite parallel to the fibres.
   Hint: you may need to use a Mohr's circle construction.

4. For a composite material with aligned fibres, show that the major Poisson ratio, ν_{12}, may be estimated by applying the rule of mixtures.
   Hint: you will need to use both the series and parallel models.

5. Derive an expression for the density, ρ_c, of a unidirectional composite in terms of the densities of the components and the fibre volume fraction.
   From the material parameters given in the table find ρ_c and E_c for the following composites:
   (a) carbon-fibre/epoxy-resin of $V_f = 0.6$,
   (b) glass-fibre/polyester-resin of $V_f = 0.65$, and
   (c) steel-fibre-reinforced concrete of $V_f = 0.025$).

Material	Density ρ, 10^3 kgm^{-3}	Young's Modulus E, GPa
Carbon fibre	1.90	390
Glass fibre	2.55	72
Epoxy resin	1.15	3
Polyester resin	1.15	3
Steel	7.9	200
Concrete	2.4	45

A rectangular-section beam of fixed width, w, unspecified depth, h, and fixed length, L, is laid horizontally on two supports at either end of the beam. A force F acts vertically downwards through the centre of the beam. The deflection, δ, of the loaded point is:

$$\delta = \frac{FL^3}{4E_c\,wh^3}$$

ignoring the deflection due to beam's own weight. Which of the composites a) to c) will give the lightest beam for a given force and deflection?

6. A composite consisting of 40 vol% of uniaxially aligned continuous fibres in a polyester resin supports a stress of 100 MPa parallel to the fibres. If the tensile moduli, E_f and E_m, respectively, of the fibres and matrix are 75 GPa and 5 GPa, and their Poisson ratios, v_f and v_m, are 0.21 and 0.35, determine the strains that will result in the longitudinal and transverse directions.

Make an estimate of the transverse modulus, E_2, of the composite, and comment on the likely validity of the estimate.

7. A thin unidirectional ply of a carbon-fibre-reinforced plastic has Young moduli of 178 GPa in the fibre direction (E_1) and 9 GPa in the transverse direction (E_2). The in-plane shear modulus, G_{12}, is 5 GPa, and the major Poisson's ratio, v_{12}, is 0.25. Experimental measurements are to be made in uniaxial tension on this material, and the accuracy of the values determined is of major importance. Calculate what angles of misalignment between the fibre direction (*i.e.* the x_1 direction) and the loading direction would cause reductions of 1% and 5% in the measured values of E.

What conclusion do you draw about the need for precision of alignment of fibres to the test direction?

8. A unidirectional lamina of glass fibres in polyester resin has a fibre volume fraction of 0.55. The elastic moduli of the fibres and matrix are, respectively, 73 GPa and 3.5 GPa. When the lamina is stressed in-plane at an angle θ to the fibre direction, the modulus, E_θ, has a value given by the equation:

$$E_1/E_\theta = \cos^4\theta + 9.5\,\sin^2\theta\,\cos^2\theta + 5\sin^4\theta$$

where E_1 is the composite modulus in the fibre direction. If eight of these laminae are combined to give a balanced symmetric cross-ply laminate of 0/90 lay-up, calculate the Young's modulus of the laminate (a) in the fibre direction and (b) at 45° to the fibre direction.

9. A composite consists of 30 vol% of aligned short fibres in a polymer matrix. The fibres are 7 μm in diameter, with a mean aspect ratio of 73, and their mean tensile strength is 4 GPa. If the polymer matrix has a strength of 68 MPa and the shear strength of the interfacial bond between fibres and matrix, τ_i, is 35 MPa, deduce the mean stress in the fibre when it fails and use this to calculate the composite tensile strength.

10. The probability, $R(\sigma)$, that the strength of any fibre in a bundle exceeds the stress, σ, is given by the modified Weibull function:
$$R(\sigma) = \exp(-A\sigma^m)$$

and the strength of a bundle of these filaments is given by:

$$\sigma_B = (mA)^{-1/m}\,\exp(-1/m)$$

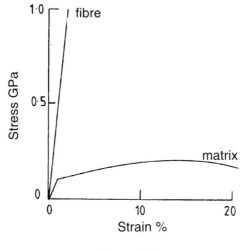

Figure 8.1

where A is a scale parameter and m is the Weibull modulus. Obtain an expression for the ratio of bundle strength to median filament strength, σ_B/σ_f, and estimate the value of this ratio for fibres having a Weibull modulus of 10.

11. An injection-moulded composite bar contains a volume fraction, $V_f = 0.2$, of short carbon fibres in Nylon. The fibres, which are roughly aligned with the length of the bar, are of average length 600 μm and diameter 5 μm, and have a strength of 3.2 GPa. The matrix polymer has a strength of 65 MPa. An experiment to measure the interfacial bond strength, τ_i, by pulling a fibre from a block of the polymer gives a value for τ_i of 30 MPa. Derive a simple force-balance equation to describe the pull-out experiment and use it to determine the mean fibre stress when the fibre fails and the tensile strength of the composite material.

12. Seventy tests on 20 mm long carbon fibres gave a mean strength of 2.1 GPa and a coefficient of variation (standard deviation ÷ mean) of 20%. Estimate a Weibull modulus for this sample, and calculate the mean strength of a group of 5mm test samples of the same material. Briefly explain the significance of this for calculations of the strength of fibre composites.

13. Stress/strain data for a brittle reinforcing filament and a ductile matrix metal are given in Figure 8.1. The modulus of the fibre is 200 GPa and that of the matrix is 100 GPa. Describe in physical terms the sequence of events when two unidirectional composites containing (a) 2% of fibres and (b) 50% of fibres are loaded to failure in the fibre direction.

Calculate (or otherwise deduce) the stress/strain curves for these two composites and draw a fully labelled graph showing the variation of strength with composition for this composite system.

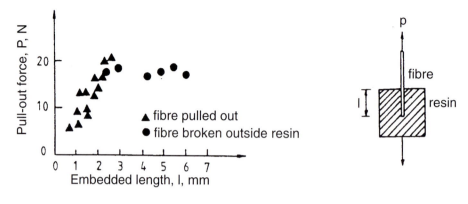

Figure 8.2

14. An experimental composite beam has been made which contains a regular array of boron fibres in an epoxy resin. The boron fibres, which are uniaxially-aligned along the beam length, are 0.08 mm in diameter and the fibre volume fraction, V_f is 0.60. In a preliminary series of experiments some fibres of various lengths were embedded in blocks of resin and the loads required to start pulling them out were determined. The results are shown in Figure 8.2.

    Calculate the minimum span-to-depth ratio of the composite beam if premature shear failure is to be avoided when the beam is loaded in three-point bending.

    The matrix tensile strength, $\sigma_m = 70$ MPa, and you may assume that in a composite where $V_f = 0.60$, the interlaminar shear strength is equal to the fibre/resin interfacial shear strength. Remember that the maximum value of the horizontal shear stress in a 3-point loaded beam is 1.5 times the mean value.

15. Unidirectional hybrid composites are to be manufactured from alternate laminae of CFRP and GRP. If the breaking strengths of the plain CFRP and GRP, respectively, are 2 GPa and 1.5 GPa, and their failure strains are 1% and 2%, construct a fully-labelled scale diagram to show the variation of strength of the family of hybrids as a function of composition. The elastic moduli of the *fibres* are 400 GPa (carbon) and 70 GPa (glass); assume that the volume fraction of fibres in each individual composite is 0.60.

16. For a carbon-fibre/epoxy-resin composite with $V_f = 0.65$ having a mean tensile strength of 2 GPa, the interlaminar shear strength is found to be 90 MPa. If the angle between the fibre alignment direction and the stress axis is θ, estimate the value of θ at which there will be a change in failure mode from fibre failure to in-plane shear failure, and sketch the approximate variation of σ_θ with θ.

Appendices

A1 ORIENTATION-DEPENDENCE OF THE ELASTIC PROPERTIES OF A SINGLE UNIDIRECTIONAL LAMINA

Composites are rarely used in the form of unidirectional laminates, since one of their great merits is that the fibres can be arranged so as to give specific properties in any desired direction. Thus, in any given structural laminate, predetermined proportions of the unidirectional plies will be arranged at some specific angle, θ, to the stress direction. In order to calculate the properties of such a multi-ply laminate, it is first necessary to know how the elastic response of a single unidirectional lamina, such as that we considered in section 3.5 of chapter 3, will vary with angle to the stress direction. In order to do this we first transform the $x_1 x_2$ axes through some arbitrary angle, θ, as shown in Figure A1-1.

The initial set of Cartesian axes (1,2 or $x_1 x_2$) are defined with respect to the orientation of the fibres in the unidirectional lamina, as shown. For the general case, we need to transform the axes from (1,2) through an angle θ to the orientation (x,y). Definitions of the stresses with respect to the old and new axis systems are as given in the diagram. The transformations can be obtained directly by resolution, as follows, or by the use of Mohr's circle, the two approaches being identical.

The steps in the analysis are as follows:

1. summing in the x direction in triangle 1, and remembering that $\tau_{12} = \tau_{21}$, etc:

    $\sigma_x (A/\cos\theta) = \sigma_1 (A\cos\theta) + \sigma_2 (A\tan\theta.\sin\theta) + \tau_{21} (A\tan\theta.\cos\theta) + \tau_{12} (A\sin\theta)$

    or $\quad \sigma_x = \sigma_1 \cos^2\theta + \sigma_2 \sin^2\theta + 2\tau_{12} \sin\theta \cos\theta$

2. summing in the y direction in triangle 2:

    $\sigma_y (A/\cos\theta) = \sigma_1 (A\tan\theta.\sin\theta) + \sigma_2 (A\cos\theta) - \tau_{12} (A\tan\theta.\cos\theta) - \tau_{21} (A\sin\theta)$

    or $\quad \sigma_y = \sigma_1 \sin^2\theta + \sigma_2 \cos^2\theta - 2\tau_{12} \sin\theta\cos\theta$

3. summing in the y direction in triangle 1:

    $\tau_{xy} (A/\cos\theta) = -\sigma_1 (A\sin\theta) + \sigma_2 (A\tan\theta.\cos\theta) + \tau_{12} (A\cos\theta) - \tau_{21} (A\tan\theta.\sin\theta)$

    or $\quad \tau_{xy} = -\sigma_1 \sin\theta\cos\theta + \sigma_2 \sin\theta.\cos\theta + \tau_{12} (\cos^2\theta - \sin^2\theta)$

Thus, the transformation matrix, T, for the transformation from the (1,2) axes to the (x,y) axes is:

$$T = \begin{pmatrix} \cos^2\theta & \sin^2\theta & 2\sin\theta\cos\theta \\ \cos^2\theta & \sin^2\theta & -2\sin\theta\cos\theta \\ -\sin\theta\cos\theta & \sin\theta\cos\theta & \cos^2\theta - \sin^2\theta \end{pmatrix} \qquad (1)$$

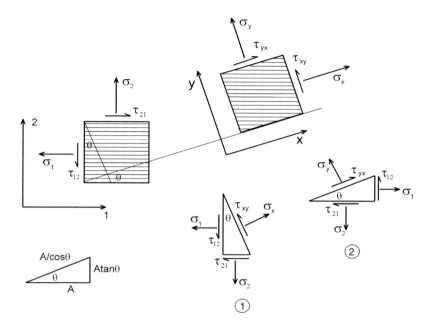

Fig. A1-1. Transformation of co-ordinates to obtain the orientation-dependence of the elastic properties of a unidirectional composite lamina.

and the transformed stresses are given by the tensor relationship:

$$\begin{bmatrix} \sigma_x \\ \sigma_y \\ \sigma_{xy} \end{bmatrix} = T \begin{bmatrix} \sigma_1 \\ \sigma_2 \\ \sigma_{12} \end{bmatrix} \qquad (2)$$

or

$$\sigma_{xy} = T\sigma_{ij} \qquad (2a)$$

Note: this derivation specifically relates to the transformation $(1,2) \Rightarrow (x,y)$, *i.e.* it concerns a change from the material properties framework to the structure framework. Other authors choose to start with the structural case (x,y) and transform to the material system $(1,2)$, in which case the transformation matrix will be T^{-1}, the inverse matrix of that in the above equation. This method appears more logical to engineers dealing directly with structures but it makes no difference which transformation is used provided that we retain consistency in our approach.

The effect of this transformation can be seen in a typical example. If we apply the initial stresses, $\sigma_1 = 800$ MPa (tension), $\sigma_2 = -200$ MPa (compression) and $\sigma_{12} = 150$ MPa (shear) to a unidirectional lamina and then rotate the $(x_1 x_2)$ axes through $180°$, the resulting variations in the stress components σ_x, σ_y, and σ_{xy}, as given by equation 1, are as shown in Figure 2.

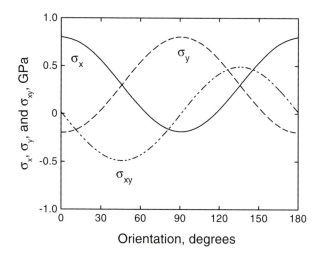

Fig. A1-2 Variation of the stress components, σ_{ij}, as the $x_1 x_2$ axes are transformed through 180°.

Strain transformations from the (1,2) axes to the (x,y) axes involve the same transformation matrix, T, but in this case it is the tensor shear strains, ε_{xy} and ε_{12} that must be used rather than the engineering shear strains, γ_{xy} and γ_{12}. As before, the transformation gives:

$$\begin{bmatrix} \varepsilon_x \\ \varepsilon_y \\ \varepsilon_{xy} \end{bmatrix} = T \begin{bmatrix} \varepsilon_1 \\ \varepsilon_2 \\ \varepsilon_{12} \end{bmatrix} \qquad (3)$$

or:

$$\varepsilon_{xy} = T\varepsilon_{ij} \qquad (3a)$$

If we wish to derive the generalised form of the Hooke's law relationship of equation 3.16 of chapter 3, between the strains and stresses in the general axes (x,y), we must substitute equations 2a and 3a into that equation. As given in chapter 3, the equation is:

$$\varepsilon_{ij} = S_{rs}\,\sigma_{kl} \qquad (4)$$

remembering that when writing equations 2a and 3a explicitly for σ_{ij} and ε_{ij} we have:

$$\sigma_{ij} = T^{-1}\,\sigma_{xy} \quad \text{and} \quad \varepsilon_{ij} = T^{-1}\,\varepsilon_{xy}$$

where T^{-1} is the inverse of the T matrix *(note: T^{-1} differs from T only in the signs of the terms containing the function $\sin\theta\cos\theta$ in equation 1)*. Equation 4 then becomes:

$$T^{-1}\,\varepsilon_{xy} = S_{rs}\,T^{-1}\,\sigma_{xy} \qquad \text{or}$$

$$\varepsilon_{xy} = T\,S_{rs}\,T^{-1}\,\sigma_{xy} \qquad (5)$$

Note that the two geometric terms T and T⁻¹ do not cancel, and that the product
T S$_{rs}$ T⁻¹ thus now represents the compliance matrix relating the strains ε_{xy} to the stresses
σ_{xy}. The full form of the generalised Hooke's law for a thin orthotropic lamina is then:

$$\begin{bmatrix} \varepsilon_x \\ \varepsilon_y \\ \varepsilon_{xy} \end{bmatrix} = \begin{bmatrix} \bar{S}_{11} & \bar{S}_{12} & \bar{S}_{16} \\ \bar{S}_{12} & \bar{S}_{22} & \bar{S}_{26} \\ \bar{S}_{16} & \bar{S}_{26} & \bar{S}_{66} \end{bmatrix} \times \begin{bmatrix} \sigma_x \\ \sigma_y \\ \sigma_{xy} \end{bmatrix} \tag{6}$$

and the equivalent set of relationships giving stresses as functions of strains can be
obtained in a similar manner:

$$\begin{bmatrix} \sigma_x \\ \sigma_y \\ \sigma_{xy} \end{bmatrix} = \begin{bmatrix} \bar{Q}_{11} & \bar{Q}_{12} & \bar{Q}_{16} \\ \bar{Q}_{12} & \bar{Q}_{22} & \bar{Q}_{26} \\ \bar{Q}_{16} & \bar{Q}_{26} & \bar{Q}_{66} \end{bmatrix} \times \begin{bmatrix} \varepsilon_x \\ \varepsilon_y \\ \varepsilon_{xy} \end{bmatrix} \tag{6a}$$

where the $\bar{Q}_{ij}$ terms are referred to as *elastic coefficients* or *elastic stiffnesses* to distinguish
them from the *compliances*, S$_{ij}$.

Except for the case when the (x,y) axes coincide with the composite orthotropic
axes (the fibre direction and the transverse direction) the compliance matrix, S$_{rs}$, now
contains six elastic constants instead of the four in the matrix of equations 3.15 of
chapter 3. In this case, since there are no zeros in the S matrix **each** component of stress
will contribute to **all** of the components of strain. A shear stress will therefore induce
normal strains and normal stresses will induce shear strains. This is referred to as *shear-
tension coupling*, and it has important consequences for designers since a thin plate of a
unidirectional composite will distort in complex fashion if loaded off-axis.

The compliance terms in the matrix in equations 6 are barred to remind us that these
are not simply material elastic constants like the S$_{rs}$ in equations 4, but that they are in
fact functions of stress, as shown by the presence of the matrix functions T and T⁻¹ in
their definitions. It can be seen that if only a tensile stress, σ_x, is applied, the strain
measured in the direction of the stress, ε_x, is related to σ_x, by the familiar Hooke's law, *i.e.*

$$\varepsilon_x = \bar{S}_{11}\sigma_x$$

and the compliance $\bar{S}_{11}$ therefore represents the reciprocal Young's modulus for this
deformation:

$$\bar{S}_{11} = 1/E_x$$

but clearly $\bar{S}_{11}$, and therefore also E$_x$ are functions of the orientation, θ. If the matrix
algebra needed to evaluate the barred compliances in equations 6 is carried out, it can be
shown that the full expression for $\bar{S}_{11}$ is:

$$\bar{S}_{11} = S_{11}\cos^4\theta + S_{22}\sin^4\theta + 2(S_{12}+S_{66})\cos^2\theta\sin^2\theta \tag{7}$$

where the compliances S$_{11}$, S$_{22}$, S$_{12}$, and S$_{66}$ are the four elastic constants for the orthotropic
lamina relative to the axes (1,2) and defined in terms of the normal engineering constants

at the end of section 3.4 of chapter 3. If those definitions are substituted in equation 7, we have the orientation dependence of the elastic modulus, which we shall call E(θ):

$$\frac{1}{E(\theta)} = \frac{\cos^4\theta}{E_1} + \frac{\sin^4\theta}{E_2} + \left(\frac{1}{G_{12}} - \frac{2\nu_{12}}{E_{11}}\right)\cos^2\theta\,\sin^2\theta \tag{8}$$

Similarly, the shear compliance component, $\overline{S}_{66}$, can be evaluated to obtain the orientation dependence of the shear modulus, G(θ):

$$\frac{1}{G(\theta)} = 4\left(\frac{1+2\nu_{12}}{E_1} + \frac{1}{E_2} - \frac{1}{2G_{12}}\right)\cos^2\theta\,\sin^2\theta + \frac{1}{G_{12}}(\cos^4\theta + \sin^4\theta) \tag{9}$$

For readers wanting to convince themselves of the validity of these derivations without the tedium of doing the algebra by hand, the commercial software package Mathcad by MathSoft is highly recommended. All of the manipulations in this Appendix have been carried out with this programme, which also enables easy graphing of the various functions.

A2 STATISTICAL ASPECTS OF THE STRENGTH OF BRITTLE SOLIDS

The strengths of brittle solids are controlled by small defects or flaws, arising either during manufacture or through handling, which concentrate stress approximately in proportion to the function $\sqrt{(a/\rho)}$, where a is the depth of a flaw and ρ the radius of its tip. The distributions of size and acuity of flaws in any sample will be difficult to characterise by any other than statistical means and the measured strength values of a batch of supposedly identical samples will thus also be statistically distributed. The strengths of brittle solids sometimes appear to be distributed in a Normal (*i.e.* Gaussian) fashion, but usually the distributions are skewed and other statistical models are therefore more appropriate.

The failure of a brittle fibre can be described in terms of a 'weakest-link' theory by supposing that it consists of a series-connected arrangement of elemental volumes. One of these volumes — one link in a complete chain — will contain a flaw deeper and more damaging than those in any other elemental volume, and that 'link' will fail first at an applied load level which, when concentrated by the flaw, corresponds to the inherent or theoretical strength of the material. A statistical model which is frequently used to deal with problems of reliability and life is that which has come to be known as the Weibull distribution (Weibull, 1951). It is one of a family of related models, of which the exponential model is another member, which are of interest in tackling extreme-value problems in engineering. The reader should consult texts by Gumbell (1958), Bury (1975), Chatfield (1983), or Castillo (1988) to find out more about the statistical background to this subject.

THE WEIBULL MODEL

The simplest statement of a mathematical function that may be described as a Weibull function gives the probability density of the occurrence of an event z as:

$$p(z;m) = m \, z^{m-1} \exp(-z^m) \qquad z, m > 0 \tag{1}$$

where m is referred to as a *shape parameter*. This expression defines a *probability distribution function,* or pdf, and it is the derivative of the *cumulative distribution function,* or cdf, which is the likelihood of the occurrence of z:

$$P(z;m) = 1 - \exp(-z^m) \qquad z, m > 0 \tag{2}$$

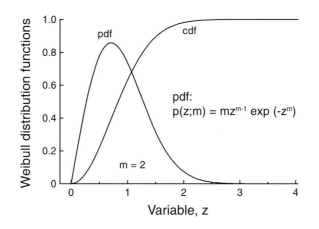

Fig. A2-1 Probability distribution function and cumulative probability function for a simple Weibull model with shape parameter, m = 2.

Plots of these functions for m = 2 are shown in Figure 1 from which it can easily be seen how the Weibull pdf differs from the familiar Normal distribution, particularly in being bounded at the lower end. If m is equal to unity, the distribution becomes exponential, in which case there is no lower bound.

In applying such a model to a physical process such as failure, it is clear that the exponential term must be dimensionless, and this may be achieved by treating z as a reduced or normalized variable, $z = x/b$, where b is another characteristic of the distribution and is called the *scale parameter*. This scale parameter is the value of the new variable x for which there is a fixed probability given by:

$$P(x; b, m) = 1 - \exp(-1) = 0.632$$

The parameter b is often treated as a *characteristic value* analogous to the familiar mean value of a Normal distribution. Applying the cdf of equation 2 to the specific case of the tensile failure of a set of samples, all of the same dimensions, of a brittle material, we may write the probability of failure, $P(\sigma)$, at a stress level σ, as:

$$P(\sigma; b, m) = 1 - \exp\left[-\left(\frac{\sigma}{b}\right)^m\right] \qquad \sigma, b, m > 0 \qquad (3)$$

This is known as the *two-parameter Weibull model* of failure. As shown in Figure A2-1, the pdf associated with this model is bounded at its lower limit by $x = 0$. In speaking of the failure of solids, it would be uncomfortable to have to accept that there was not some finite level of stress below which there was no possibility of failure, and it is more usual to shift the distributions of equations 1 and 2 to the right along the z axis so that there is some lower limit, a, to the distribution. The reduced variable is then defined as:

$$z = \frac{x - a}{b}$$

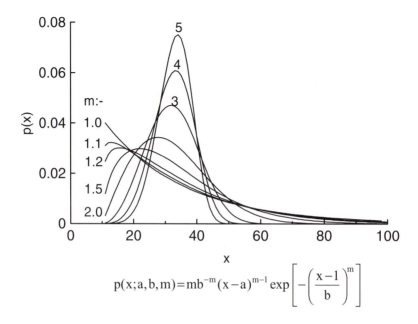

$$p(x;a,b,m) = mb^{-m}(x-a)^{m-1} \exp\left[-\left(\frac{x-1}{b}\right)^m\right]$$

Fig. A2-2 Variation of the shape of the Weibull probability distribution function, p(x;a,b,m), with the value of the shape parameter, m. The pre-selected values of the location parameter, a, and the scale parameter, b, are 10 and 25, respectively.

and the parameter *a* is called the *location parameter*. Shifting the distributions in this fashion makes no difference to their shapes. The Weibull cdf may then be written as:

$$P(\sigma;a,b,m) = 1 - \exp\left[-\left(\frac{\sigma-b}{b}\right)^m\right] \tag{4}$$

This is referred to as the *three-parameter Weibull model* and it is often argued that this is the only possible valid Weibull function that can be applied to physical processes like tensile failure of brittle solids, or the fatigue failure of engineering materials. The appearance of the distribution varies widely with changes in the value of *m* from the exponential form for m = 1 to shapes resembling Normal distributions for higher values of *m*, as illustrated by the family of pdfs shown in Figure A2-2.

The shape and scale parameters are of great importance in that they provide a means of characterizing the Weibull distribution in the same way that the average, variance and standard deviation are used to characterise Gaussian distributions. Some of the parameters of a Weibull pdf are:

i.   The expected value of x, or arithmetic mean (called the *first moment* of the distribution)

$$b\Gamma\left(1+\frac{1}{m}\right)$$

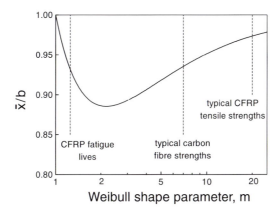

Fig. A2-3 The effect of the value of the Weibull shape parameter, m, on the ratio between the arithmetic mean of a distribution, $\bar{x}$, to the Weibull scale parameter, b.

ii. The variance (called the *second moment* of the distribution)

$$b^2\left[\Gamma\left(1+\frac{2}{m}\right)-\Gamma^2\left(1+\frac{1}{m}\right)\right]$$

iii. The median (*e.g.* the value for which there is 50% probability of failure)

$b(\log_e 2)^{1/m}$

iv. The mode (most probable value)

$$b\left[\frac{m-1}{m}\right]^{\frac{1}{m}}$$

where Γ is the gamma function. From these relationships, it is easy to show that for values of the shape parameter, m, of the order of 10, there is a useful approximate relationship between m and the familiar coefficient of variation (cv = standard deviation divided by arithmetic mean) of the form:

$m \approx 1.2/cv$

As noted earlier, the characteristic value, b, is often used as a measure of a property like the fracture strength of a brittle solid in place of the more usual mean value. The difference between the two is often small, but depends on the value of m through the gamma function $\Gamma(1 + 1/m)$, as shown above, for which the argument will usually lie between 1 and 2. The variation of this function, which thus displays the ratio of the mean to the characteristic value, $\bar{x}/b$, is shown in Figure A2-3.

When the Weibull distribution is used to deal with weakest-link problems like brittle fracture and fatigue failure, the parameters b and m have to be estimated from available experimental data. There are several methods of doing this, of greater or lesser degrees of complexity, and it is by no means easy to be sure, for given distributions, which method will give the most satisfactory estimates. A discussion of these methods is beyond the scope of this book, but we mention two that are familiar to users of the Weibull model. The first, referred to as the *method of moments,* uses the conventionally defined values of the arithmetic mean and the variance, deduced from the data, to solve equations 1 and 2 above. This method is not universally well-regarded.

The method most familiar to materials scientists is that of plotting the probability function against the measured parameter so as to analyse equations 3 and 4 graphically. The experimental data are first ranked from lowest value, i = 1, to the highest, i = N, where N is the total number of data values. The likelihood of failure associated with i = 1 is small while that for i = N is great, and a method is needed for assigning a probability, P(x), that reflects this range. There are several common methods, but perhaps the most familiar, the mean-rank assignment, gives the probability as

$$P = \frac{i}{N+1} \tag{5}$$

where *i* is the rank (*i.e.* the position in a list of ascending order) of a given datum value. For a distribution of measured strengths, σ_f, then, to obtain the distribution parameters from equation 3 the double logarithmic function ln[-ln(1-P)]is plotted against $ln\sigma_f$ (Gumbell, 1958). Provided the points fall on a straight line, the slope and intercept then give the required estimates of *m* and *b*. If the resulting line is curved, the two-parameter Weibull model is not valid and it may be that the three-parameter Weibull model is more appropriate. A non-linear curve fit of P versus σ_f will, with care, permit the estimation of all three distribution parameters. It must be admitted that these simple methods tend to be ill-regarded by statisticians. More sophisticated treatments, such as the use of *maximum likelihood estimators*, can be found in the texts referred to below.

In the foregoing discussion of strength, we have assumed that all samples were of equal dimensions. A weakest-link model must be able to account for volume effects, however, since it is the dimensions of flaws that we are dealing with in evaluating the strength of brittle solids. The form of the three-parameter Weibull model that is most commonly invoked in discussions of strength, therefore, is one which includes the volume of the sample, V:

$$P(\sigma) = 1 - \exp\left[-V\left(\frac{\sigma - a}{b}\right)^m\right] \tag{6}$$

REFERENCES

K.V. Bury: *Statistical Models in Applied Science*, J. Wiley and Sons, London,1975.

E. Castillo: *Extreme Value Theory In Engineering*, Academic Press, Boston/London, 1988.

C. Chatfield: *Statistics for Technology: a Course in Applied Statistics*, 3rd edition., Chapman and Hall, London, 1983.

E.J. Gumbel: *Statistics of Extremes*, Columbia University Press, New York,1958.

W. Weibull: *Journal of Applied Mechanical Transaction ASME*, **73**, 1951, 293-297.

A3 SOME APPLICATIONS OF COMPOSITES

This is a brief listing of current and proposed applications of composite materials in various branches of industry. It is not intended to be comprehensive or all-embracing, but merely to give an indication of the range of possibilities for designers.

AEROSPACE

A wide range of load-bearing and non-load-bearing components are already in use in both fixed-wing and rotary wing aircraft. Many military and civil aircraft now contain substantial quantities of lightweight, high-strength carbon, Kevlar and glass-fibre composites, as laminated panels and mouldings, and as composite honeycomb structures with metallic or resin-impregnated paper honeycomb core materials. They are used in air frames, wing spars, spoilers, tail-plane structures, fuel tanks, drop tanks, bulkheads, flooring, helicopter rotor blades, propellers, pressured gas containers, radomes, nose and landing gear doors, fairings, engine nacelles (particularly where containment capability is required for jet engines), air distribution ducts, seat components, access panels, and so forth. Many modern light aircraft are being increasingly designed to contain as much lightweight composite material as possible. For elevated-temperature applications carbon-fibre-reinforced carbon is in use. Concord's disk brakes use this material, rocket nozzles and re-entry shields have been fashioned from it, and there are other possibilities for its use as static components in jet engines. Rocket motor casings and rocket launchers are also frequently made of reinforced plastics. A particularly interesting (and important) application of composites is in its development as a means of repairing battle damage (patching) in metal aircraft structures.

Space applications offer many opportunities for employing lightweight, high-rigidity structures for structural purposes. Many of the requirements are the same as those for aeronautical structures, since there is a need to have low weight and high stiffness in order to minimize loads and avoid the occurrence of buckling frequencies. Dimensional stability is at a premium, for stable antennae and optical platforms, for example, and materials need to be transparent to radio-frequency waves and stable towards both uv radiation and moisture.

AUTOMOTIVE ENGINEERING

There is increasing interest in weight reduction in order to permit both energy conservation and increased motoring economy. Reduction in the weight of an automobile structure achieves primary weight-saving and if carried to sufficiently great lengths enables the designer to use smaller power plants, thus achieving substantial secondary improvements in fuel economy. The majority of automotive applications involve glass-reinforced plastics because the extra cost of carbon or aramid fibre is rarely considered to be acceptable in this market. Even so, the cost of using GRP is usually being weighed against the lower cost of pressed steel components, and the substitution is often rejected on purely economic grounds, leaving aside the question of energy saving. A wide range of car and truck body mouldings, panels and doors is currently in service, including complete front-end mouldings, fascias, bumper mouldings, and various kinds of trim. There is considerable interest in the use of controlled-crush components based on the high energy-absorbing qualities of materials like GRP. Leaf and coil springs and truck drive shafts are also in service, and GRP wheel rims and inlet manifolds have been described in the literature. Selective reinforcement of aluminium alloy components, such as pistons and connecting rods, with alumina fibres is much discussed with reference to increased temperature capability.

BIO ENGINEERING

Carbon-fibre-reinforced plastic and carbon components are in use for prosthetic purposes, such as in orthopaedic fracture fixation plates, femoral stems for hip replacements, mandibular and maxillary prostheses (jaw remodelling, for example), and for external orthotic supports in cases of limb deformity *etc*. Pyrolytic carbon is used to manufacture heart valve components, and the substitution of a carbon/carbon composite is not unlikely. There have also been developments in the use of particulate hydroxyapatite as filler in a thermoplastic composite for bone remodelling or replacement.

CHEMICAL ENGINEERING

A substantial amount of GRP is currently in use in chemical plant for containers, pressure vessels, pipe-work, valves, centrifuges etc. These may be filament-wound or moulded components for containment of process fluids.

CIVIL/STRUCTURAL ENGINEERING

Again the bulk of composites used in this field are glass-reinforced plastics. The low inherent elastic modulus of GRP is easily overcome in buildings by the use of double curvature and folded-plate structures: thin GRP panels also offer the advantage of translucency. Glass-reinforced cement (GRC) products made with CemFil (alkali-resistant glass fibres) are gradually being introduced as structural cement-based composites, but these GRC are still regarded with some suspicion by architects who prefer to consider only non-load-bearing applications for glass-reinforced cement. Development of suitable highly-drawn polymer fibres and net-like polymeric reinforcement has made it possible to produce stable polymer-reinforced cement for a variety of purposes. But concrete is the cheapest engineering material available, and it requires very little in the way of expensive reinforcing filaments to be added to it to make it uncompetitive. The answer is usually to use more concrete! But GRC is perhaps likely to attract the more adventurous designer with light-weight concrete structures in mind (thin shell structures for example). A good deal of GRP is used in this industry for folded-plate structures, cladding panels, decorative 'sculptured' panels (like those in the doors of the Roman Catholic cathedral in Liverpool), services mouldings and ducting, racking, pipework, rainwater mouldings, domestic and industrial water tanks, form-work for concrete, and complete small structures like foot-bridges. Lightweight composite panelling for partitioning and similar applications have also been tried. CFRP have been less used until recently because of the cost, but are increasingly being considered for building lightweight structures, including a number of bridges.

In recent years there has been a major surge of interest in the use of structural composites in civil engineering infrastructures. At least two major composites journals have run Special Issues on the subject in 1996/99, and the 1998 European Conference on Composite Materials, ECCM8, ran a series of well-attended sessions on civil engineering uses for the first time. Current applications include the use of pultruded FRP shapes as individual structural elements and shear stiffeners for concrete structures, as reinforcing bars for concrete, components in composite/concrete structures, as externally applied impact-containment supports, and as patches for damaged concrete bridgework. A good deal of upgrading of existing bridge structures by externally applied composite components is being carried out.

DOMESTIC

Injection-moulded reinforced thermoplastics and polyester moulding compounds are perhaps the most common composites used in consumer items for the domestic market, and the range is vast. Mouldings of all kinds, from kitchen equipment of all kinds to casings for the whole gamut of domestic and professional electrical equipment, motorcycle crash helmets, television and computer casings, and furniture.

ELECTRICAL ENGINEERING

Typical applications are radomes, structural components for switch gear, power generator coolant containment and large-diameter butterfly valves, high-strength insulators (*e.g.* for overhead conductor systems), printed circuit boards, and casings for electronic equipment. The majority of applications in this field again use GRP, although the use of composites which are more thermally stable and more moisture-resistant is increasingly predicated for sensitive, small-scale electronic components. Many prototype and practical wind-generator designs incorporate GRP or hybrid blading.

MARINE ENGINEERING

Marine applications include surface vessels, offshore structures and underwater applications. A vast range of pleasure craft has long been produced in GRP, but much serious use is also made of the same materials for hull and superstructure construction of passenger transport vessels, fishing boats and military (mine-countermeasures) vessels. Sea-water cooling circuits may also be made of GRP as well as hulls and other structures. Off-shore structures such as oil rigs also make use of reinforced plastics, especially if they can be shown to improve on the safety of steel structures, for fire protection piping circuits, walkways, flooring, ladders, tanks and storage vessels, blast panels, and accommodation modules. High specific compression properties also make composite materials attractive for submersibles and submarine structures, both for oil exploration and for military purposes, and for towed transducer arrays for sea-bed sonar mapping.

SPORT

Perhaps the most visible development in the use of composites has been in the sports goods industry. Manufacturers have been quick to seize on the potential advantages of new materials like carbon and boron fibre composites over conventional wood and metal for sports equipment of all kinds, but whether the average sportsman (and perhaps even some of the above-average ones) who have been inveigled into buying this more expensive composite equipment in the hope that it would improve their game have been able to demonstrate genuine improvement remains uncertain. GRP vaulting poles were perhaps the earliest of the composite sports gear, but one can now obtain tennis rackets, cricket bats, golf clubs, fishing rods, boats, oars, archery equipment, canoes and canoeing gear, surf boards, wind-surfers, skateboards, skis, ski-poles, bicycles, and protective equipment of many sorts in composite materials of one kind or another. In an industry that is often less directly subject to controls exercised in other areas of engineering there is often a tendency to dupe customers with the use of names which incorporate the words 'carbon' or 'graphite' to describe expensive, black-coloured items which may at the present time legitimately contain little or no carbon fibre.

Index